P9-AOJ-172

$\mathscr{P}resented\ to:$

Butler Area Public Library

In Memory of
Jim Walker

Donor
Bass Buddies

A SHEARWATER BOOK

STRIPER
WARS

23
12/06

233 8/55

Butler Area Public Library
218 North McKean St.
Butler, PA 16001

STRIPER WARS

AN AMERICAN FISH STORY

799.17
RUS

BY DICK RUSSELL

ILLUSTRATIONS BY ANTHONY BENTON GUDE

⬤ ISLANDPRESS / SHEARWATER BOOKS

Washington • *Covelo* • *London*

A Shearwater Book
Published by Island Press

Copyright © 2005 Dick Russell

All rights reserved under International and Pan-American Copyright
Conventions. No part of this book may be reproduced in any form or by any
means without permission in writing from the publisher: Island Press, 1718
Connecticut Ave. NW, Suite 300, Washington, DC 20009.

SHEARWATER BOOKS is a trademark of The Center for Resource Economics.

Library of Congress Cataloging-in-Publication data.

Russell, Dick.
  Striper wars : an American fish story / Dick Russell.
    p. cm.
  Includes bibliographical references and index.
  ISBN 1–55963–632–7 (cloth : alk. paper)
  1. Striped bass fishing—United States—History. I. Title.

SH691.S7R87 2005
799.17'732—dc22

                 2004029347

British Cataloguing-in-Publication data available.

Printed on recycled, acid-free paper ✪

Design by Maureen K. Gately

Manufactured in the United States of America

10 9 8 7 6 5 4 3 2 1

To Mel and Jessie, who taught me how to fish

Contents

1   *Prologue*
      ORIGINS OF A FISH STORY

13   *Chapter 1*
      A TALE OF TWO FISHES

31   *Chapter 2*
      STORM OVER THE HUDSON

49   *Chapter 3*
      THE CONSCIENCE OF A LURE MAKER

71   *Chapter 4*
      A MAN NAMED MENDONSA

91   *Chapter 5*
      HOW THE STRIPED BASS STOPPED A HIGHWAY
      AND ELUDED THE MOB

114   *Chapter 6*
      HOW RHODE ISLAND CHANGED THE WORLD

129   *Chapter 7*
      SHOWDOWN AT FRIENDSHIP AIRPORT

146   *Chapter 8*
      REVOLT OF THE BIOLOGISTS

165   *Chapter 9*
      STRIPER MAGIC

185   *Chapter 10*
        THE DOUBLE-EDGED SWORD
        OF "FULL RECOVERY"

200   *Chapter 11*
        THE MYCO MYSTERY

217   *Chapter 12*
        THE TOWN THAT MENHADEN BUILT

235   *Chapter 13*
        UPRIVER

254   *Chapter 14*
        CALIFORNIA STRIPERS

277   *Chapter 15*
        KEEPERS OF THE RIVER

295   *Chapter 16*
        STRIPERS FOREVER?

317   *Acknowledgments*

322   *Notes*

349   *Index*

Prologue

Origins of a Fish Story

Ⅰn the summer of 1973, I first "wet a line" seeking striped bass in waters off the island of Martha's Vineyard, Massachusetts. I'd not gone fishing since I was a boy living in the Midwest, during an occasional summer outing with my parents and brother on Lake Michigan. I grew up with little interest in recreational fishing, and certainly no natural instincts for it. Then a friend who happened to be a master fisherman invited me to visit him on the Vineyard. I'd rarely tasted anything as sweet as a striped bass. And I'd rarely known an experience as exciting as heading out to sea in the middle of a star-filled night, trolling the rip in a twenty-three-foot boat off Gay Head's treacherous shoals.

My mentor didn't sell his catch, taking home only enough for his family to eat. Nor did he use live bait on his rod-and-reel, which would have almost guaranteed hooking into a fish, and likely a big one. He was a purist who used plastic, metal, or wooden artificial lures designed to look like food a bass would enjoy. It took quite a bit more casting skill to make a Rebel, an Atom, or a Lupo plug move the way a baitfish does below the surface, or to make a Popper dance invitingly across the waves.

You simply couldn't tell about striped bass. They were the most confounding of fish. You might figure to find them feeding on a flood tide, but on this particular night they'd choose an ebb tide. You might guess they'd go after the same lure you'd successfully employed yesterday, only to have the fish be completely oblivious to its charms today. Once in a while you might gun the boat toward a flock of birds hovering above a school of bass in a seeming feeding frenzy, only to have every last fish disappear into some twilight zone as you came within casting distance. If you hooked into a bass that had made that mistake before, it would dive deep and quickly attempt to wrap your line around a rock in order to break loose.

Even after I'd practiced my backyard casting long enough to avoid getting a snarl in my line *most* of the time, it took many a fishing trip before I finally landed a striped bass. I still remember the first time vividly. Ask any striper fisherman, and he will tell you his story.

At midnight, standing at the bow of the boat, a lighthouse beacon from the Gay Head cliffs was my only external guide. The moment the fish struck, I knew. This was not the hammering strike of a bluefish or the furiously darting pull of a bonito. The feeling was more subtle, the leap of my heart more intense. "It's a bass!" resounded inside me, though I dared not utter it aloud.

The fish ran and line flew from my reel. Anything could happen, I realized. Conceivably, the fish could roll and cut the line with its razor-sharp gill edges. It might turn sideways and hang in the current, like a sail against the wind. It might simply jerk free. No fish, I already knew, is so wise as a striped bass.

But there was no time to think of these possibilities, nothing to do but follow into the fish's invisible realm. The feeling of each moment was all that connected the thin thread I was holding to the surge of life on the opposite end.

Try to remember all you've learned. Bring the rod-tip high, reel down on her, not too fast, not too slow, oh God have I lost her? No, still there, I can feel her again, breaking the surface, oh look at that!, gone under again now. Where is she?, close maybe, a friend's voice echoing "Bring the fish around this side," slow turn, moon's behind a cloud, nothing to go on but this feeling inside and my heart's in my throat and there is only a flash where the net meets the fish and . . .

I stood by the bow-rail staring down at the deck. A whisper: "You caught a bass, Dick. You caught a bass." If time had passed as I slowly reeled the fish to the boat, I'd had no awareness of it. But when the bass was in, and I saw the struggle and the joy reflected in the eyes of my partners, I believed the fish had somehow belonged to me the whole time.

Probably no near-shore fish, not even the prized Atlantic salmon, arouses so much emotion as striped bass—the biggest of which have been known to reach lengths of five feet, weigh more than a hundred pounds, and live for more than thirty years. To several million sport-fishermen like me along the Atlantic coast, it is the premier gamefish to pursue: intelligent, crafty, the ultimate challenge for a rod-and-reeler. To thousands more commercial fishermen, striped bass are a lucrative source of seasonal income. To hundreds of restaurateurs, they are the favorite item on a summer menu.

But it is more than the striped bass's size, mettle, and flavor that gives them such mystique. They are an intrinsically American fish. It might even be said that striped bass are the aquatic equivalent of the American bald eagle. Without the sustenance this fish provided, the Pilgrims would not have survived their first harsh winters in the Plymouth Colony. Two decades later, the fish was the impetus behind America's first conservation law, when the Massachusetts Bay Colony's general court ruled that striped bass were too valuable a resource to be used as fertilizer by the maize growers and squash planters. Along the coastline of the original thirteen colonies, the fish traced its migratory path and provided food for the table. Later, striped bass joined the westward pioneers, being transplanted from New Jersey out to California on the early transcontinental railroad in 1879.

Presidents from George Washington to George W. Bush have pursued this most wily and tasty of creatures. So have millions of less renowned mortals, for the striper is readily accessible to the common man (or, as the case may be, the uncommon woman). You don't need a boat to go after striped bass, unlike tuna or tarpon. They can be sought from shore, dock, or jetty; on a rock, a rip, a channel, a drop-off, or a reef. They can be fished deep or shallow. They can be fished using lures—swimmers, jigs, umbrella rigs, spoons, and such—or live bait such as herring, mackerel, and eels. They can be taken on flies with names like Lefty's Deceiver, Skipping Bug, and Surf Candy. They can also be snared in gillnets, fish traps, haul seines, and bottom trawls.

With primary spawning habitats in the Chesapeake Bay and the Hudson River systems, striped bass have prompted the development of an entire boat and tackle industry around their migration along the East Coast. They are also the most sought-after gamefish in the San Francisco Bay–Delta estuary. They are pursued in the Southern river systems of North Carolina, South Carolina, and Georgia. They have been stocked in dozens of landlocked lakes and streams along the Gulf Coast, from the Suwannee River in Florida westward through Alabama, Mississippi, Louisiana, and Texas. You can now fish for stripers in the Mississippi and the Colorado rivers, or in Lake Texoma, Oklahoma, and Lake Meade, California. Freshwater striped bass tournaments are now held in at least thirty-eight states, with prizes as high as $70,000 awarded to the winner of the Raytown Lake, Pennsylvania, event.

Many striped bass are dependent upon food-rich, sandy shorelines during portions of their life stages. Like we humans, a striper seems to love a good beach. But the fact that they live in such close proximity to people, who can utilize so many different means of catching them and whose practices on land can affect them in myriad ways, makes the striped bass as vulnerable a creature as it is adaptable.

This vulnerability is what has made the striped bass a flash point for conservation struggles over the last half century. It was over striped bass that, in the 1960s, citizens established their right to sue a government agency in order to protect natural resources. A study of striped bass became the precursor for the first Environmental Impact Statement (EIS). This set the stage for development of the National Environmental Policy Act of 1969, now the cornerstone of American environmental law, requiring all federal agencies to prepare detailed EISs to ensure that environmental effects are taken into account along with economic and technical considerations. The fight for the long-term survival of the striped bass was also crucial to the establishment of the Natural Resources Defense Council, which employs legal action as a vital arm of environmental activism, and to spawning the national Riverkeeper movement, a loose alliance of citizens' groups that have come together to protect their local waterways. Yet while striped bass have been a watershed species in terms of successful environmental action in the past, their story today vividly illustrates the need for an expanded view: If we want to preserve species, we can't do so one by one; rather, we must look at the entire ecosystem of which they—and we—are a part.

When I first began fishing for them, striped bass were abundant along the Eastern seaboard. In 1970, the largest "dominant year class" ever had been recorded in their primary spawning grounds of the Chesapeake Bay. This phenomenon, first studied in depth by a Yale University researcher named Daniel Merriman, occurred periodically —often once every half-dozen years, occasionally longer—when female fish encountered optimal natural conditions for reproductive success. If the especially large numbers of females from this year class entered a good nursery environment and then received protection from fishing pressure, they would similarly replenish the fishery when they reached spawning maturity, about the age of six years. The annual young-of-the-year survey, conducted in the Chesapeake's river systems each summer and fall, had recorded an average of more than thirty juvenile striped bass every time the marine scientists hauled in their seine nets to catalogue the various fish before letting them all go. By 1973, many of the three-year-old females from this bounteous hatch had joined the coastal migration. Commercial landings of striped bass from Maine to North Carolina would reach a record high of 14 million pounds that year. Knowing nothing about all this at the time, I figured the vast numbers available to fishermen were the norm.

Then, in the early 1980s, the puzzling steep population decline of the striped bass was among the first indications that something had gone terribly wrong with our oceans and their piscine inhabitants. The Atlantic population of stripers was so depleted that the fish seemed destined to be soon added to the endangered species list. A fishermen's campaign on behalf of the striper, of which I became an integral part, ensured that that did not happen, however. For the first time, the U.S. Congress intervened to protect an inshore fish whose territory had always been regulated—or misregulated—by the states. For the first time, too, pivotal states declared moratoriums on possession or sale of a prized commercial and recreational fish species. A striped bass population estimated to contain about 4.6 million fish in 1982 would reach a historic peak estimated at an astounding 56.7 million fish in 2004.

The fish's remarkable comeback has become part of modern conservation lore. Today it is once again among the most popular saltwater fish in the United States. As Carl Safina wrote in *Scientific American* in 1995: "The resurgence of striped bass along the eastern coast of the U.S. is probably the best example in the world of a species that was allowed to recoup through tough management and an intelligent rebuilding plan."

What follows is the behind-the-scenes story of how this came to pass, of what I call the striper wars, along with the strange history of the striped bass since then, a story about a magnificent fish and those of us who fought against commercial interests and government bureaucrats to bring it back from the brink. It is a case study in environmental activism, one suggestive of lessons that might be applied to today's critical questions of how to govern other fisheries, at a time when so many species are in grave jeopardy. And it's also a story about the dilemmas that face us as the need to move beyond protecting single fish stocks, and toward taking into account entire ecosystems, becomes the focus of the future.

For today, a new set of circumstances has again put the striped bass's future in peril. While its primary source of food—the menhaden—are apparently being overfished in the Chesapeake Bay region, for example, a chronic and debilitating disease has struck a majority of the striper population in its Chesapeake Bay spawning habitat. No longer is it simply a matter of preventing overfishing, as it was in the past; now the struggle involves the life cycle of the fish and the realm it inhabits. The striped bass still has something to teach us.

As winter recedes in the Atlantic waters around Cape Hatteras, North Carolina, a female striped bass stirs from her lethargy. Born a decade and a half ago just above the salt wedge inside the Chesapeake Bay estuary, she is at the southern apex of her migratory range. She has been in near-dormancy, eating little since arriving here several months earlier, but now, as the water warms, it is time to move north. Immensely fecund, she is laden with eggs. Like the salmon, she will return to spawn in the river where she was born. But while a salmon will home in on a precise location, sometimes swimming upriver for hundreds of miles, a striped bass will seek out merely the same subestuary or estuary. Striped bass are classified as one of the anadromous fishes—born and returning to spawn in tidal freshwater, while many spend the majority of their lives in salt water.

Passing Cape Charles, she turns to enter the Chesapeake Bay, a 195-mile-long stretch of serrated peninsulas and veinlike tributaries. The largest and most complex of America's 130 estuaries, the bay's watershed embraces about 90 percent of Maryland, 60 percent of Virginia, and a considerable part of eastern Pennsylvania. The bass might enter any of three rivers in Virginia's sector of the lower bay—the James, the York, or the Rappahannock—or, farther north, the Potomac or Patuxent rivers along the bay's western shore. Or she might journey up the Eastern Shore, where three large rivers—Maryland's Nanticoke, Choptank, and Chester—and several lesser tidal ones are drained by the Delmarva Peninsula. Some of her sisters will travel into the upper bay, to spawn either there or in the Sassafras, Bohemia, Elk, Northeast, or Susquehanna rivers.

On a late afternoon in early May, when the waters reach an optimum temperature just shy of 65 degrees Fahrenheit, this particular fish moves swiftly into the spawning reaches of the Choptank. The blood supply to her eggs shuts down, and they now have a limited time to be successfully fertilized. Mature male striped bass have preceded her by days or even weeks, waiting, schooled up in a strong current above the river bottom. The female gives off a scent to attract them. The water in the shallows begins to boil as ten males—it can be as many as fifty—surround and engage her in a pre-spawning courtship ritual. They thrash, they splash; the males sometimes even turn on their sides and skid across the surface, as if wounded. This is sometimes called a "rock fight." In the Chesapeake region, striped bass are more commonly known as rockfish, for their propensity to hide and forage around large boulders in the tributary rivers. (The *saxatilis* part of their scientific name, *Morone saxatilis*, means "rock dweller" in Latin.)

Then, a quietude. As she slowly sinks below the surface, pressed upon still by perhaps half a dozen males who nudge her belly with their snouts, she begins to expel a pale river of eggs. From 10,000 to 5 million eggs may be broadcast in a matter of hours, each one a translucent green sphere covered by an amber globule of oil, the size of tiny pearls. The males swim back and forth through the cloud of eggs, releasing their milt to mix with it. And so it continues, new males moving in to replace their spent brethren, continuing to prod the female until all of her eggs are released. Only then will both sexes meander off downriver to rest and eat, leaving the fertilized eggs to sweep back and forth in the brackish currents.

Within two to four days, a mouth forms on our newborn bass. Shortly thereafter, the larva disperses in the water column and begins feeding on microscopic zooplankton. Soon its diet will progress to tiny crustaceans, including copepods and cladocerans (water fleas). When the new bass reaches fifty to eighty days old, its prey will advance to little mysid shrimp, scavenging crustaceans called amphipods, and even smaller larval fish.

Nursery areas of striped bass are located in littoral shoal habitats in freshwater and brackish tidal regions of estuaries, where the salinity is higher. At about thirty millimeters, our baby bass becomes a juvenile. With fins now completely developed, this highly mobile and gregarious animal gathers in schools often numbering several thousand other bass along sandy near-shore bottoms. A keen watch must be kept for predators amid the spiked marsh grasses and waterlogged riverbanks: adult bluefish, weakfish, and other piscivores that would enjoy a good meal of baby bass.

By late summer of its first year, our "fry" has grown to between two and three inches long. As a "fingerling," it will remain in the estuary over the winter, reaching between five and seven inches by the following spring. During its second summer, our young striper will begin to forage on what oil-rich menhaden it can find, bay anchovies, and other small fish. By the time it is two, it will have attained a size of as much as twelve inches.

Customarily, striped bass remain in the Chesapeake Bay for between two and four years. A majority of the males, in fact, will never migrate into the Atlantic at all. But many females leave the Chesapeake system at the beginning of their third year. In the ocean, they will join older fish that have been wintering offshore and are returning to their summer feeding grounds.

They mingle as they follow the coastline northward, traveling about twelve miles a day, sometimes with speed-bursts as fast as twenty miles per hour when chasing bait-balls of menhaden. An adult female from the Chesapeake may cover as many as 2,000 miles in a year's roundtrip journey. Beyond Delaware to the north, the springtime migration finds younger fish joined by much larger and older ones along the beaches of New Jersey.

On they journey, past Sandy Hook, eastward along the coast to Long Island. Some of the Hudson River–spawned fish—abounding near the United Nations, Ellis Island, and the Statue of Liberty—now

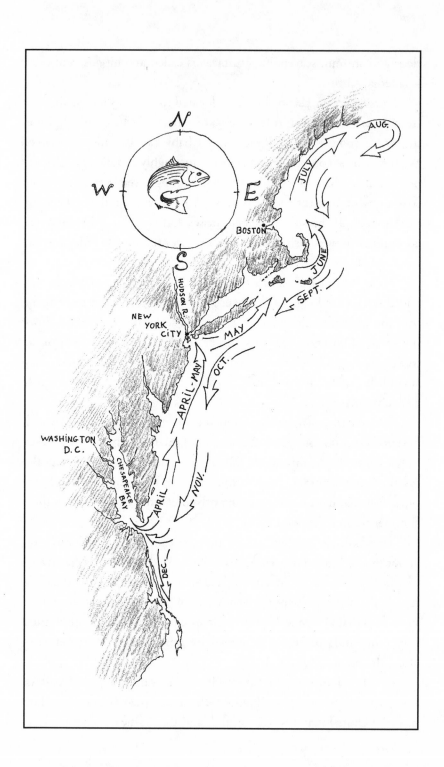

merge with their Chesapeake relations. They travel together along the Connecticut coast, then onward to Rhode Island's Narragansett Bay and the shorelines from Sakonnet Point to Watch Hill. Many keep going, into Cape Cod Bay and the waters around Martha's Vineyard. And some journey beyond the Gulf of Maine to summer in the waters off Nova Scotia in Canada.

Striped bass, of course, do not swim in one vast parade from south to north, then back again in the fall when the waters grow colder. They will pause, and often stay, where the summer's food supply is good and the water temperature satisfactory. Other fish constitute more than 95 percent of their diet. The bass do have favorite meals—with menhaden ranking at the top in nutritional value—but they will devour almost anything in their path: sand worms and sand lances, silversides, eels, squid, lobsters, crabs, shrimp, soft-shelled clams. They've even been discovered with Portuguese man-of-war, a poisonous jellyfish, in their stomachs.

Mike Laptew, a diver/filmmaker who makes underwater videos, has described observing a striper feeding "blitz" as akin to "watching a pride of lions work a herd of zebra." Laptew writes: "The majority of bass would circle the silversides, sealing off their escape, while several individuals slowly made their way to the center of the school. The tiny baitfish didn't panic as long as the bass moved slowly; they simply maintained a fixed distance from the bass, creating a clear zone around each fish, as if they were projecting some type of invisible forcefield. However, when the time was right, the bass would suddenly charge into the wall of silversides, gulping down several fish at a time."

The larger, older bass (the age can be determined by counting the rings in their ear bones, or otoliths) are usually solitary feeders. Most active soon after dark and shortly before sunrise, they often establish territory behind boulders or rocky outcroppings, facing up-current to await the scent of prey that may be flushed through the inlets.

Watch the fins move as food approaches—the twin dorsals, the pectoral, pelvic, anal, and caudal fins fluttering ever so slightly beneath the scales that cover their entire frame. The dorsal and anal fins act as anti-roll stabilizers; the tail, pectorals, and pelvic fins are used for maneuvering and braking. Their bodies are shaped like torpedoes, their heads nearly as long as the depth of their bodies, their tails also large and slightly forked. The bass have extremely strong tails that, when they desire, can propel them through the water with incredible speed and grace. In addition, like many other predatory fishes, the bass possess a very efficient gill system for extracting oxygen from the water and a swim bladder that selectively adjusts buoyancy at a variety of depths. The latter is extremely important, for it allows the bass to choose a particular depth and then travel in a plane there with relatively little effort.

The waters that bass prowl are often turbid, and their sense of smell is believed to be among the finest any fish has ever developed. While most species have a single pair of nares (the term for nostrils in fish), the bass are possessed of a double pair, two on each side of the head. Their ears are buried deep inside their heads, finely attuned like radio directional finders as the fish turns. And their large eyes, while myopic, are set high and are beautifully placed in order to see whether prey is ahead, below, or above. Indeed, there is a human quality about those eyes, which are able to adjust quickly and easily to varying intensities of light.

The sense of taste is equally acute. The taste buds exist not just on the tongue, but around the roof of the mouth and in the throat, with even the end of the snout and the lips containing an array of external taste buds. Thus can a striped bass taste its food before opening its mouth. And, like us humans, the bass can tell whether something is sweet or sour.

Striped bass have been making this journey every year, since long before the first European settlers arrived in America. But that journey today traverses what has become the most densely populated and highly developed coastline anywhere in the world. And *that* has come to pose a considerable challenge, in a variety of ways, to the strength and sinew—indeed, the very viability—of the striper population.

Chapter One

A Tale of Two Fishes

Captain John Smith, the first white man to explore and map the Chesapeake Bay, felt that "Heaven and earth have never agreed better to frame a place for man's habitation." The "fruitful and delightsome land" Smith encountered bore great woods of oak and walnut that sheltered abundant deer and wild pig. Corn was plentiful, as was fruit, including grapes, mulberries, and strawberries. And the rich waters—which would eventually produce more seafood per acre than any others on earth—were teeming with striped bass. "I myself at the turning of the tyde have seen such multitudes," Smith wrote of them in 1614, "that it seemed to me that one mighte go over their backs drisho'd [dry-shoed]."

For centuries before Smith, striped bass had been caught and dried in prodigious numbers by Native American tribes along the Atlantic coast. According to Roger Williams, the Narragansett Indians called the bass *missuckeke-kequock*, which meant "much fish" or "great fish." A Dutch commercial agent named Isaack De Rasieres wrote in 1623 of the Hudson River tribes: "It seems the fish makes the Indians lascivious, for it is often observed that those who have caught any when they have gone fishing have given them, on their return, to their women, who look for them anxiously."

The first Thanksgiving might as easily have served striped bass as turkey. Local Algonquins alerted the Pilgrim newcomers to the qualities of the fish, and the colonists noted its resemblance to the sea bass of the British Isles, except that the North American variety bore a series of seven to eight close-set charcoal stripes that ran laterally along the upper sides. These were beautiful creatures indeed, the stripes and scales reflecting like brass in the autumn waters. The fish's back, or dorsal, surface came in colors ranging from steel blue to a dark olive-green, paling on the sides to a white belly. The lean, white flesh had a delicate, succulent flavor—seeming to combine the sweetness of the fresh water where it spawned, the saltiness of the ocean, and the meatiness born of muscling its way down the Eastern coastline. The fish were robust, sometimes as thick around as your waist, and plentiful. In the summer of 1623, the Plymouth settlers used their last, leaky little boat and a single net to catch enough striped bass to provide them sustenance well into the autumn. As William Wood noted in the *New England Prospect* (1634): "Though men are soon wearied with other fish, yet they never are with basse."

Rules were early laid down around the fish. Colonial bond servants, for example, could not be fed more than two meals a week of striped bass. The promulgation of America's first fishing regulation, by the Massachusetts Bay Colony in 1639, stated: "And it is forbidden to all men . . . to imploy any codd or basse fish for manuring of ground." Another distinction was an act passed by the Plymouth Colony in 1670 that required that all income accrued annually from the fisheries at Cape Cod for striped bass, mackerel, or herring be set aside for a free school. As a result, the first public school in the New World was made possible in part through moneys derived from the sale of striped bass.

George Washington, upon arriving at Mount Vernon in 1759, would write a letter about the Potomac River being "well-stocked with various kinds of fish," including bass "in great abundance" in the spring. Washington employed a small net designed to be hauled ashore by hand and ordered his overseer to admit "the honest poor" to partake of fishing privileges at one of his shores. In 1776, the legislatures of New York and Massachusetts took time out from raising revolutionary armies for General Washington to enact one of our first fisheries management efforts: laws prohibiting wintertime commercial sale of striped bass.

The fish were, in fact, then abundant in nearly every major river that fed into the Atlantic Ocean, from Canada's St. Lawrence River all the way to Florida. When newcomers settled in North Carolina, part of its attraction was the bountiful anadromous fishes—especially striped bass—that thrived in the Roanoke River and other tributaries of Albemarle Sound. From his first years as a congressman, Daniel Webster is reported to have pursued striped bass along the Potomac, "eminent among the celebrated fishermen of the day," according to Spencer Baird, secretary of the Smithsonian Institution and later America's first fish commissioner as well as a devotee of bass fishing.

Shortly after the Civil War, recreational fishing for striped bass began in earnest, initially as a pastime for wealthy sportsmen. They founded clubs to fish for stripers with long cane rods adapted from European salmon rods, often from wooden piers specially built out into the surf. As one historian described the world of Massachusetts's Cuttyhunk Striper Club, "robber barons had their hooks baited with lobster tails and called in their stock-market trades to the mainland via carrier pigeon." (One technique, known as "walking the lobster," found fishermen baiting their hooks with baby crustaceans, which would move slowly but tantalizingly along the bottom until a striper located them.)

As the nineteenth century drew toward a close, an intensified wave of concern about overfishing and habitat surfaced on the striper's behalf—a wave that would wax and wane for the next hundred years.

By 1870, Robert B. Roosevelt, a renowned fisherman-conservationist who served as a role model for his nephew Theodore Roosevelt, was warning that "the insatiable maw of the New York market" was cutting drastically into the Northeast's striper population. In addition, dams built to power the mills of the Industrial Revolution were preventing the customary upriver spawning migrations of striped bass in New England.

Yet heavy fishing continued to be rampant along the Eastern seaboard. One newspaper account from May 1896 reported that 38,000 pounds of striped bass had been landed in a single seine haul near Norfolk, Virginia. Some of these fish weighed upward of a hundred pounds, while others were as small as six inches. The market became so glutted that the fish were selling for a mere penny a pound.

By the early 1900s, when Presidents Theodore Roosevelt and William Howard Taft came to cast for stripers from the "bass stands" of Cuttyhunk, the fish's population had been reduced to extremely low levels. The Cuttyhunk club would close its doors by 1910, and for most of the next thirty years, striped bass were few and far between.

Then gradually came a shift in the fish's fortunes. Because the streams in Maryland, Virginia, and North Carolina were comparatively slow and meandering, they had not been harnessed for power. In the major Chesapeake tributaries—as well as in the Hudson River—the bass typically spawned below the first dams. The Chesapeake fish, emanating from so many diverse and productive tributaries across a wide area, would come to comprise up to 90 percent of the Atlantic coastal stock.

In 1919, the federal government purchased the Chesapeake and Delaware Canal. The canal had been little more than a ditch since being opened ninety years before to connect the Chesapeake Bay eastward to the Delaware River. Now it was decided to make the waterway larger. For striped bass at the northern end of the Chesapeake system this meant they could take a much shorter route to the Atlantic, through the canal and down into Delaware Bay. The expansion also allowed greater water flow into the bass's primary spawning habitat in the Chesapeake. Some have speculated that, at the nadir of the fish's population in 1934, this helped result in a "baby boom."

A "great awakening of public interest in maintaining the supply of this fish at a high level of abundance coincides with one of the most remarkable returns to high levels of yield that the striped bass fishery has ever experienced," William C. Neville of the U.S. Fish and Wildlife Service told the first meeting of the Atlantic States Marine Fisheries Commission (ASMFC) in 1942. He was referring to formation of a remarkable fifteen-state compact "to coordinate the conservation of management of near-shore fishery resources," inspired by concern over the striper's future viability.

Led by Massachusetts, most of the coastal states began implementing a sixteen-inch minimum size limit to allow more of the small fish to reach spawning age. In 1947, Massachusetts also outlawed any netting of striped bass within its three-mile territorial boundary. While regulations remained inconsistent along the bass's migratory route, soon the fish were flourishing as never before and continued to do so for several decades. The 1960s were the halcyon years, a time when a fisherman camped alongside the Cape Cod Canal could be awakened by the slapping of thousands of tails as an endless school of bass headed toward the open sea.

But fishing was again beginning to take a devastating toll on the species. New power boats equipped with radar, short-wave radios, powerful engines, and sonar devices for detecting and locating fish schools—often referred to as "killing machines"—were partly responsible. In 1971, commercial draggers working offshore of Virginia had discovered a vast horde of big striped bass wintering over, for example, and netted them all. And on a single day in January 1972, two crews of haul seiners at Cape Hatteras pooled their nets and brought in more than 100,000 pounds of stripers, most of them roe-laden females weighing between twenty and fifty pounds. Gillnetting of fish entering the Chesapeake's tributaries to spawn also took a heavy toll on large females. Recreational angling increased exponentially in the 1970s as well, targeting smaller "panfish."

Suddenly the flood-tide of bass began to recede. Nobody knew whether the decline was a result of overfishing, or increased pollution within the Chesapeake Bay spawning grounds, or a combination of both. But within less than a decade, the population had declined to such an extent that scientists were questioning whether the fish could survive at all as a coastal resource. An Emergency Striped Bass Study was mandated by Congress. Maryland scientists observed incredibly low numbers of eggs, larvae, and adult spawners in the Potomac River and upper Chesapeake Bay. By 1981, the young-of-the-year survey of surviving newborn bass in the Chesapeake had hit an all-time low: an average of 1.2 juvenile fish taken per seine haul.

All I knew at the time was that catching, or even seeing, a striped bass had become a rare occurrence. For three years in the late 1970s, I'd worked as a feature writer in the Hollywood bureau of *TV Guide*. By the time I returned to the East Coast, the plentiful summer striper fishing I'd known so briefly and beautifully was over.

Still, I couldn't stop trying. On a crisp late summer day in 1981, I remember marveling at the majestic red clay cliffs along Gay Head as my friend cut the engine and let our boat drift toward shore on the incoming tide. The sea was dark and churning after a September storm, the way striped bass liked it best. Suddenly I glimpsed the outline of a fish on the crest of a wave. The torpedo-like shape was unmistakable. Then came a second wave, a second fish. They were huge, at least 30 pounds each—swinging their strong, broad tails, their gills protected by a tough and bony plate. Now dozens appeared, their iridescent colors outlined against the light green wave tops. For an instant a pair of bass came tail to tail, forming a wondrous mirror image inside a single wave.

Into this rainbow of sun and sea and bass, we could not even cast. Especially given how few bass we had been seeing lately, it was a spectacle too precious to disturb. Finally we journeyed on along the island's coast. After a spin through the riptides offshore to pick up a few blue-fish, we returned for one last look. But in the same spot below the cliffs, there was now an intruder: another fishing boat had anchored in the path of our drift.

Rapidly and with ruthless abandon, several fishermen were hauling in bass after bass, clubbing each fish as it came over the side and toss-ing it into a box. They were "chunker fishing," tossing hunks of men-haden overboard and then casting out hooks fitted with more of the irresistible bait. This school of bass that had danced for us in the twi-light was being wiped out. It was too painful to speak as our boat returned to Menemsha Harbor that evening.

A few days later, at a dinner party hosted by a small group of my friends, some of the island's finest fishermen were on hand. These weren't just weekend anglers; among the recreational fishing communi-ty, over the years they had established a reputation for being able to find, and successfully catch, striped bass anywhere, anytime. Over sev-eral glasses of wine and a wonderful meal, we had a long discussion about what we had all been observing—and had been the subject of much dockside talk over the summer. As we compared notes, it was clear this went well beyond fishermen's disgruntled palaver about "the one that got away." They confirmed that seeing fewer and fewer stripers was a common experience.

"Spider" Andresen, an associate publisher at *Salt Water Sportsman* magazine, paused at the kitchen door on his way out. "So long," he said. "I don't know how we can deal with this now. It's impossible to regulate a migratory species, just can't be done. Eventually, our best hope is that they'll be fished almost to extinction. Striped bass will become an endangered species. And then the regulations will come."

For a long time thereafter, his words resounded. But for my friends and me, a line had already been crossed. We could no longer passively watch the species decline. There was nothing to do but take a stand.

Starting in 1982 in my home state of Massachusetts, that is precise-ly what we did. We even brought our children into the streets to help collect signatures on petitions calling for stronger regulations to protect the striped bass. Years later, hearteningly, one of these young activists, Lincoln Lyman, would write me that this experience had imbued him with the belief that "if you want the world to change, you had better roll up your sleeves and get to work."

Certainly my world changed. Journalism became a tool that I put to use as an activist. I began speaking out on radio and TV. Eventually I ended up testifying before Congress, organizing a national conference, and forming a new organization to bring attention to the plight of the striped bass. For several years, my life was dominated by this effort as I traveled up and down the Eastern seaboard, finding allies in each state where the striper had touched the shoreline and the hearts of people. They, too, knew that this majestic fish was worth fighting to save.

You will meet many of them in these pages—from a Massachusetts lure maker to a Rhode Island postman to a jeweler from Maryland. Had the striped bass not brought us together, from so many walks of life, we would never have known what kindred spirits we were or what we could do together. I've been to see them all again while working on this book, some for the first time in years—to probe their memories about what happened then, and to talk about what they are seeing now, as threats to the striped bass mount once again.

On a gray late October day in 2003, rain clouds were on the horizon. It was about two in the afternoon as the four of us loaded our fishing rods and tackle boxes onto *Diamond Jim,* Jim Price's sleek twenty-eight-foot Bertram. From the marina in Oxford, a watermen's enclave on Maryland's Eastern Shore of the Chesapeake Bay, it was maybe a twenty-minute run to reach the mouth of the Choptank River.

Price climbed a ladder to the flying bridge, sat down at the wheel, and started his new, 650-horsepower twin engines. In the stern stood Joe Boone, readying a bucktail jig for his spinning rod. He had to be approaching seventy, but retained the same lanky frame, ruddy complexion, and reddish hair that I remembered from years earlier. A fisheries scientist, Boone had supervised Maryland's annual young-of-the-year fishing surveys for thirty years. Now retired to a farm about a two-hour drive from here, Boone still made a half-dozen trips every summer to go fishing with Price. Standing next to Boone was Jim Uphoff, his longtime colleague on the surveys, still employed as a marine biologist by Maryland's Department of Natural Resources.

I'd flown in from Boston for this reunion. Twenty years earlier, Boone and Uphoff had put their scientific careers on the line for the sake of the striped bass. Price had abandoned his charter-boat business to lobby for the bass's designation as a threatened species in Maryland and started a nonprofit organization known today as the Chesapeake Bay Ecological Foundation.

As we left the dock, I remembered the last time I'd been on the water with Boone and Uphoff, in 1983. That was the year the Atlantic coast's commercial striper landings fell to their poorest level since record-keeping began: 1.6 million pounds, a drop of nearly 90 percent in a decade. We'd then waded in past the tall reeds lining the Bohemia River, one of the Chesapeake tributaries, holding onto a 100-foot-long knotted nylon net. We'd made a wide circle along the shallow shoreline. The net had a quarter-inch mesh and a single heavy lead line on the bottom. Slowly the circle closed and the seine was pulled in. Boone charted the count while Uphoff and another biologist sifted through schools of little menhaden and silversides, as well as an occasional carp or catfish. In the haul, there was a single juvenile striped bass about three inches long. In a normal year, the net would have held at least eight. "If this is an indication of what's going on up here," Boone had said, "we're in bad shape."

I'd been on the water with Price one other time too, the following summer of 1984, when newborn stripers had become even scarcer. We'd motored down the Choptank River then in a much smaller boat, as he described being part of the fifth generation of fishermen in his family. His father made a living setting pound nets and gillnets for "rock" and other species in the bay. Price started helping out before he was ten, then found that he most enjoyed hook-and-line fishing. A jeweler by trade, Price had ended up using his business to found and support the organization he first called the Chesapeake Bay Acid Rain Foundation (since he believed at the time that acid rain was largely responsible for the bass's decline). "Every so often I think I'm wasting my time," I recalled him saying. "But this fish put me through school and always put food on our table, so I feel like I owe them something."

The four of us in the 1980s had been among those in the forefront of fighting commercial interests and government bureaucrats to save the bass. Now, two decades later, we found ourselves—literally and figuratively—in the same boat. For all that was ultimately achieved on the striped bass's behalf, the fish was once again in potential jeopardy.

Price first realized something was wrong in the autumn of 1997, when he hauled in a bass covered with red sores. "I'd never seen anything like it in my life," he recalled. "It sickened me." Since then, an outbreak rarely found in wild fish has been spreading. Recent surveys have found that at least half of the Chesapeake's striper population is infected with mycobacteria, a chronic "wasting" disease that attacks an animal's organs and can prove fatal. While the infections are often internal, nearly one-fifth of the fish exhibit external lesions. Many of these fish are also emaciated.

So today on the Choptank River, we would not be looking to catch "keepers" for dinner. No, we were in search of relatively small fish, ones exhibiting signs of disease or malnutrition. Price would then deliver these to marine biologists at a fisheries lab back in Oxford. "It's so disheartening to see the striper deteriorate in physical condition," Joe Boone was saying. "And it's simply from lack of food. There are a lot of things you can't do anything about, but this is manageable. But who's going to do it?"

No longer is it the bass themselves who are the victims of too much fishing pressure. Now it's their food of choice: the Atlantic menhaden. Customarily, these small, silvery, herringlike fish have comprised between 70 and 80 percent of the diet of the Chesapeake's adult striped bass. Although they have no value whatsoever as table fare, menhaden comprise the largest commercial fishery in the United States. In 2001, some 1.7 billion pounds of menhaden valued at more than $100 million were landed in the Gulf of Mexico and the Atlantic. While some are sold as bait to fishermen, about 85 percent of the oily little fish taken are ground up, dried, and made into a high-protein feed for pigs, chickens, cattle, and aquaculture operations as well as a growing market in fish oil for human consumption.

Since the menhaden fleet has now been prohibited from most state's inshore waters, the industry has lately been forced to concentrate its efforts in Virginia and North Carolina. Out of Reedville, Virginia, the Omega Protein Corporation operates ten factory-sized vessels. Focusing their fishing near the Maryland line (they are prohibited by Maryland law from crossing it), these vessels deploy smaller boats with purse-seine nets that encircle and haul in tons of menhaden. While it's not an easy figure to calculate, some scientists estimate that the numbers of menhaden in the Chesapeake have declined in recent years by 75 percent.

Boone continued: "The industry has used the argument that the stripers are taking all the menhaden, not the purse-seiners. I was hearing this same song thirty and forty years ago. Anytime something's declined, it's blamed on the natural predator—never the fishermen."

Boone could hardly be called an environmental elitist. He was an avid lifelong hunter who'd spent weeks the previous summer pursuing caribou in the Canadian Arctic. He was not a man to pull his punches, though, which never endeared him to his managerial superiors in Maryland.

The bass have done their best to cope in these changed circumstances, he explained. They've turned to eating smaller, less nutritious bay anchovies, whose population has now also diminished. They've started eating large quantities of blue crabs, much to the dismay of local watermen who depend upon the crabs for a substantial portion of their livelihood. A booming bass population is being observed, but at well below their customary body weight and suffering from a disease that's believed to be stress-related.

By the time we reached the mouth of the Choptank River, the rain was falling slowly but steadily. The four of us donned slickers as the *Diamond Jim* drifted with the tide and we bounced our jigs off the river bottom. We'd brought in a few weakfish, but no bass. Every waterman knows that, to find fish, you follow the birds. But they didn't seem to be here either.

"You used to see 200 loons in the fall chasing the schools of menhaden, which were also being pursued by bass," Price commented. "How many loons did we see today? Two!"

Price, whose elongated and somewhat orbicular countenance is not unlike that of the magnificent creature he's fought to preserve, continued: "If I'd have known in 1983 what I know today, I would not have filed for threatened species status for striped bass. I'd rather have seen them fewer in numbers and healthy."

Uphoff shook his head in disagreement. "We were at the point in 1983 of pushing 'em right over the brink. If we hadn't done what we did, they'd be gone."

I felt a slight tug at the end of my line. Reeling in, I couldn't tell if anything was on there or not. Momentarily, a juvenile striped bass, five inches long at best, was struggling to free itself from my hook. As Uphoff deftly returned it to the water, Boone chuckled. "That bass was smaller than the jig!" he exclaimed. "Now how in the world did it expect to eat *that*?" There was nothing to do but laugh.

"Consider then the dark forces it would take to extinguish the striper's fires, to end the survival of this most determined survivor," conservationist John Cole noted in his 1978 book *Striper*, during the beginnings of the striped bass crisis. "What would be the message for the men who have assigned to this fish the qualities of courage, the virtues of bravery and strength? If this fish were to vanish, how much time would be left to the men who extol it?"

Through the devotion of the Boones, Uphoffs, Prices, and many others, the fish did not vanish. Far from it. In 1989, commercial landings of striped bass totaled a little more than 7,400 fish. A decade later, the commercial take had soared to more than 1.1 million fish. Recreational fishermen landed another 1.3 million fish—and are estimated to have released an additional 12.5 million. That's the biggest recreational catch-and-release reported for any marine fish.

The tale of how this dramatic turnabout came to pass contains lessons that may help us address the desperate need to keep other species from falling off the brink. Consider what has been happening to other fish populations at the same time that the striped bass came roaring back.

In the United States, where commercial fishing is a $25 billion industry and 9.4 billion pounds of fish were landed in 2002, the National Marine Fisheries Service can only assure us that 22 percent of the stocks under federal management are being fished in a manner to ensure their future productivity. Government officials note that eighty American fish stocks have serious problems. It's estimated that 65 percent of fish populations already considered depleted continue to be overfished. By 1989, the numbers of New England cod, haddock, and yellowtail flounder were at historic lows; soon, the New England groundfishery had to be shut down altogether. While some areas have been reopened, those stocks are far from recovered. West Coast fish are faring even worse. Scientists say that some species, including the Pacific rockfish and grouper, are at risk of extinction. The prestigious journal *Nature*, in May 2003, published a paper estimating that industrial fishing has reduced the numbers of large fish—including tuna, swordfish, and marlin—to but 10 percent of their abundance a century ago.

In most of these instances, it's a case of too many fish being taken by too much high-tech fishing gear, notably longlines, bottom trawls, purse seines, and gillnets. The pattern is that once a prized species is decimated, fishermen move on to other, less valuable species. And the chain of destruction continues. Modern fisheries take a mere ten to fifteen years to bring any new fish community they encounter to one-tenth of its previous population.

In June 2003, the Pew Oceans Commission released a major report calling for reform in the way U.S. fisheries are governed. A success story they pointed to was that of the striped bass: "In the midst of crisis, there are expressions of hope. . . . Striped bass, severely depleted along our Atlantic shores, made a striking comeback when given a chance."

But the question today is, will this success story yet prove to be a transient one? The current dilemma of the striped bass brings into focus problems that were seldom considered during the years of come-back. The menhaden situation, for example, cries out for management of the entire Chesapeake ecosystem rather than simply individual species. Aside from the effect their decline has had on striped bass, menhaden are also important filter feeders—they improve water quality by removing algae and stifling, oxygen-depleting algal blooms. Their decline comes at a time when an overload of nutrients remains a terrible problem in the Chesapeake Bay—the result, in part, of waste from huge poultry farms that, ironically, are being fed by the menhaden reduction fishery.

Climate change that may be linked to long-term global warming is also beginning to play a role in the health of the fish. "Striped bass is a marginal cool-water species or temperate species. The big fish migrate north to get out of the hotter conditions," Jim Uphoff observes. Though their reproductive success appears to have benefited from some wet, cool springs in the Chesapeake, a series of uncommonly mild winters in the 1990s may have exacted a toll on their nutritional health. Warmer-than-normal surface waters increase the energy requirements of fish, which must eat more to survive.

In the spring, the deep waters of the Chesapeake start to lose their oxygen due to excess nitrogen and phosphorus that emanates from agriculture, atmospheric deposition, and human and animal waste emptying into the rivers. That's what causes algae to bloom wildly. When this mass of organic matter eventually exhausts its nutrient supply, it sinks to the bottom and decomposes, eating up oxygen and killing grasses and tiny organisms that larger fish feed upon. As U.S. Environmental Protection Agency (EPA) scientist Kent Mountford explains, "This really denies a lot of habitat to the bass. They can hold their breath, scoot in, and grab a clam, but are not capable of really functioning down there." So the bass are forced back toward the surface, where summer temperatures are hotter than they are accustomed to.

Nor, despite their escalating numbers, are the bass now immune from too much fishing pressure. Warns Gary Shepherd of the National Marine Fisheries Service: "In terms of total number of eggs, the spawning stock is now probably as large as it's ever been. But the big fish are getting cropped off. Most of them are gone by the age of about fifteen, and stripers can live to be thirty. So we're limiting their life span to about half."

And those large fish include the females bearing the most eggs, whose survival is crucial to future reproductive success. Meanwhile, in the Chesapeake, it's the males—who don't generally migrate around the age of three, as the females do—who are suffering the most from undernourishment and disease.

"Nine o'clock, just about 90 degrees straight off the port right now, two of 'em breaking!" Uphoff called out from the deck. Above us, Price headed the *Diamond Jim* in that direction. The rain was coming harder, but in a moment we all had our lures in the water. Soon Price's "Jackie jig" started hauling in one small bass after another. "Finally catching some damn fish! The fishing gods are shining on me now!" he exulted, then added, in quieter tones: "These are just exactly the size I want— for the lab to examine and see if they show any abnormalities."

There was no overt evidence of disease, but the fish were clearly all on the skinny side. Price explained that most of the bacterial infections he'd been observing were in bass eighteen inches long and above. That's the size when they start searching for menhaden. By looking at smaller fish, marine scientists are trying to figure out just when the disease strikes—and even if, perhaps, the mycobacterial infection might be passed on through the egg.

Now a smaller boat came speeding toward us, the *Miss Erin G.* Its three fishing partners were all members of the Coastal Conservation Association, a national organization founded in 1977. They were bringing over two small bass for Price to take back for testing. "Chartreuse jigs have been working the last few days," Sherman Baynard informed us, as he handed the fish across the stern. "Well, no chartreuse for me!" Price shouted back. "My Jackie jig's working just fine!"

I asked Baynard what he'd been seeing lately of the striper population. "When the moratorium was lifted in 1990, there was an abundance of large bass," he replied. "We caught those up, but there were still plenty of smaller fish. You could catch a hundred fish a day around the shorelines. If you had fifteen locations you felt were good to fish, you'd get striped bass in all fifteen. That's no longer the case. Now you probably only find them in three of those same areas."

Darkness closed in. We reeled in our lines, stowed our gear, and silently headed back to the marina.

That evening, sitting in Jim Price's den where large striped bass were mounted on both walls, the skipper skimmed through a pile of papers and brought out some grisly photographs and a report that he was submitting to various governmental agencies as well as two congressional committees. Titled "Chesapeake Bay Forage Base Collapse," it was chock-full of comparisons between what's been happening recently to striped bass and the simultaneous overharvesting of menhaden.

"Even this small bass, look at how his tail is eroded away," he said, pointing to one of the pictures. "This is a freshly caught fish. Here's a blow-up of the spleen—isn't that awful? And look how skinny! This one, no bad spleen, but you can see it's flat as a pancake. If this was a mammal on land, you'd see all kinds of protesters complaining."

Price paused, then went on: "At this point, the menhaden fishery in the bay probably should be stopped completely. Give us a few years so these fish can spawn at least once before you start harvesting them."

Price was not alone in calling for the same type of moratorium on menhaden that had once preserved the bass. The National Coalition for Marine Conservation had circulated a petition calling for a halt to all industrial-scale menhaden seining in the Chesapeake. Meanwhile, in April 2003 the Omega Protein Corporation had committed to building a new 100-metric-ton-per-day menhaden processing plant, designed to triple the existing capacity and vastly expand production of a "heart-healthy" fish oil containing omega-3 fatty acids.

This is a political hot potato. The parent company of Omega Protein is the Houston-based Zapata Corporation—founded initially by George H. W. Bush and now controlled by Malcolm Glazer, billionaire owner of the Tampa Bay Buccaneers football team.

The morning after our fishing trip, I traveled with Price to a hotel conference room near the Baltimore airport, where the Menhaden Stock Assessment Committee of the Atlantic States Marine Fisheries Commission (ASMFC) was meeting. Twenty years earlier, the number-crunchers made their presentations on a chalkboard. Today, it was all computer-driven PowerPoint. There were new models, or "tuning indices," as well. The old method started from the present and then "back-cast to the past," as one modeler put it. It had been replaced by models that projected current populations into the future ("the difference between a Model T and a BMW," ASMFC technical director Matt Cieri proudly informed us). Looked at this way, the picture looked bright. Rather than what appeared to be a substantial decline in the Chesapeake's menhaden population, the new model estimated that there were millions more along the Atlantic coast than previously thought.

As Cieri angled his laser pointer toward a fishing mortality graph that used happy and sad faces, he concluded: "We are slightly above the happy face in terms of population fecundity." Ken Hinman, president of the National Coalition for Marine Conservation, questioned how the diets of predators such as striped bass fit into these models. It turned out that such considerations hadn't been factored in.

"The coastwide menhaden stock is not overfished, and overfishing is not occurring," Cieri insisted. That was as far as he was willing to go. It simply wasn't possible to calculate numbers for the Chesapeake alone. After lunch, from the end of the conference table, a burly fellow challenged Hinman and Price. "Why do we need catch limits on menhaden? If the stocks are fine and it's not overfishing, there is another problem. You know what it is? Too many striped bass! Just lying around and eating menhaden!" I'd learn later that this was Jule Wheatley of Beaufort Fisheries in North Carolina, the only other menhaden operation besides Omega Protein on the Atlantic coast. It wouldn't be the last time, I was sure, that I'd hear this argument raised. As Price was allowed to make a brief presentation, Wheatley shouted, "That ain't on the agenda!"—and stormed out of the room.

After the meeting, Ken Hinman and I adjourned to the coffee shop. He'd been another ally in the first fight to save the bass. "When striped bass came back," he said, "it was to a very different environment than the last time they were around in such numbers. In the Chesapeake Bay, I think the water quality is worse, despite all the well-intended cleanup efforts. Today it's not kepone or PCBs, it's the runoff, the nutrient overload. Whole areas are almost void of fish. There are stories of crabs actually crawling out of the water to get oxygen.

"What we're getting now are warning signs with striped bass that something is very wrong. While the stripers were absent, we were fishing out their food supply. A decision has to be made about menhaden—do we want an industry at this level, or do we want to make sure there are enough menhaden to support striped bass and other fisheries?"

Concern about the striper's ecosystem—and how it was being affected by a different kind of industry—had a remarkable modern-day antecedent. It had played out in the bass's other crucial Atlantic spawning habitat, the Hudson River, and would be a forerunner of landmark national legislation and nonprofit litigation. Striped bass, it turns out, were not only present at the creation but were also central to the founding of the modern environmental movement.

Chapter Two

Storm over the Hudson

The tallest mountain and northernmost point in the majestic Hudson Highlands is Storm King, some fifty miles upriver from New York City. Looming 1,355 feet above the water, Storm King Mountain dominates the rugged river gorge near Cornwall. Its often cloud-veiled peak captivated landscape artists of the Hudson River School. In the 1800s, the mountain received its name as "the most sure foreteller of a storm"—an integral part of a uniquely American wilderness that inspired many of the nation's earliest writers, including James Fenimore Cooper, Washington Irving, and Nathaniel Hawthorne. "It rises like a brown bear out of the river, a dome of living granite, swelling with animal power," professor Vincent J. Scully Jr. of Yale University once said of the mountain. "It is not picturesque in the softer sense of the word but awesome, a primitive embodiment of the energies of the earth."

Beneath Storm King, the Hudson is more than a river. It's an estuary where the tides of the Atlantic cut through the Appalachian chain at sea level. Just to the north, across from a gap between Storm King and Breakneck Ridge, river currents and winds are treacherous enough to capsize sailing vessels around sentinel-like Bannerman's Island. Perched above a bluff two miles to the south is the U.S. Military Academy at West Point.

Storm King Mountain was to prove the locus of the spark that led to a fight to save the bass and the region, making it one of the cradles of the modern environmental movement. In 1955, just below West Point, two young marine biologists discovered the existence of "the principal spawning area" of the Hudson River's population of striped bass. With the river's only dam nearly a hundred miles above this area, the Hudson at the time not only retained a sizable local bass population but also contributed considerably to the migratory one. Fully 88.8 percent of the striper eggs collected by researchers Warren Rathjen and Lewis Miller were found along a seven-and-a-half-mile stretch between Highland Falls and Denning Point, with Storm King Mountain precisely in the middle. Based on their counts of eggs, larvae, and juveniles, the scientists concluded that a large percentage of the total spawning occurred within a relatively short segment of the Hudson, once the fish passed above the saltwater portion of the tidal river and into fresh water.

These findings were published in the January 1957 issue of the *New York Fish and Game Journal.* There they remained in obscurity for almost eight years, until Robert H. Boyle happened upon an old copy that a colleague at *Sports Illustrated* had given him. By that time, something very different than spawning was being planned for the striped bass's favored habitat around Storm King Mountain. Consolidated Edison, the power company that supplied electricity to New York City fifty miles away, wanted to build a huge pumped-storage hydroelectric facility at the base of the mountain.

As Boyle would later write in his landmark book *The Hudson River: A Natural and Unnatural History,* "the plant would be able to suck up to six billion gallons of river water during a single daily operating cycle. This water would be pumped from the river through a two-mile-long tunnel, forty feet in diameter, to a storage reservoir. When Con Ed wanted to generate electricity, the plug on the reservoir would be pulled, so to speak, and the water would then surge down the tunnel to turn turbines before flowing back into the Hudson. The plant would be able to generate two million kilowatts of electricity, and it would be the largest of its kind in the world."

It would also kill countless numbers of striped bass. For them, the power plant would be a daunting gauntlet indeed. Should the fish's eggs and larvae not succumb after being initially sucked into the tunnel, they would be unlikely to survive in the storage reservoir because they require a current to keep them buoyant. And even if the hardiest should remain alive, as Boyle put it, "they would then be subjected to whirling turbine blades on the trip back down the tunnel, which Con Ed has exultantly described as 'a "fall" six times higher than Niagara.'"

None of this was yet considered when certain Highlands residents formed the Scenic Hudson Preservation Conference to challenge the power plant's designs on Storm King. They feared the project was the opening wedge in the industrialization of the valley. Theirs were aesthetic and recreational concerns—the defacing of the mountain and the riverfront, the lost hiking trails, the ugly power lines destined to spring up nearby—legitimate concerns that were nonetheless dismissed by the utility as the "self-centered complaints" of a "few local dreamers." New York Governor Nelson Rockefeller shrugged that if the wealthy Highlands denizens were so offended, why didn't they buy the mountain? After all, this was a $162 million project.

Early in 1964, two hearings on the proposed plant were scheduled by the Federal Power Commission (FPC). The only testimony about fish life came from Alfred Perlmutter, a marine biology professor at New York University and a consultant for the power company. Perlmutter said he could "almost guarantee" that the Storm King plant would have "little effect" on fish such as the striped bass. The last study on the Hudson River, Perlmutter testified under oath, had occurred in 1938 and had proven that the striper's primary spawning grounds were "much farther upriver."

Not surprisingly, that March 1964, the hearing examiner recommended to the FPC commissioners that Con Ed's license application be approved. But they had not yet reckoned with the allure of the striped bass, the Hudson fishermen, and Robert H. Boyle.

Nearly forty years later, Boyle is having coffee at a little table in his colonial farmhouse near Cooperstown, New York. "Fish Stories Told Here" reads a plaque above his kitchen door. He is recalling how he spent days trying to track down the two marine biologists whose work on where striped bass spawned directly contradicted the testimony of Con Ed's witness. Finally reaching one of them by phone in Alaska, Boyle brought up Perlmutter's testimony. As it turned out, none other than Perlmutter—employed in the mid-1950s by New York's Department of Conservation—had hired Rathjen and Miller to conduct the striped bass study that he'd so conveniently forgotten about.

Boyle couldn't help but smile, remembering Storm King's "smoking gun," then exclaimed: "The bullshit being pulled at the time was unbelievable!" Now in his mid-seventies, Boyle retains the bulldog demeanor that characterized him as a writer-activist during his thirty-two-year career at *Sports Illustrated*. For a brief time, we had been colleagues at the magazine, though Boyle was already a grand eminence and I but a rookie reporter who had yet to learn about either striped bass or Storm King Mountain. Only later, during my own fight to save the fish, did we strike up a long-distance relationship.

Boyle had grown up in New York City, but spent summers in New England. "I guess I was four years old when I first went fishing," Boyle says. "I loved to be in brooks and streams turning over rocks and seeing what was underneath. When I came back from the country in the summer, my mother would meet me at Grand Central and I'd have cans filled with salamanders, turtles, and aquatic insects."

After serving a stint in the Marine Corps and earning a master's degree in history and international relations from Yale, Boyle went to work for United Press and, after selling a story to *Sports Illustrated,* joined the magazine in 1954, when it was only four issues old. He spent several years as a correspondent in Chicago and San Francisco, then returned to the East in 1960, bought a house in Croton overlooking the Hudson Valley, and "started fishing again. And that's when I ran into striped bass."

What fascinated Boyle was the fish's "absolute unpredictability." You never knew when they might strike, or what would interest them. "They move around to different environments, they go from fresh water to salt water, they'll get onto a particular feed or they won't, who knows? The other thing is—I've seen it on tagging returns—somebody will catch and release a fish, and then find the same fish in the same place a year later! Do the fish live there *permanently?*"

Boyle's round face fills with wonder. He begins to recall fondly the fishermen and scientists he came to know, who helped him collect specimens for museums and taught him everything they knew about the river. Then, soberly, like a lawyer building his case, he recounts what two of these men told him in 1964. Art Glowka was a pilot for Eastern Airlines who'd written about striper fishing for *Outdoor Life,* and Dominick Pirone was a consulting biologist for the Long Island League of Salt Water Sportsmen. They'd been hearing rumors about huge fish kills of mature striped bass at an existing Con Ed facility fifteen miles south of Storm King and decided to investigate for themselves. Indian Point's daily withdrawal of river water by the millions of gallons, to cool its nuclear reactors, offered a portent of what might occur elsewhere.

On June 12, 1963, Glowka, Pirone, and two others surprised the utility by rowing a boat right up to the reactor site. They were horrified. "We saw 10,000 dead and dying fish under the dock," Pirone told Boyle. "We learned that Con Ed had two trucks hauling dead fish to the dump when the plant was in operation." Subsequently, Pirone saw photographs of the dumping ground, showing flocks of crows feasting on heaps of rotting bass piled twelve feet high, covering an area bigger than a city lot.

Boyle later wrote: "Apparently the fish were attracted by hot water discharged from the plant after it had been used to cool the condenser. Swimming in to investigate the hot water—many fishes have a knack for discovering such currents, especially in cold water—they were then trapped underneath a pier which had sheathing partly down the sides to keep debris from the nearby intake pipe. . . . Some swam inside the bowels of the plant to meet their death; others milled around hopelessly, crowding all the closer as new recruits swam in under the boarding. Many apparently became enervated and diseased. This went on for at least six months, and Con Ed operated a huge wire basket elevator to remove the dead and dying."

The utility proceeded to erect a guarded fence around the dump. And by the time Boyle went looking for the photographic evidence, it had "disappeared" into state files. But the furor over the Indian Point kills—eventually estimated at a staggering seven million fish—did not abate. When state legislative hearings on Storm King convened, Boyle testified about the striper slaughter and predicted that Con Ed's new project, twice the size of Indian Point, would kill millions more. He also persuaded John Clark, a biologist from the U.S. Fish and Wildlife Service, to come forward. Perlmutter's calculations were in error, Clark testified; the waters around Storm King remained a vital striped bass spawning area, just as the Rathjen-Miller study had indicated.

Early in 1965, despite a growing chorus of protest, the Federal Power Commission issued Con Ed a license to proceed anyway. Around the same time, Boyle's detective work led him to a former deputy game warden who'd allegedly taken the incriminating pictures of the bass fish kill at Indian Point. Boyle still delights in the story of what ensued. The fellow, George Yellot, was now a bank officer in Peekskill. When Boyle confronted him, Yellot lifted his shoulders and explained that a higher-up in the Conservation Department had ordered him to turn over the photos. "So I gave him the pictures," Yellot said, adding that the same official had returned two weeks later asking for any duplicates. "So I gave him the duplicates," Yellot said.

"Then I watched Yellot reach into a bottom drawer of his desk," Boyle remembers. "He was smiling as he produced this sheaf of Polaroid photos and handed them across to me. I'll never forget George Yellot saying: 'I knew a reporter would come along some day. You see, they never asked me about triplicates.'"

Boyle's exposé, "A Stink of Dead Stripers," which included one of the gut-wrenching photographs, appeared in the April 26, 1965, issue of *Sports Illustrated*. The article concluded: "Perhaps, ironically, the Hudson River, the living river, may yet be saved by dead fish long thought buried in an obscure dump and by pictures buried in Conservation Department files." The state's Conservation Department was not pleased. It even checked employee phone bills trying to find out, as one official is said to have put it, "Who squealed to Boyle?"

That July of 1965, Scenic Hudson filed suit in the Federal Circuit Court of Appeals in New York seeking to have the FPC's licensing decision on Storm King reversed. Its attorney was Lloyd Garrison, a descendant of the famed nineteenth-century abolitionist William Lloyd Garrison. The FPC, not wanting to open the floodgates to future Scenic Hudsons, argued that the citizen's group had no standing to sue because, under existing jurisprudence, in order to bring a federal court action the plaintiffs needed to prove that they would suffer tangible economic harm. Garrison countered that the FPC failed to develop a full and complete record. Under the Federal Power Act passed by Congress in 1920, he argued, all factors in the public interest needed to be considered—specifically, the beauty and historical significance of Storm King Mountain and what the power plant would do to the striped bass.

On December 29, 1965, the Court of Appeals handed down a landmark ruling. It held that citizens no longer needed an economic interest to be granted "standing" in such cases, entitling Scenic Hudson to judicial review of the FPC's ruling. The court reversed the licensing decision and ordered the agency to hold new hearings. The judge criticized the commission for "seemingly placing great reliance on the testimony of Dr. Perlmutter" (whom the FPC had hailed as an "outstanding ichthyologist") and instructed it "to consider the fisheries question before deciding whether Storm King is to be licensed."

So, with the striped bass as a centerpiece, what became known as the Storm King doctrine would establish the right of citizens' groups to sue a government agency in order to protect natural resources and scenic beauty—the precedent for "environmental standing" in court cases ever since. The decision also mandated, for the first time, that a federal agency undertake a complete environmental review of a project. Thus the stage was set for Congress to pass one of the country's most important pieces of environmental legislation, the National Environmental Policy Act (NEPA) of 1969, which required all government agencies to consider the full environmental impacts—including social costs and potential alternatives—of proposed projects before any major decisions were made.

The battle of Storm King was still far from over, but Boyle had also found time to introduce the concept of environmental reporting to Time, Inc., the parent company of *Sports Illustrated*. His pieces for the magazine focused not just on the Con Ed dispute, but also on the appalling condition of the Hudson. "The abuse I saw was intolerable," Boyle recalls. "It had been written off as an 'industrial river.'" Raw sewage had turned much of the Hudson's 315-mile-long stretch into one big septic tank. Old tires floated above debris that included discarded automobiles, a grand piano, and even a dead giraffe!

In researching an article, Boyle had come across two long-forgotten laws: the New York Rivers and Harbors Act of 1888, and the Federal Refuse Act of 1899. "Basically, both those laws said that no one should put anything into any navigable waters other than what flows through a city sewer in a liquid state." Both statutes also had a clause, Boyle noticed, that granted anyone who reported a violator the right to half of any fines that were assessed. He wrote to Dominick Pirone, who'd helped blow the whistle on the Indian Point fish kills. "I said, 'We've

got to attack these problems, and bring science and law to the river,'"
he remembered. That was when Boyle formed a group called the
Hudson River Fishermen's Association.

They decided to ferret out the polluters, one at a time. "The most
flagrant one was the New York Central Railroad, which had been
dumping oil into the Hudson since 1929. It took us three years before
we could get the U.S. Attorney to take action, and the railroad wound
up being fined $4,000. We collected $2,000, which we used to fight
Con Ed at Storm King and Indian Point. We also printed up 10,000
copies of the Refuse Act on 'Bag a Polluter' postcards. People could fill
these in with polluters' names and mail them prepaid to the
Fishermen's Association office."

As the organization developed successful actions against other
companies, its bounties increased, and membership over its first three
years grew to some 300 sports and commercial fishermen, bait-shop
owners, and marine scientists. The Fishermen's Association also joined
with Scenic Hudson to continue the fight to stop Con Edison at Storm
King. Once again, it would be the striped bass that turned the tide.

The court-ordered new round of FPC hearings dragged into 1969.
Con Ed's scientists still contended that the Storm King plant would
have minimal impact on the stripers, resulting in mortalities of less
than 3 percent as the eggs and larvae floated by the intake pipes. This
was enough to convince the FPC commissioners, who voted 2-1 to
approve construction, a decision upheld in 1971 by the Court of
Appeals.

"It looked like we were finished," Fishermen's Association attorney
Al Butzel would recall. Butzel was nevertheless able to persuade the
state's Department of Environmental Conservation to open new hearings for public comment. The Clean Water Act, passed by Congress in
1972, required any federally licensed facility that planned to discharge
into a water body to first obtain certification from the state. As it

turned out, there were some major scientific problems with Con Ed's calculations about the effect on striped bass.

Boyle had had Con Ed's testimony analyzed by John Clark, the scientist who'd testified about the striper spawning ground below Storm King. Clark quickly realized that the utility's study was fundamentally flawed. Its low mortality estimate for the bass was based on the faulty assumption that the fish's eggs and larvae would go by the intakes only one time. "They had treated the Hudson like a one-way river flowing downstream," as Butzel put it, "whereas in fact it was tidal at Storm King." Boyle recalls: "The text accompanying the mathematical equation mentioned the tide, but the equation *omitted* the tide. It was a real con job!" Every little fish, on average, would float past the plant not once but *ten times* before finally making it to safe haven in the lower river, Clark suggested, increasing the likely mortality to approximately 35 percent.

When Clark couldn't make it to the hearing, Butzel put Boyle on the witness stand before the Honorable Emmanuel Bund. "While Con Ed's lawyers and pseudo-scientists smirked and laughed," says Boyle, "he dismissed as hearsay my testimony that the Hudson was tidal at Storm King." Told he had no expertise, Boyle was ordered to step down. Al Butzel remembers "going to Bund's place in Riverdale afterwards to deliver him our brief, and he basically said to me, 'You know, I can only decide this one way.' He had been given his marching orders from higher up."

By this time, the Storm King case had captured attention far beyond the Hudson River. It had been a time of rising discontent with the way the nation's environment was being transformed, indeed trampled upon, by powerful interests—a discontent that culminated in the first Earth Day, in 1970. Thousands of people from across America had sent contributions to Scenic Hudson; some of Con Ed's stockholders even signed over their dividend checks. In 1969, a Wall Street attorney named Stephen Duggan, whose wife was active in Scenic Hudson, had come up with the idea of forming a nonprofit environmental law organization to help fight the power plant. That marked the beginning of the Natural Resources Defense Council (NRDC), which grew to become one of the leading environmental law firms in America. The NRDC's first client was the Fishermen's Association.

While the Fishermen's Association was appealing Bund's decision on Storm King, the fish kill situation at Indian Point came to the fore again. To emphasize the problem, Boyle plopped a twenty-five-pound striped bass onto a congressman's desk during a hearing in New York City. With a second reactor unit up for licensing by the Atomic Energy Commission (AEC), Boyle and the Hudson River Fishermen's Association succeeded in having the potential impact on the striped bass included as a consideration in the decision. If Indian Point II withdrew cooling water from the river as it planned, Clark testified before joint hearings of the AEC and the recently created Environmental Protection Agency (EPA), there would be massive deaths of young stripers. The AEC therefore made the licensing conditional. Con Ed would first need to install an expensive closed-cycle cooling system. This would cost the utility an estimated $240 million, but it would reduce its usage of river water—and its fish kills—by some 95 percent.

That decision's consideration of the welfare of the striped bass population meant the handwriting was on the wall for Storm King as well—especially after May 1974, when the Second Court of Appeals ordered the FPC to conduct additional hearings on the fishery issue. By then, Con Ed was also facing a dilemma, prompted by the Clean Water Act, as to where they could legally landfill their excavations of the mountain. Meanwhile, two young scientists, Phil Goodyear and John Boreman, who had been enlisted by Boyle's compatriots, began looking at cumulative impacts on striped bass from the various power plants. These, they concluded, were already considerable; Indian Point I alone was estimated to have killed another 1.1 million fish in 1971.

Boreman, who today runs the Northeast Fisheries Science Center at Woods Hole, Massachusetts, recalls of Boyle: "He has a very inquisitive mind, reads a lot, understands a lot. Because he loves the river, the fish, and fishing, he would do anything to protect the resource. Bob always had the knack of asking the questions that should be asked, the ones that we as scientists should know the answers to and don't."

All of the testimony impressed Administrative Law Judge Thomas B. Yost, who spent three years taking evidence in the EPA hearings called to hear an appeal by Con Ed of an EPA order requiring cooling towers but, under the bizarre rules of practice in existence at the time, had never been authorized to render a decision. At one point, the judge summarized the Con Ed attorneys' "recipe" for court success: "Combine 1 large grain of salt with 2 cups of self-interest. Stir vigorously with tame scientist until slightly thick. Adjust seasoning to taste."

On the witness stand, the scientists who'd done Con Ed's study of Storm King were eventually forced to concede that, having corrected for the error on tidal fluctuations, at least 35 percent of the striper eggs and larvae could be killed. At that point, Con Ed asked to sit down and negotiate with Boyle's Fishermen's Association, Scenic Hudson, and the NRDC. Russell Train, a former EPA administrator, was brought in to mediate.

After a seventeen-year fight and twenty months of negotiations, the Hudson River Peace Treaty was announced in December 1980. There would be no power plant at Storm King; instead, the land would become a state park. At Indian Point, Con Ed agreed to protect the striped bass by installing devices to prevent the bigger fish from being trapped on the intake screens and also to shut down the plant during the spawning season so that eggs and larvae wouldn't get sucked into its maw. These steps would spare the utility the huge cost of building cooling towers, at least for the time being. And, through Boyle's impetus, Con Edison agreed to endow a new Hudson River Foundation with an initial outlay of $12 million. The foundation would study the river's fisheries resources and advise fishery managers based on its findings.

Years later, attorney Butzel would say: "I can't tell you how many times I thought this case was lost. At best, it was a stalemate. In fact, the fish saved Storm King. And striped bass *were* the fish."

Not long after the Peace Treaty was signed in 1980, the fight to pre-serve the Hudson molted into a new organization called Riverkeeper, yet another brainchild of Boyle's. In his book *The Hudson River*, Boyle had foreseen enlisting someone to be "out on the river the length of the year, nailing polluters on the spot"—something he felt was needed "on every major river in the country." John Cronin was hired as the first Hudson Riverkeeper in 1983, and Robert F. Kennedy Jr. became the group's chief prosecuting attorney in 1984. Today the organization has a staff of more than twenty, and it has inspired at least 130 similar pro-grams across the United States and internationally.

On the Hudson, Riverkeeper continues to fight against the impact of power plants on striped bass and other species—because the Peace Treaty, it turned out, did not stop the fish kills. In return for not build-ing Storm King, all through the 1980s Con Edison had been allowed to continue operating at Indian Point under its old permit. "The New York Department of Environmental Conservation [DEC] was in bed with the power companies," Boyle maintains. "Once they sent out a notice about fish kills: 'After you read this, destroy it so Riverkeeper doesn't get it.'"

Finally, in the early 1990s, the DEC agreed to assess whether Indian Point was meeting requirements for using the best available technology. However, the power plant was allowed to do its own administrative review. After lengthy and fruitless negotiations over how to mitigate fish kills without costing the power company too much money, Riverkeeper initiated a series of lawsuits. The latest round of legal action charged the DEC with violating the Clean Water Act by failing to force the state's power companies to obtain new permits. Meanwhile, Indian Point continued to draw up to 2.5 billion gallons of water a day from the Hudson in order to cool its two nuclear reac-tors—despite a study showing that Indian Point, between 1981 and 1987, caused the yearly destruction of more than 1.2 billion eggs, lar-vae, or hatched fish of several aquatic species. When you factor in three more power plants within that same thirty-mile radius, the water with-drawal doubles to up to five billion gallons of water daily and the mor-tality estimate stands at more than two billion fish. It's been calculated that these five facilities continue to kill approximately 40 percent of the Hudson's young striped bass annually, when the eggs, larvae, and juve-niles get sucked into the cooling systems. And today the Hudson is believed to provide perhaps as much as 25 percent of the coastal migra-tory stock of striped bass.

Pressed by the Riverkeeper lawsuits, on November 12, 2003, the New York DEC at long last ordered the Indian Point plants either to replace the older technology (known as once-through cooling) with closed-cycle cooling systems—thus dramatically reducing water with-drawals and decreasing fish kills by as much as 95 percent—or to shut down operations when their licenses expire in ten years. The DEC also ordered forty-two-day shutdowns and reductions in water usage dur-ing spawning seasons to minimize the harmful effects on striped bass and other fish. This, of course, is something the power plants had promised to do in the settlement agreement more than twenty years before—but perpetually delayed implementing. Then, in February 2004, a suit initiated by Riverkeeper brought a U.S. Court of Appeals ruling that upheld the EPA's regulation mandating closed-cycle cool-ing as the national minimum technology for new power plants and factories. Alex Matthiessen, the son of nature writer Peter Matthiessen and the current executive director of Hudson Riverkeeper, says he is "hopeful Riverkeeper will finally prevail against the wanton slaughter that's been happening for more than forty years."

The vision conceived by Robert Boyle during the Storm King battle—of a community of people fighting to protect their local water-ways from pollution and harmful development—has come to fruition with the Riverkeeper movement. Besides investigating and successfully prosecuting more than 300 environmental lawbreakers, the Hudson Riverkeeper group that Boyle chaired for two decades is credited with continuing the river's restoration and with forging a landmark agreement that protects New York City's drinking water. Often called the "conscience of the [Hudson] river," Boyle must too be regarded as a pioneering force behind protecting the habitat of the striped bass—and in defending America's rivers, lakes, and streams.

It was also Boyle—and the striped bass—that brought the problem of PCB contamination into the nation's consciousness. Oily polychlorinated biphenyls, or PCBs, are among the most stable chemical compounds known. They came into industrial use in the 1930s in a wide range of products. Because of their heat-resistant properties, the 200-plus related compounds were used extensively as insulating agents in electrical equipment such as transformers and power capacitors; as plasticizers; in adhesives for tape and envelopes; and much more. PCBs are also long-lived, don't dissolve in water, and easily bioaccumulate in animal and human tissues. (Bioaccumulation is commonly defined as the measure of uptake over time of toxic substances that can stay in a biological system.)

The first detection of PCBs in the environment came in 1966, when a Swedish chemist investigating DDT and other chlorinated pesticides in human fat and in wildlife stumbled upon them. Subsequent research by Robert Risebrough of the University of California at Berkeley on the harmful effects of PCBs led Monsanto Chemical—since 1930 the major manufacturer of PCBs—to issue a denial that these were "highly toxic" chemicals. Risebrough responded early in 1970: "The possible PCB hazard, like so many environmental hazards, is one of long-term, low-level exposure and perhaps of effects from its combination with other poisons." Today we know that PCBs can not only cause cancer but also disrupt human endocrine systems during fetal development. Studies conducted in recent years, in both the United States and the Netherlands, have established that children who are exposed in the womb can suffer impaired intelligence as well as a greater susceptibility to infectious diseases.

That spring of 1970, John Clark—then assistant director of the U.S. Marine Gamefish Laboratory at Sandy Hook, New Jersey, and

Boyle's key witness in the Storm King hearings—approached Boyle at a scientific meeting. Clark had learned from a biologist with the Michigan Department of Natural Resources that DDT had been detected in coho and Chinook salmon in Lake Michigan. Despite Rachel Carson's landmark 1962 book on the dangers of DDT contamination, *Silent Spring*, this marked the first time the pesticide was known to bioaccumulate in freshwater fish. Clark suggested to Boyle that perhaps *Sports Illustrated* would fund a study looking at coastal gamefish as well, because "it would take the government several years to get off its ass."

Boyle got the magazine to enlist its correspondents on the Atlantic, Gulf, and Pacific coasts to gather fish samples and ship these to a laboratory at the Wisconsin Alumni Research Foundation (WARF) Institute. This was done under strict protocols; the fish had to be wrapped in aluminum foil and dry ice, and air-expressed. While the sports editor of the *San Francisco Chronicle* arranged to capture striped bass from the Bay-Delta region, Boyle—who already had a license from New York State to collect stripers for the American Museum of Natural History—recalled seining on the Hudson with a fisherman named Charley White. "I think he'd have been lost in the daytime, but at night Charley knew right where to go. We headed for a cove over by the FDR Veterans Hospital. Charley said, 'The bass'll be in here right now, the moon's in the right phase.' We put out a net and caught so many bass that they broke right through it. I'm talking about twenty-, thirty-, and forty-pound fish."

About a month later, Boyle happened to read Risebrough's article describing the discovery of polychlorinated biphenyls in birds on the West Coast. Boyle had never heard of PCBs before, but he persuaded the WARF Institute to check for those compounds in addition to DDT and other better-known chemicals. The institute found PCBs in the flesh of every coastal fish species tested, with particularly high levels in striped bass from the Hudson and from California. The Hudson stripers contained an average of 11.4 parts per million (ppm) of PCBs in their eggs and 4.01 parts per million in the flesh. Boyle published these findings in October 1970, in a pathbreaking *Sports Illustrated* article headlined "Poison Roams Our Coastal Seas."

"The state of New York secretly began testing after my article, but they never admitted it. The cover-up went on for five years," says Boyle. It would later be revealed that the DEC had indeed started specifically analyzing fish for PCBs in 1972, finding high levels in largemouth bass, northern pike, sturgeon, white perch—and striped bass. The following year, of twenty-two striped bass collected from the Hudson near the Tappan Zee Bridge, eighteen were found to have excessive residues of PCBs, as high as nearly fifty parts per million. Four out of six stripers sampled in 1975 also showed high PCB levels. None of this, however, was revealed to the public.

"I had no idea where the PCBs were coming from, none at all," Boyle recalls. "I thought probably some nut in Jersey City. I had a list of every polluter of the Hudson River. General Electric's manufacturing capacitor plants, in the foothills of the Adirondacks at the head of the canalized Hudson about 180 miles north of Manhattan, were discharging something they called Pyranol. Well, it turned out that this was really Aroclor—which was Monsanto's trade name for PCBs. Once GE got it, they would rebaptize it and call it Pyranol. It was like a shell game."

Between 1940 and 1976, those two old GE factories discharged about 1.3 million pounds of PCBs into the Hudson. The story finally broke into the headlines in August 1975, after a new DEC commissioner named Ogden Reid was stunned to read the report of a federal EPA investigation that "revealed the largest source of PCBs into the aquatic environment" to be GE, in its pollution of the Hudson. A front-page article in the *New York Times* had Reid warning the public against eating striped bass from the Hudson because, in samples taken from their flesh, fully 90 percent contained levels of this potentially carcinogenic compound that were above the Food and Drug Administration's standard, which had been set at five parts per million. Boyle wrote another piece for *Sports Illustrated* headlined "The Spreading Menace of PCB."

In 1976, the same year that the U.S. Congress outlawed the manufacture, sale, and distribution of PCBs, commercial fishing for stripers in the Hudson was banned by New York officials. As it turned out, the no-sale edict would prove a blessing in disguise for the species's survival. But that is getting ahead of our story.

In the early 1970s, Boyle decided to start a new, Atlantic coast–wide conservation group devoted specifically to protecting the striped bass. He called it the Striped Bass Fund. The man Boyle asked to run it was a Long Islander named Al Reinfelder. "He used a fishing rod the way an orchestra conductor would use a baton," Boyle recalled. "If I had to name the five purest souls I've known in my life, Al would be one of them." Reinfelder was also a fine writer who penned articles for fishing publications on pollution, power plants, water diversion, and unregulated netting—"rousing the troops," as Boyle put it. "We had the damnedest people show up at our meetings," Boyle said—including actors Richard Boone and William Conrad, who turned out to be striper fishermen. From Martha's Vineyard, where Reinfelder spent his summers, to the Jersey shore, citizens were being enlisted into the striper cause.

On May 14, 1973, in a letter to readers of the *Long Island Fisherman*, Reinfelder wrote: "The task of uniting fishermen from Maine to the Carolinas on the East Coast . . . is considerable. And the enemies of the fishery we face are many and powerful." Tragically, four days later, on a float trip down the Delaware River that he had fished since his youth, Reinfelder's canoe capsized and he was not able to make it to shore. He was only thirty-five.

Without Al Reinfelder's charismatic leadership, the Striped Bass Fund foundered. Boyle remembers that "guys on Long Island and people in the Chesapeake said, 'You just want to rope us in to fight *your* battles.' I told them, 'You're going to have your own battles—this is not stopping here. I don't know what's coming down the line, but there are all sorts of threats going to hit you guys.'"

Over the ensuing decade, Boyle's words were to prove prophetic. For me, the battle to save the striper would begin in a place called Buzzards Bay.

Chapter Three

*The Conscience
of a Lure Maker*

Ｔhe early Massachusetts colonists
named Buzzards Bay after the numerous "fish buzzards" (actually
ospreys) that fed and nested along the shoreline. The Massachusetts
Maritime Academy located today on its shore is about an hour's drive
from Boston. And it was here, on a cool evening in mid-March 1982,
where some friends and I showed up in two carloads for a public hearing on new proposed regulations for the striped bass.

Since witnessing the awful "chunker fishing" scene that wiped out
an entire school of big fish six months earlier off Martha's Vineyard, I'd
begun doing some homework on what was happening to the striper
population along the Atlantic coast. Over the course of the past decade,
the fish's numbers had been plummeting. Coastwide, commercial
striper landings had fallen from nearly 15 million pounds in 1973 to
3.4 million pounds in 1979. The young-of-the-year survey of reproductive success in the Chesapeake Bay had gone into a tailspin—from
a record-breaking average of more than 30 juvenile striped bass found

every time marine biologists hauled in the sampling net in 1970 to an all-time low average of a mere 1.2 fish in 1981.

I'd never been much of an activist, nor did I expect to become one now. It simply seemed important to support Massachusetts's strengthening of a minimum size limit to protect the younger fish until they'd had an opportunity to spawn. That, however, didn't appear to be a popular view in the auditorium at Buzzards Bay. The moment we walked in, I sensed that we were greatly outnumbered. Several hundred commercial fishermen in T-shirts and caps had packed the place. Their livelihood would be affected by any tighter restrictions on their catch. My friends and I sat down in the back of the room and waited to see what might transpire.

I reviewed my notes on what the experts were saying—and doing—about the plight of the striped bass. In 1978, Ben Florence, then chief of the tidewater-finfish section of Maryland's Department of Natural Resources, had closed the entire Chesapeake Bay system to bass fishing during the striper's two-month-long spring spawning season. "Official Warns on Demise of Striped Bass in 5 Years" was the headline of a *New York Times* column about Florence's speech to a Long Island group called Save Our Stripers. "People are looking for magic answers and there aren't any," Florence said. "The only answer is that there are too many striped bass being fished." But the closure was vehemently attacked by Maryland's powerful commercial watermen's lobby. The next year, the springtime fishing ban was lifted, and Florence was shunted off to a less controversial fisheries job.

In 1979, Rhode Island Senator John Chafee read *Striper*, written by his former Yale roommate, John Cole. The book's warning of impending doom for the bass galvanized Chafee to push a bill through Congress establishing a three-year, $4.7 million Emergency Striped Bass Study. "It's a great sportfish, like a few special animals in Africa," Chafee said. "It would be a pity if this great fighting fish were lost."

The Atlantic States Marine Fisheries Commission (ASMFC) was already charged with drawing up a new striped bass management plan that coastal states from Maine to North Carolina were supposed to implement. The trouble was that while some states had scarcely any limits at all, among the remaining states there were already six different minimum size limits for fish caught and two different maximum size limits. The conflicting regulations, designed to encompass the varying types of gear used to take the fish, made the management of striped bass more complex than that of any other inshore fish species.

The situation was only exacerbated by the fact that, as the supply of stripers dwindled, the market price had soared, and commercial fishermen were in some cases doubling their money. In 1970, the value of the commercial catch was $2,528,000. By 1980, this figure had risen to $4,901,000, even though landings had declined drastically. Where a fisherman might get as little as a few cents a pound for his bass catch during the boom years, in 1981 he could earn as much as $3.65 a pound at dockside. For the consumer, striped bass had become a luxury item—up from a dollar a pound in supermarkets in 1978 to a standard $6 a pound in many locales.

In the past the striper fishery had been replenished by periodic bountiful hatches, or "dominant year classes," of young fish, as mentioned earlier. These had occurred in 1956, 1958, 1961, 1964, 1966, and 1970. But there hadn't been a dominant year class in the Chesapeake for the past dozen years, while commercial and recreational fishermen continued to catch between 40 and 60 percent of the migrating stripers every year (in addition to natural mortality estimated at 10 percent annually). Most of those fish were the large female breeders. At the same time, Chesapeake watermen were snapping up many of the males that didn't customarily migrate.

The only even close-to-average recent year class had been in 1978, and most of these females had now joined the coastal migration and so were extremely vulnerable to fishing pressure. They wouldn't be returning to spawn until they'd reached the age of five or six and achieved a size of approximately twenty-four inches. If the 1978 class wasn't protected, it could spell doom for the Chesapeake stripers. To meet this threat, the ASMFC's proposed management regime called for an increase in the minimum size limit for striped bass caught along the coast—from sixteen inches to twenty-four inches. Massachusetts, always in the forefront of striper conservation efforts, had in fact already promulgated that regulation in 1981, six months before the ASMFC made this standard a formal recommendation.

The trouble was, Massachusetts fisheries officials had allowed a loophole. Its rod-and-reel fishermen had been allowed to catch, and keep or sell, up to four fish a day between sixteen and twenty-four inches long. Some fishermen, unhappy about the stronger sanctions to begin with, simply continued to market undersized fish. They claimed that a friend had caught them or that they had caught four bass before midnight and four bass after midnight.

To close the loophole, in 1982 Massachusetts proposed making it illegal to hang onto *any* striped bass smaller than twenty-four inches. That was why all these commercial fishermen were out in force in Buzzards Bay. Their seasonal income would be reduced. I jotted down some of their comments as they berated the managers. "Look, the striped bass have gone through natural cycles before; this decline is just another one. . . . The small fish are the best eating! . . . We're rod-and-reel only; why aren't the rest of the states? . . . I'm just releasing my small fish to wind up in somebody else's nets."

They had a good point, at least judged by what the ASMFC had put forward for the Chesapeake region. There, the commercial fishery used primarily nets to catch the available fish—smaller ones that hadn't yet migrated out of the rivers—so a twenty-four-inch minimum size limit would put most watermen out of the bass business. The ASMFC's interstate management plan proposed a fourteen-inch minimum inside the bay and its tributaries, coupled with a seasonal ban on bass fishing during spawning time. This was intended to strike a compromise that would permit both Chesapeake and coastal fishermen to continue catching a fair share of the resource, while taking some of the pressure off the species.

A new report I'd read by a Maryland fisheries biologist named Joe Boone, however, indicated that a change in the size limit probably wasn't going to be enough to maintain the fish population. "If a lesson has been learned over the past five years," Boone wrote, "it is that striped bass are a valuable and irreplaceable natural resource of limited quantity, which must be carefully husbanded from now on. The colonial concept of relatively open and unhampered fish harvest is over. This magnificent fish can no longer cope with modern demands. Salvaging some semblance of traditional striped bass fisheries necessitates a pragmatic interstate management effort responsive to the needs of the species rather than the desires of man."

I planned to read Boone's eloquent comments aloud when my turn came to speak on behalf of those who favored the no-exception, twenty-four-inch minimum size limit. It took more than an hour for the commercial fishermen to finish airing their grievances. Then, looking ridiculously out of place in a three-piece suit—and making even more of a spectacle of myself by walking up to the microphone instead of standing up where I was—I became the first of the minority to talk.

I should have expected the acrimonious reception I got, but somehow it took me by surprise. When I brought up the subject of overfishing, they yelled, "It's all pollution!" When I said we had to do something for the fish's sake, they hollered, "It's not our fault! It's those net fishermen in Rhode Island and Maryland!" In anger and frustration I found myself shouting back: "I didn't interrupt when you talked! Why don't you shut up and let me finish?" They booed and laughed and hooted. I'd rarely felt so alone.

Then, from the back of the auditorium, a fellow I'd never seen before rose to my defense. He stuttered when he talked, but the feeling behind his words was powerful enough to silence the crowd. The striped bass desperately needed more protection, he said, and everyone should at least have the courtesy to listen to that viewpoint.

More than two years would pass before I saw the man again. One evening off Martha's Vineyard, our boat was coming home from a fishing trip about twenty miles offshore. As it passed No Man's Island, a desolate figure loomed up from the beach. His boat had crashed into

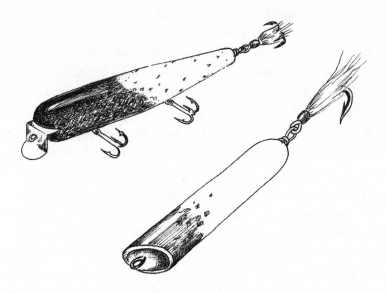

the rocks the night before, in a fog so thick you couldn't see more than a foot away. For twenty-four hours he'd been marooned, trying in vain to contact the Coast Guard. Badly shaken, he wept as he came on board. This turned out to be the very man who had added his voice to mine that lonely night in Buzzards Bay. Fate works in mysterious ways. Dr. Alan Cordts, a physician, would become another ally in the fight.

Two other men in the audience that night at the Maritime Academy would come to play significant roles in the striper wars. One was John Boreman, who'd first become involved in striped bass research during the Storm King controversy. Now he was a chief scientist for the congressionally mandated Emergency Striped Bass Study. He was here strictly to observe the proceedings. But he seemed badly shaken by all the confrontation. When we were introduced briefly after the public hearing, he said something about never wanting to be in a room as volatile as this again. At the time, I doubted Boreman would be of much help.

The other man also received his share of catcalls at Buzzards Bay: Bob Pond. A tall, handsome man in his mid-sixties with chiseled features and thick, wavy white hair, he took the microphone and began,

"I'm the man who's probably responsible for killing more stripers than anyone else alive." It was a statement to make everyone take notice, and I listened intently as Pond spoke of having invented the Atom plug soon after World War II. For many years, this had been the artificial lure most attractive to striped bass. Its creation had, Pond admitted, made him a wealthy man.

Then, in 1965, in the midst of apparent plenty, he'd been the first to sound the alarm that the striped bass resource was in potential trouble. He'd formed the initial organization strictly devoted to conserving the fish, Stripers Unlimited. Now, staring intently at a largely disdainful hall of fishermen, Pond cited overfishing, pollution, and habitat destruction as the three reasons for the striper's impending demise.

"We have lost in my lifetime the St. John's River population, because they built a dam above it in Nova Scotia," Pond intoned in a resonant baritone. "The Delaware River population is completely gone; you can't find an egg. The Southern one has vanished—killed off by farm chemicals, except for one area."

The man seemed to be a walking history of striped bass. After the 1958 dominant year class that made him rich, the rivers were full of bass, he continued. Wintering off New Jersey, schools of young bass had stretched fifty miles long and thirty miles wide. Then, over the next few years, a million pounds were wiped out in gillnets and by powerful cherry bomb firecrackers thrown into the water—what Pond described as "a money crop worth a fortune taken by a few opportunists." When the remaining breeders returned to the Chesapeake to spawn in 1964, net fishermen stacked them helter-skelter on the docks, without enough trucks to cart them to market or ice available to keep them from spoiling.

That was why Pond had felt compelled to start Stripers Unlimited. That was why he was here tonight, warning the fishermen that unless they bit the bullet, the striped bass would soon be history.

When the commercial fishermen called for a vote that night on the twenty-four-inch limit, our side lost overwhelmingly. But I knew that Bob Pond was someone I had to see again.

In South Attleboro, Massachusetts, we were sitting together inside an old ice house that Pond had turned into a three-story tackle manufacturing shop. Above the molding room, the assembly area on the top floor had formerly been a jewelry factory and, before that, housed the Silver Moon Dance Hall. Now fishing rods, reels, line, lures, and stacks of scientific papers surrounded us. Every dollar of profit Pond made was devoted to saving the very fish that his pioneering Atom plug had been designed to catch. "We still turn out about 65,000 lures a year," he said. "Mostly they're for bluefish now."

Pond smiled as he began recalling the mysterious source of what had become his life's crusade. Born July 4, 1917, the son of an architect who designed homes around suburban Larchmont, New York, he had spent his boyhood fishing the waters off Long Island. He'd attended the Syracuse University School of Forestry and spent the first years of World War II as an inspector of ammunition for the British government. Schooled in critical observation, he found he had a knack for solving knotty problems.

By the summer of 1944, at twenty-seven, Pond was in New England looking for work, and it was the Cape Cod Canal region in particular that seemed to beckon him. "The bass were everywhere," he remembered, "though not easy to catch. In those days, nobody'd even thought of using a wooden lure. People figured you could only catch bass at night, bouncing the bottom with a bucktail jig. One afternoon I was sitting on an embankment under the Bourne Bridge, watching these big fish playing and thrashing around. Then all of a sudden, one came rolling and tumbling to the surface. I saw the tip of a rod emerge from behind a catwalk—and there was an old man fighting this bass on a tall surf rod! I was transfixed with excitement. After the old man brought the bass ashore, I ran down to find out what he was using. But when he saw me coming, he threw a big Turkish towel over the fish.

You don't ask a guy a question when it looks like he isn't going to answer it.

"But I was obsessed with what I'd seen. It just didn't happen! A few weeks went by. One day I was standing down on the big breakwater, almost in the same location, still thinking about it. I looked down. And there in the water was a handmade piece of colored painted wood circling around. I picked it up. It was a battered old Creek Chub type of plug, the kind people used to catch giant pike in the Great Lakes. But never striped bass. Well, I tied it to the end of my line, started casting—and caught fourteen bass!"

Pond's eyes sparkled as he recalled tossing one fish up onto the bank and trying to stop it from falling back into the water, while fighting yet another fish. "I knew it was the old man's lure. Later I found out who he was and tracked him down. I said, 'I've got something of yours.' 'What you got?' I said, 'I got a plug.' He said, 'Give it to me.' I said, 'No way.' I've still got it hanging in my shop, right over there."

He pointed with a forefinger at a nondescript-looking piece of weathered wood, then went on: "Anyway, that's what gave me the pattern. I worked away on it from September all through that winter. I enlarged it big enough to use on heavier surf gear, made 400 of 'em, and started selling 'em. Some I delivered to Red Top Sporting Goods over in Buzzards Bay, and they held a contest to name them. People started catching so many bass on those plugs that somebody said, 'This has got to be an atom bomb!' This was 1945, right after Hiroshima. So that's how it got the name."

An Atom lure, I remembered, had enticed my first striped bass. Now, I finally realized why my mentor had urged me to use it.

The first of hundreds of saltwater swimming plugs that would be created for the burgeoning striper fishery, Pond's Atom made a wiggling motion like a crippled baitfish on top of the water. It could also be jerked or popped to make a splash and would prove especially effective seeking big stripers feeding on menhaden, herring, mullet, or other fish whose behavior it mimicked. Through the ensuing years, the Atom probably brought in more big bass—in the thirty- to sixty-pound range—than any other plug. By 1949, Pond was manufacturing plastic ones and later went to Styrofoam, a considerable improvement over the Ponderosa pine variety that often grew waterlogged. He followed with Junior Atoms and Striper Swipers and Reverse Atoms, the last rigged to move backward and look like a frantic live squid.

Traveling along the Atlantic coast to promote his products and to gather information on the bass so as to improve his lures, Pond ran across Daniel Merriman, the Yale fisheries biologist who'd conducted the pioneering studies in the 1930s that revealed the phenomenon of the dominant year class. "He gave me the basic patterns of the bass's movements, where it would be and where it spawned and what it did." Merriman also gave Pond a copy of his doctoral thesis, "The Striped Bass of the Atlantic Coast." "There is no necessary connection," Merriman had written, "between the numbers of eggs produced in a particular spawning season and the number of fry that survive. . . . Anything that affects the rivers in which the eggs hatch and larvae develop is worthy of consideration."

It was food for thought. Fisheries management, Pond noticed, had not even begun to approach the concepts he'd learned in forestry school—the careful planning, planting, harvesting, and recruitment of a resource. In fact, fisheries management had become so politicized that Merriman told Pond he'd put aside his striped bass studies and moved on to less controversial biology. Pond, already active by the mid-1950s in the Cape Cod Canal Salties club, presciently warned his fellow fishermen that "the state-by-state approach to protecting a migratory species seemed to be asking for trouble." He made a motion at one meeting to petition the federal government to put the striped bass under its jurisdiction, only to be vehemently attacked for being against states' rights.

When he decided to form Stripers Unlimited in 1965, Pond conceived of it as a "service organization" for striped bass fishermen. It had five main goals: to develop a fraternity of fishers willing to help each other, to stop indiscriminate netting of stripers, to develop legislation of benefit to sports fishermen, to develop more public access areas, and to curb pollution in waters the bass inhabited.

The more he had followed the striper's coastal odyssey, the more of an avid naturalist Pond had become. He was particularly fascinated with the biological uniqueness of the Chesapeake strain of striped bass. "By the 1950s," Pond continued, "scientists had determined that the Chesapeake bass spawn on the tide line where fresh and salt water meet. It's the only strain with a buoyant, floating egg. All other varieties of stripers have a heavy egg that drifts a long way before it hatches and swims. But the Chesapeake bass didn't require long fluvial rivers; the spawning and nursery grounds were in exactly the same area."

By the mid-1960s, a dominant year cycle that had started in 1958 had run its course. A relative scarcity of striped bass led Pond to the University of Rhode Island (URI), where he discussed Merriman's work and its implications with Saul Saila, head of the Fisheries Department. "His solution to the problem was a suggestion to break the six-year cycle [of dominant year classes] by starting new areas of reproduction in rivers that once held spawning populations. This project seemed to be within our scope and limited budget."

Pond knew that several important breakthroughs had been made in hatching and rearing techniques. At South Carolina State University, Robert Stevens had developed a hormone treatment for female striped bass that enabled them to ripen their eggs and spawn in captivity. When Pond questioned Stevens about the feasibility of taking fish from the Chesapeake and restocking New England rivers with their offspring, he received not only encouragement but also a textbook of techniques that the biologist called his "striped bass cookbook." Meanwhile, at the Edenton National Fish Hatchery in North Carolina, techniques for rearing fry and fingerling striped bass had been developed.

"A river can only take care of one major spawning population at a time," Pond realized. "So we had to treat the striped bass as you would the salmon, which spawn every three years. We figured we'd alternate our new sites for stocking every year, so that dominant year hatches might occur at intervals in the different rivers."

In the spring of 1970, accompanied by a graduate student from URI, Pond headed for Maryland's Nanticoke River in a pickup truck. With the help of Maryland biologist Joe Boone and some commercial fishermen, they hauled in the nets. Pond put several ripe males and one ripening female they had netted into tanks and made the long drive to Edenton, where the superintendent was waiting to inject the females with hormones. The spawning seemed to work. Pond shook his head, remembering: "We were very hopeful of success. But the eggs would hatch and just die. Oh, that was a heartbreak."

The next spring, Pond set up a hatchery on the banks of the Nanticoke River itself and tried the same spawning procedure right there, to avoid subjecting the fish to the stress of long travel. While curious churchgoers watched from shore on their way to sunrise Easter services, Pond gathered the newly hatched eggs in buckets and raced them to his car for transport home to Massachusetts. "We were able to raise a few and watch them develop. Certain odd behavior patterns became evident. The fry became lethargic, started to spiral in the tanks, and failed to develop a visible swim bladder."

Figuring this might still be an anomaly due to poor conditions in the hatchery, Pond persisted. He also took on a partner, Avis Boyd, a plucky younger woman who began traveling everywhere with him and "made English" out of his hastily scrawled notes. In the late spring of 1971, they journeyed to New York to enlist the help of Robert Boyle and the Hudson River Fishermen's Association. The first time they tried collecting stripers for breeding, the big carp in the river tore the net to bits. Eventually, though, male and female striped bass from the Hudson were carted back to URI, where a graduate student suggested having the eggs analyzed for chlorinated hydrocarbons. The behavior of the fry, he thought, indicated symptoms of DDT poisoning, which was already known to damage nerves and inhibit reproduction in Great Lakes fish.

Pond was familiar with DDT, the powerful pesticide that came into use across America in 1946, designed to kill insects and heralded as the greatest boon to mankind since the mousetrap. In the 1960s, scientists had discovered that DDT was sterilizing birds such as pelicans and hawks by thinning the egg shell. After the 1962 publication of Rachel Carson's landmark book *Silent Spring*, the chemical had eventually been banned. But Pond knew its impacts could be long-lasting, and an analysis by URI's Dr. Charles Olney did indeed find high levels of DDT in the fish eggs from Pond's Hudson River samples.

In July 1972, Pond managed to raise a small supply of Maryland striper fry to fingerling size. He planted these in the Parker River on Massachusetts's North Shore, where he knew Yale biologist Merriman had once had some success. "At first they looked normal and healthy, but as they grew larger, half weren't swimming properly. Most fish swim level, but these swam upward all the time, never resting. They apparently didn't have the air they needed in their swim bladders."

Bob and Avis decided to start their own hatchery, renting the basement of an old dairy next door to the tackle factory. Here they would raise and sell tropical angelfish to help defray Stripers Unlimited's expenses. With striped bass, Pond set out to do something novel. He fashioned a small "aquarium" out of microscope slides, by running wire through rubber tubing on three sides to keep it water-tight. Then he would place a couple of striper eggs at a time into the contraption. Sitting in front of a 35-millimeter still camera with a bellows and a magnifying lens, Pond photographed the eggs as they developed. Others had taken pictures of striped bass from the top down; those, Avis would recall, "looked like a fried egg. As far as we know, Bob was the first to look at them from the side."

In the beginning, it was tremendously exciting. Unlike salmon, striped bass in their earliest stages are transparent. Viewed through the magnifying lens, one egg filled the entire frame. Pond could see the heart beating, the blood flowing, the food passing through them. Then, as he would write in a special edition of the *Stripers Unlimited Research Guide:*

> We recorded on film the weakening of the egg membrane to a degree where rapid movement of the eggs caused the weight of the developing larvae to tear the tissues. Photographs also showed the eggs stretching into an hourglass shape and finally rupturing; in other eggs small holes appeared in the membranes, allowing the yolk material to escape. We lost about 85 percent of our eggs during the first twenty-four hours due to these causes.
>
> At about thirty hours, while the larvae was still in the early stages of development, the outer chorion [a membrane], which protects the small developing cells, ruptured and exposed the immobile, helpless larvae to the outside elements. The few which survived these insults and lived for seven days died when their swim bladders failed to inflate. This occurred in 99 out of 100.

As a result of Pond's findings, two federal projects were initiated. A monitoring program by the U.S. Fish and Wildlife Service found extremely high levels of contamination in all the striped bass eggs tested from the Chesapeake. These turned out to be unusually susceptible to chemicals that concentrated in a large globule of oil inside the egg, which otherwise served to make the eggs buoyant. DDT had been heavily sprayed during the 1940s and 1950s on the marshy, mosquito-infested wetlands along Maryland's lower Eastern Shore, Pond learned. And earlier pioneering researchers, especially Romeo Mansuetti from the University of Maryland, had observed some striped bass eggs developing strangely in the '50s. Pond feared that the abnormalities were being passed down from generation to generation among the fish.

There were, of course, other potential manmade culprits that had invaded the bass's territory. With the advent of no-till farming methods in the late 1960s, vast amounts of agricultural pesticides and herbicides began washing into the Chesapeake rivers. The bay started losing most of its submerged aquatic vegetation, formerly a place where young stripers could hide from their predators. Evidence of chlorine—a by-product of sewage treatment, but often fatal to aquatic life, including newborn bass—was also increasing dramatically as more and more people purchased land by the bay. Acid rain had not yet entered the American consciousness, but it was there nonetheless, like an unknown parasite whose presence served only to increase the effects of the contaminants.

Then there were PCBs. The Hudson River wasn't the only place where these compounds were starting to be recognized as a problem. For thirty years, several Massachusetts manufacturers of electronic capacitors had been dumping chlorinated hydrocarbons by the ton into the Acushnet River estuary, which flowed into New Bedford Harbor. Some of the harbor's sediments would eventually test out at an astounding level—one part PCB for every three parts of sand—the worst such contamination on the planet. From there the chemicals drifted out along Cape Cod and into Vineyard Sound and Rhode Island Sound, where migrating fish such as striped bass bioaccumulated PCBs in their flesh.

When Pond sent fatty tissue samples from the eggs to a lab in Pennsylvania, one of the technicians replied: "Did you soak these in PCBs?" At the time, Pond had never even heard of the compounds. But he learned quickly what was known of their impact on reproductive systems.

What Pond's ongoing photographic observations revealed, though, was a puzzling phenomenon. In all the literature on striped bass, the oil globule inside the egg started to diminish in size only when the newborn fish had almost absorbed the yolk material. The oil provided floatability, but for some reason the fish appeared to be excreting it through the gut and living on the protein in the yolk alone. "How the fish knew the poison was in the oil I don't know, but they didn't use it," Pond continued. This apparently weakened the fish's tissues. Some lucky ones nonetheless survived.

Pond's theories were not published in academic journals, nor peer-reviewed. So he was merely an amateur in the eyes of most pedigreed scientists—though some, like Joe Boone, were already beginning to recognize his talents. "What amazed me was the energy he had," Boone would recall. "Bob would stay up twenty hours a day, taking the eggs out and putting them under a microscope, getting pictures of fish at every stage of their cell growth, saying, 'This oil globule is not attached; this yolk is not big enough.' He was so inspired, he never seemed to get tired."

Pond still needed to keep peddling his lures in order to keep Stripers Unlimited and his conservation efforts afloat. Robert Boyle dropped in to see him in Attleboro one day and asked, "How do you keep selling those plugs?" Pond replied without missing a beat: "Well, I just tell 'em that, in order to be a sportsman, you need to use a lighter test line."

"Of course!" Boyle recounted with a hearty laugh. "Then they'd *lose* them and need to buy more!"

As Pond traveled the Eastern seaboard, he preached a heretical gospel to fishermen: the striped bass were in big trouble. Will Barbeau, later to become active in striper conservation in Rhode Island, would remember: "In the early '70s we're in striped bass up to our eyeballs, and Bob shows up at our club full of doom-and-gloom, and we think he's crazy. All these egg stories! Well, biology is a difficult subject at best, and especially when a lot of the professionals aren't inclined to believe him, fishermen are thinking, 'Why should we?' So we were not too attentive. But a year or two passes, and all the things that he's been predicting start happening. We're not getting fish, tournaments are not producing the way they should. And us little spoiled children start believing Bob Pond."

Similar reactions faced Pond wherever he went. Charles Witek, later to be a founder of New York's Coastal Conservation Association, joined Stripers Unlimited after encountering Pond in the tackle shop where he worked summers as a student. "He was a classic Cassandra," Witek would remember, "speaking the truth while nobody believed him." Charley Soares, today a prominent writer-lecturer about striper fishing, first encountered Pond talking conservation at his Linesiders Bass Club in Fall River, Massachusetts. "Pond was the guy who first gave me an awareness," Soares would remember. "I brought this up to the biologists and they told me, 'Don't worry about it.' Bob was looked upon as an oddball, a Chicken Little, 'sky is falling down' type. But he was right."

Pond's focus was primarily to the south and to a deteriorating Chesapeake Bay, where in 1976 the federal EPA authorized a $27 million study. The bay was being overwhelmed by algal blooms from high levels of nutrient overload, coming from both municipal and agricultural waste. High concentrations of toxic chemicals and metals were accumulating along the river bottoms. Large sections of the bay were devoid of oxygen, which striped bass and other fish need to survive.

Given a tremendous striper hatch in 1970—indeed, a record-setting young-of-the-year count that summer—another boom should have been expected when those females reached spawning maturity by 1976. But in making his annual pilgrimage to the Chesapeake that spring, Pond observed what appeared to be a shortage of male fish when the females arrived in one of the rivers to spawn. Many females never released their eggs. When caught by sports fishermen later that summer, many were still carrying green roe when they were gutted. No new dominant year class occurred. In Pond's mind, there was no longer any doubt that the natural cycle of the bass had broken down. He petitioned the federal government that fall to have the Chesapeake striper placed on the endangered species list.

"They told me I was out of my mind. They said, 'The bass aren't in danger now, or in the perceivable future. There are 11 million pounds out there!' I said, 'That doesn't make any difference; 11 million pounds of parents are not properly replacing themselves.' But the government never seems to function until there's an absolute crisis."

Now, in 1982, sitting with him in the tackle shop where it all began, I could not have had a more inspirational guide than Bob

Pond in fighting for the striped bass population. Their situation had only deteriorated further as the early '80s dawned. Now, Pond said, the critical first step was protecting that 1978 Chesapeake year class, letting them grow to twenty-four inches before anyone could keep or sell them along the coast, making sure they made it back to the spawning grounds in 1984. By midsummer, Pond added, he'd be traveling to the Chesapeake to accompany Joe Boone on the young-of-the-year survey; maybe I'd like to come along.

In the meantime, it was crucial, he said, that my friends and I do everything in our power to persuade Massachusetts to establish that twenty-four-inch minimum size limit before the 1982 summer fishing began. The vote by the Massachusetts Marine Fisheries Commission was coming up in two weeks, in early April. I said I'd do whatever I could.

Back in Boston, I started placing phone calls to local media to alert them to the striped bass's plight. The *Good Day Show*, a popular morning talkfest, expressed interest in a debate about the size limit controversy—if I could find a commercial fisherman who would appear alongside me. I got the number of a spokesperson for the commercial fishermen at the Buzzards Bay meeting, a big redheaded fellow named Kenny Baker, whom I suspected had probably been among my louder hecklers at the public hearing. After some hemming and hawing, Baker agreed to do the show.

I called the producer at *Good Day*. Well, she asked, why did I think this segment would be of interest to housewives? I had to think fast. "Because the bass are a great-tasting fish that've become real expensive!" I blurted out. She said she'd get back to me. To my surprise, she called a day later to say they'd plan for a seven-minute spot, Russell versus Baker.

I hadn't been on TV since my dad, an advertising executive, plopped me down in the *Howdy Doody Show* audience back in the

'50s. I was scared to death. I studied the *Good Day* hosts; one of them, I thought, wanted to be Barbara Walters. After the producer called to give me the questions I'd be asked, I spent the rest of the day rehearsing.

Tossing in my sleep that night, I dreamed that I was out fishing and put my lure on backward. Early the next bleary-eyed morning, I found myself waiting in the "green room" with the other scheduled guests—including John Ehrlichman of Watergate fame, who'd just had a novel published. I wondered whether striped bass would be his opening act or the curtain-closer. (To this day, I can't remember.) I'd decided to adopt a casual look—wearing an open-collar paisley shirt that my wife had embroidered. Kenny Baker, wouldn't you know, showed up in a three-piece suit.

Once we were on camera, those seven minutes flew by. The bass were simply in a natural down cycle, Baker argued, so why should Massachusetts fishermen have to bear the brunt of tougher regulations? I spoke about overfishing and pollution and the need for urgent action, and they gave me the last word on the need for a stronger size limit. On my way out, the woman who answered the phones looked up and exclaimed: "I don't know what you guys were talking about, but our switchboard is going crazy!" I didn't reveal that I'd asked just about everyone I knew to watch and then call the station to say how important this was.

When my friends and our children took to the streets that weekend by the New England Aquarium, armed with petitions calling for a twenty-four-inch minimum catch rule, a number of people we encountered turned out to have seen the program. We'd tried to imbue our kids, who ranged in age from six to twelve, with a sense of the magic of the bass and of what was at stake. They proved to be not only tenacious (chasing more than one reluctant signer down the street) but also eloquent. When six-year-old Bing Guerin encountered a "mean man" who responded that he couldn't care less about the bass, Bing was so outraged that he went over to a snowbank and started kicking the bejesus out of it. But more often than not, our children's crusade captured the imagination of the local citizenry. Our stack of petition pages filled with signatures mounted higher and higher.

A musician friend, Jim Kweskin, added new lyrics to a song called "Fishin' Blues," performed it on the radio, and even brought his guitar to a meeting of the Massachusetts Striped Bass Association. Afterward, the association voted to adopt a resolution supporting the new regulation (and Kweskin won a rod-and-reel in their raffle!). I spoke at a saltwater fishing seminar back in Buzzards Bay. At the New England Environmental Conference I cornered U.S. Representative Ed Markey, who over lunch suggested I speak with his Massachusetts colleague Gerry Studds (a crucial piece of advice, I later realized). Another friend, Owen de Long, the maitre d' at a fancy Boston restaurant, raised the striper subject after finding a table for Boston mayor Kevin White. The mayor agreed to sign a letter if Owen would write it for him, which he did.

The day before the scheduled vote by the Massachusetts Marine Fisheries Commission, I carted petition pages bearing more than 6,000 signatures down to the state's fisheries officials. They'd apparently been trying to get the attention of Mayor White for years, so his long letter recommending the twenty-four-inch limit was also highly impressive. "From Hizzoner himself? Do you guys have any spare time on other issues?" assistant director Randy Fairbanks marveled. His boss, Phil Coates, expressed his own surprise, then added: "I saw the mayor down at Monument Beach one day last year with a pretty good haul of striped bass. You must have aroused his guilty conscience!"

Another letter that caught the attention of the commissioners came from Anthony Benton Gude. He was only eighteen at the time. I'd known him since he was a boy. He, too, was a summer fisherman on Martha's Vineyard, and we'd shared the same mentor. Grandson of the renowned American artist Thomas Hart Benton, and an up-and-coming painter himself, Anthony also had a passionate way with words: "I have fished for bass all my life, and have great respect and admiration for them. Over the years I've watched their numbers declining, seen their schools diminishing, until now, in the areas which used to abound with bass, I find an occasional few or none at all. They are in trouble, as I'm sure you are well aware.

"Personally, I feel very strongly about the bass. Unlike most bass fishermen, I have even gone under the water and swum with them. I observed them in their own world and strangely enough, they observed me. The younger bass, from three to ten pounds, were very interested in what I was. Many times they came circling around me within an arm's reach. The older ones, the bass around twenty pounds and over, seemed to know and kept their distance. I was swimming among an entire school of them, all basking quietly on the bottom. As I swam along I came upon them in pairs, one very large bass and beside it a small one. They weren't spawning, for it was July. Why they were in pairs I could only guess. The larger fish acted very protective of the little ones, so much so that I came to believe the older bass were actually looking after the young in the school. . . .

"They are not just an ordinary species of fish. They have a mystery about them and a sense of greatness that we can feel when they are on the line. . . . We have in our waters a fish unlike any other in the world . . . let's do our best to protect them."

On April 1, 1982, the eight Massachusetts commissioners—some of them commercial fishermen themselves—voted unanimously to adopt a strict twenty-four-inch minimum size limit for anyone fishing state waters. We'd given them enough public support to do what was needed for the fish. A proviso was attached, however, mandating a review if neighboring states did not act in a similar fashion.

Rhode Island was the crucible. Unlike Massachusetts, where only hook-and-line fishing was permitted, Rhode Island allowed traditional trap-net and gillnet fisheries. Those fisheries focused on the smaller bass for their income, and a lot of commercial money was at stake.

A single trap company had scooped up more than 100,000 pounds of the 1978 year class within a two-week period the previous fall. The owner sat on the Rhode Island Marine Fisheries Council, along with two of his cronies. No matter what the rest of the Atlantic states did, forcing Li'l Rhody to go beyond a sixteen-inch minimum size limit, I was told, was going to be just about impossible.

That's when I became a carpetbagger.

Chapter Four

A Man Named Mendonsa

George Mendonsa (pronounced Mendoza) had liked it a whole lot better when the Rhode Island state legislature was the overseer of his trap-net fishery. "There was a steak house where we would all get together for dinner in the wintertime," Mendonsa would recall fondly. "Then afterwards, I had a friend who owned a nightclub. Dancing girls around the bar. The booze flew, the doggie bags all went home. Cost us a few bucks, but that's when we were most successful with fishing regulations. We won every time. No names. The Marine Fisheries Council was the worst thing that ever happened to the industry."

When that council was created in 1977, it came as no surprise that Mendonsa was among the several commercial fishermen appointed to the new regulatory body. His family, he was proud to proclaim, had been "settin' traps out of Newport since 1910, when the old man come over from Portugal." George and his three brothers grew up on the water ("we were raised like ducks"). He started working summers with his father at the age of seven, and, during the winter, family members

71

all joined forces to hand-make the enormous new cotton nets for the
next round of fishing. His mother never did learn English, and George
himself "could speak Portuguese with the best of 'em."

They were a tough crowd. His mother's brother was "the only trap
fisherman I ever heard of that drowned," while pursuing conger eels.
His father saw the body floating on the bottom, reached down with a
boat hook, stuck it through Uncle Joe's neck, and hauled him up.
George's brother Arsen was "a horse, the strongest man that ever lived
around here. He got in a barroom fight with some sailors one time; the
first thing he did was lock the door so nobody could get out." In the
1930s, before all the fishing areas were controlled with licenses, the various trap-net companies tried to beat one another to the best fishing
locations. Legendary feuds ensued. "The word was that the Coggeshalls
put a big sickle on the bottom of their boat, and they just run right
through everybody, cutting the nets all to pieces." George didn't want
to say what happened in return.

Dropping out of high school after his junior year, Mendonsa joined
the Navy soon after the United States entered World War II and eventually became helmsman for the USS *The Sullivans*. He successfully steered
the warship through the Philippine Sea, to Iwo Jima, the Marshall
Islands, and more—avoiding Japanese suicide bombers and rescuing the

crew of the aircraft carrier *Bunker Hill.* "We put Ernie Pyle ashore at Okinawa, with twenty-five other journalists, and he got killed there."

When the war ended, Mendonsa was celebrating in Times Square when he grabbed a nurse and planted on her what became known as "the smack seen round the world," captured in a famous Alfred Eisenstadt photograph that graced the cover of *Life* magazine. "He was nothin' but a bum until that happened! I made him!" George believes. Though the magazine would never acknowledge Mendonsa as the sailor—he eventually sued because *Life* was profiting by selling prints for $1,600 apiece—today there is little doubt that he's the man. Few other people on the planet have hands as big around as their legs, while a tattoo with the initials "GM," visible on his left arm, sealed the case for a photographic expert from Yale.

Shortly after Mendonsa returned home, his family purchased the Tallman & Mack fishing company. He and his crews of between fifteen and twenty-five men (depending on the season) would set ocean traps for scup, butterfish, bluefish, bonito, sea trout—and what Mendonsa called "the smartest fish in the sea," the striped bass. The idea was that migrating fish would strike the leader, which is like a twine fence, and instinctively head for deeper water, thus entering the floating trap. First into the wings and heart, or "kitchen," and then into the box, or "parlor," where the fish are harvested.

Everybody in the Newport area knew that George, like his father before him, was the canniest trapper around. He seemed to possess a sixth sense about where the most fish would be. As the years went by, technology improved as well. The new nylon nets could last for years and now were pulled not by hand, but by big winches. Where once it took shoveling the fish off the boats and dumping them into 100-pound buckets, now the whole dockside process was mechanized. The day's catch would be unloaded from conveyors into ice-filled boxes, then trucked down to the Fulton Fish Market in New York City. The Fulton folks controlled "the whole price for the Atlantic coast," Mendonsa recalled. They would pay you for the fish, yet at the same time would reload some of your choicest product into boxes of their own. And your driver, "if he talks too much, he'll end up floatin' down the Hudson."

One thing hadn't changed, as far as Mendonsa was concerned. "Even the old man was always fightin' the sports fishermen. Way back, before I was born. Over the striped bass. I can remember as a tiny kid goin' up there with my father to the legislature, the fisheries committees, arguing about striped bass. Every winter it went on, that battle. The nets used to come right to shore in the fall of the year, right to the rocks, and the sports fishermen always wanted us to move them. They howled so much in Massachusetts, they eventually put the netters out of the bass business. That's always been their whole idea. The rod-and-reelers have been hidin' behind this conservation thing for almost a hundred years!"

Bob Pond was the first to clue me in that Rhode Island's eight-man Marine Fisheries Council was dominated by this fellow Mendonsa, who owned a $250,000 boat and a $100,000 fish trap. Massachusetts's director of marine fisheries, Phil Coates, added that in the fall of 1981, George Mendonsa and two counterparts (council members all) had hauled in 160,000 pounds of small bass—thousands of fish—in two weeks. The net fishermen were capitalizing on the striper year class from 1978 making its first migration. And that year class was the only one remaining in any numbers, the last best hope for the fish's future if they were allowed to reach spawning size before being caught.

I phoned John Cronan, the Ocean State's director of fish and wildlife, hoping to get a better sense of whether the state expected to adhere to the new coastal management plan and raise the minimum size on catchable striped bass to twenty-four inches, as Massachusetts had done. I didn't see how Rhode Island could justify allowing its fishermen to keep fish at the pre-spawning size of sixteen inches. These small bass did tend to be "frosting on the cake" for the trap-netters

(Mendonsa called them a "Christmas bonus" for his crew), Cronan replied, but the gillnet fishery also needed these smaller fish in order to stay afloat. Gillnets, I knew, were a particularly cruel form of commercial fishing. They're constructed to entrap fish as they unwittingly swim into the net, then entangle their gills in the tightly knit webbing as they seek escape. Slowly the fish suffocate from lack of oxygen. If left untended, or lost in a storm, the gillnet continues to drift around, "ghost fishing" underwater—killing and wasting hundreds of fish.

Cronan added that I needed to remember that Rhode Island's striper harvest was paltry compared to that in states such as New York and Maryland—which, he hinted, were also unlikely to change their minimum size limits anytime soon. When I suggested that wiping out the 1978 year class might be the final curtain for the striped bass, Cronan politely responded that I was "crying wolf."

Clearly, unlike in Massachusetts, not much help would be forthcoming from Rhode Island's fisheries officials. Allies were needed. Bob Pond gave me two names. One was Jim White, a postman in Coventry, Rhode Island. When I arrived at his house late one afternoon, White had just completed his day's round of mail. He was a tall, lean fellow in his early thirties with a dark mustache, and I instantly felt him to be a kindred spirit.

Both his great-grandfather and his grandfather had been commercial fishermen in Portugal. White, equally impassioned, was otherwise about as far from the Mendonsa tradition as someone could get. His father, a truck driver, was a dedicated recreational surf fisherman who refused to fish from a boat. "He was an immigrant, and from the old school," White remembered. From the age of about three, "he would let me hold the rod and feel the fish bite, but he would never let me set the hook. Not until I was about six years old. It was like torture." Still, it made White appreciate the value of fishing like nothing else could have. On weekends he'd pile into the car with his parents and head for the Cape, where his mother cooked dinners on a Coleman stove and they'd all sleep in the old Nash Rambler.

Of all the fish, there was one that completely captivated him. Once, White was gone so long that his father sent a policeman out looking for him. When the officer found the boy a few miles away at a beach, White announced: "Tell my father the stripers are running at Bristol Cove, and I'm not coming home till they're gone!" The policeman raced to get his own fishing rod, and joined him. White didn't stop fishing for thirty-six hours straight.

As a teenager, camped on warm summer nights with his friends along Cape Cod, White would keep his rod beside him and dream about casting into the waters at dawn. When the sun broke through, he would be gently awakened by a wondrous sound. It was the splashing of a huge school of striped bass, their tails churning the water as they swam atop the waves—a rainbow of shimmering green and gold fins outlined against the rising sun.

"That was only twenty years ago," he was saying now. "But that sound is like a memory from another world. Now when I take my little boy out in our boat, we're lucky if we see more than a dozen stripers the entire summer. There are so few around that it's hard to bring yourself to fish for them at all. It makes me feel as if a huge part of myself is dying with them."

White had first encountered Bob Pond in the early 1970s, not long after he got back from Vietnam, when he was attending night school at the University of Rhode Island. White had decided to do his college thesis on the striped bass, and Pond already worked closely with two of the researchers who were assisting him. "I already knew about his Stripers Unlimited, because both my father and I were members since it started," White recalled. "Bob taught me that you don't necessarily need a degree in something to be an intelligent person on the subject. He could tell you more about striped bass than anyone."

So the idea of now forming a united front with me, Pond, and other Massachusetts fishermen was a natural to White. I told him what had recently happened in Texas. Commercial netters, many operating illegally, had pillaged the redfish and spotted sea trout fisheries almost out of existence. Then the Gulf Coast Conservation

Association, a fishermen's organization, pushed the state legislature to pass a bill removing those two species from the marketplace. The sports fishermen succeeded by launching a campaign that stressed economics—specifically, the money contributed to the Texas economy by the 800,000 of them, versus the approximately 300 full-time netters whose dockside catch added up to about $3 million.

White shook his head. The trouble was, he didn't know where to begin. Forty fishing clubs, including his own, belonged to the Rhode Island League of Federated Sportsmen. The striper sports fishermen could have some clout if they weren't so apathetic. Hardly anybody even bothered to show up at the Marine Fisheries Council meetings, a lack of participation that played right into the hands of Mendonsa and company.

We had to start somewhere, I said. The next council meeting was coming up in a week in Wakefield. It was an opportunity to start lobbying for the twenty-four-inch minimum size limit. Striped bass weren't on the agenda. But we could put them there.

Huddled at one end of a long conference table, several council members tried to size up my friends and me, while George Mendonsa refused to look in our direction. He was an archetypal figure, a character right out of *On the Waterfront* or *Captains Courageous*, craggy-featured and square-jawed with a mane of wavy white hair. He was pushing sixty, but at six-foot-two and probably 220, those massive hands and arms looked like they could still haul an anchor with the best of them.

Most of the meeting was devoted to discussion of gillnetting regulations, and it was evident that self-interest, not conservation, was the real topic. At the end, under "old business," they let me speak. My approach was diplomatic, until Mendonsa interrupted. "You can't tell me there's no bass around!" he bellowed. "I caught more last year than my family ever has—130,000 pounds!" My waiter friend Owen de Long—who had enlisted Boston mayor Kevin White into our

Massachusetts campaign—shouted back that *Mendonsa* thus bore responsibility for killing off the remaining spawning stock. Big George stood up. In the audience, so did Jim White and I and a few others. Fisheries director Cronan cried out to restore order, and somehow everyone cooled down. They'd take up the size-limit question at the next meeting in May, Cronan said, and adjourned the meeting.

One of my main advisors was George Peper, a longtime fishing partner and photographer who'd worked with me on a number of articles. He was in Los Angeles at the time, but we talked by phone on an almost daily basis. He was adamant that, besides building an organization, we needed to apply political pressure from the top down, seek out influential legislators, give the fisheries council something to be scared of. It couldn't go on operating in a vacuum. Wasn't anyone concerned that Mendonsa was voting for his vested interest?

As it turned out, Rhode Island had a little-known sector of its bureaucracy called the Conflict of Interest Commission. Will Barbeau, another protégé of Bob Pond, knew of it and agreed to help. He had stood up with us against Mendonsa and served as vice chairman of the state's striped bass citizens' advisory committee. In mid-April 1982, Barbeau and I paid a visit to Rae Condon, the woman in charge of the Conflict of Interest Commission. I recorded in my diary: "It was clear a state investigation into the trappers could be launched by simply filing a complaint." It didn't take much arm-twisting to enlist Barbeau to do so.

Fortuitously, around the same time, Congress had voted to renew the Anadromous Fish Conservation Act of 1965, and Gerry Studds of Massachusetts had attached an important amendment to it. Any member state of the Atlantic States Marine Fisheries Commission (ASMFC) refusing to comply with the new striped bass management plan by the end of fiscal 1983, the amendment read, would have all its federal fisheries funds cut off. This would finally give teeth to the ASMFC. Fisheries would move from provincial to federal management—a paradigm shift from a resource management point of view.

It was educational, the way politics worked. When I'd attended a town meeting with Studds near New Bedford a few weeks earlier, he couldn't even remember the status of the Emergency Striped Bass Study that he'd cosponsored with Senator Chafee. But after being badgered by me and other fishermen to press his fellow members of Congress to do something more about the bass, Studds had followed through.

Now the ASMFC was having a meeting of its Striped Bass Board in Washington, D.C., and I flew down for it. The board, composed largely of fisheries commissioners from the Atlantic states, was discussing each state's effort to comply with the new management plan it had voted for in 1980. Especially now that the Studds amendment had been approved by Congress, Rhode Island commissioner Cronan was very much on the defensive. After complaining that he didn't see how his state could punish the netters by raising their size limit (and so comply with the management plan), Cronan huffed: "Well, this Studds amendment means we won't be able to get any money to monitor the stocks that we're all worried about. That doesn't make any sense!" Alan Peterson, Northeast regional director of the National Marine Fisheries Service, responded furiously: "You're ducking the issue, Mr. Cronan!"

The next day, I got an appointment to see Gerry Studds. He was a genteel, erudite fellow who hardly looked like he spent much time on the water, but fishermen made up a good part of his constituency and he'd risen to be the ranking Democratic member of the House Subcommittee on Fisheries. Studds explained that his amendment was "very much in the spirit that recognizes this is not directly a federal management jurisdiction, since the striped bass spends most of its life in state waters, but that this is also a fish which doesn't have a great deal of respect for state lines. The bass obviously will be endangered if the states don't get their act together, and this is a federal encouragement to do that."

Studds's means of enlisting the support of other House members had been both witty and engaging. He'd offered them a description of "what the striped bass has to go through—given that map which confuses everybody—how the hell do you get out of Chesapeake Bay? Then up and around New Jersey and New York, out past the tip of Long Island, in order to do what any intelligent fish would want to do—get to Cape Cod and the Gulf of Maine! By the time we came to the vote," Studds added, "a number of the members had grown quite fond of this fish they'd never met before."

He figured that uncooperative states would face losing anywhere from a few thousand to as much as half a million dollars in federal fisheries money. He specifically applauded the "courageous" stand of Representative Claudine Schneider from Rhode Island, "who said we'd better help persuade our states to act in what is in everybody's best interest. I'm sure," Studds added wryly, "that there are some strong forces on the other side within her own."

Back in New England I suited up again to lobby Rhode Island legislators to put pressure on the fisheries council, arrange meetings with local media, and, donning my fishing cap with a silver striper pin, address the Federated League of Sportsmen's Clubs. I talked about the state's conflict-of-interest laws every chance I got. A small army began to gather to force stronger regulatory action on striped bass, in anticipation of the council's May meeting.

I phoned John Boreman, the marine scientist I'd met at Buzzards Bay. The claim by Rhode Island officials that 93 percent of their bass catch was by sports fishermen was dubious indeed, he said. Of Mendonsa's alleged 130,000-pound take in 1981, Boreman added, "If that figure is true, then it's six times Rhode Island's entire commercial catch of 1980—and equal to the entire commercial catch they reported for 1978, 1979, and 1980."

I drafted a long memorandum to hand out at the council meeting —"Why Rhode Island May Lose Its Federal Fishing Funds"—and, on May 4, again made the two-hour drive from Boston to Wakefield

accompanied by several carloads of carpetbaggers. Bob Pond was waiting to greet us. All three Rhode Island TV stations had sent camera
crews. Jim White, who was petrified at the thought of speaking in public, found himself being interviewed. "Some people ask me, why so
much uproar over a fish?" he said. "It's not a question of economics, it's
a feeling way down deep inside. How can I explain it? The striped bass
are a part of the soul of this country. And they're a part of my soul,
too." Why, the postman was positively eloquent!

Outdoor writers from the *New York Times* and the *Boston Globe*
were on hand, along with curious Rhode Island legislators. Outside the
building, several musician friends played guitars and sang "Fishin'
Blues" with Jim Kweskin, while others held up protest signs. Will
Barbeau carried petitions bearing more than a thousand signatures, as
well as the unanimous endorsement of the Federated League of
Sportsmen's Clubs for a stronger size limit.

Inside, there was a standing-room-only crowd of about 150 people
when W. Edward Wood, director of Rhode Island's Department of
Environmental Management (DEM), called the meeting to order.
With his handlebar mustache, I thought he looked a little like the villain in a grade-B movie. Wood began by saying he'd never experienced
such a whispering campaign against honorable men before, finding it
"abysmal" behavior to use "innuendo" to attack certain council members. It was a "catch-22 situation," he claimed: the fisheries council was
mandated by law to be a panel of experts, while the state's conflict-of-
interest law prohibited a person from holding a position from which he
or she might gain financially.

Wood then turned things over to Saul Saila of the University of
Rhode Island, the lone scientist on the council, who introduced
motions on three options to send out to public hearing, including the
twenty-four-inch minimum size limit. It was all carefully rehearsed,
right down to five of the eight council members—three trap-netters,
a dragger, and a sports fisherman (who admitted occasionally selling
his striped bass)—abstaining from the vote due to their "potential

conflicts." A public hearing was set for June, with a decision to follow in July.

Mendonsa couldn't help but have the last word. "I'll have to lay off thirty men after August if this twenty-four-inch regulation passes!" he railed. "What I'd catch won't be worth putting $100,000 worth of gear in the water for! You can't knock my industry out of the bass business!" On the way out the door, Mendonsa's crew tossed epithets and issued challenges to our Massachusetts contingent. Police cruisers patrolled the parking lot for potential incidents against out-of-state vehicles. We formed a phalanx and headed toward our cars.

The *Providence Journal* followed a few days later with an editorial endorsing the higher minimum size limit, arguing: "Neighboring states have been putting Rhode Island to shame. Commercial netters can go after other fish to tide them over until the striper's future is assured." In a rebuttal op-ed, DEM director Wood branded us "single-issue zealots" and said he found it "disappointing" to see "a simplistic regurgitation of statements made by Massachusetts lobbyists." Wood's column concluded by bemoaning the "unwarranted hand-wringing of a few 'true believers' from our neighboring state to the north." Apparently we had touched a nerve.

Wood instructed the five vote-abstaining council members to go to the Conflict of Interest Commission for an advisory opinion. On May 27, one was rendered: George Mendonsa, specifically, ought *not* to deliberate and vote on regulating the taking of striped bass. That same day, the DEM's Wood was kicked upstairs by the governor to run Rhode Island's Department of Transportation.

The very next day, something else transpired that buoyed our spirits, this time on Martha's Vineyard. It was as if the fishing gods were adding their blessing.

As fog began closing in on the island, eighteen-year-old Anthony Benton Gude approached the surf off Squibnocket Point with his mother, Jessie, and her husband, Richie Guerin. About a hundred yards from the beach, a center console fiberglass boat with three men aboard killed its engine and dropped anchor. Something about their movements seemed suspicious. The boat was momentarily forgotten as Anthony spotted myriad pogeys being chased up against the shore. In his hip waders, he found a nearby rock and cast a surface swimmer lure into the gathering dusk. His first cast brought a strike, and a large striped bass whirled over in a wave about ten feet away. Anthony cast again and hooked it.

When the trio onshore glanced again at the nearby boat, the men aboard averted their gaze, hauled anchor, and headed rapidly toward the other side of the point. "I think I'll follow them down," Anthony said and headed overland. As the boat approached the rocks along a onetime jetty known as the Bass Stand, Anthony noticed the fishermen throwing something overboard. He could make out some corks, then the green of a net that the fishermen anchored at one end. It was, he quickly realized, far too large a net to be simply taking bait.

Arriving home, Anthony looked up the description of a gillnet in McLean's fishing encyclopedia. It matched perfectly. Massachusetts had outlawed any netting of striped bass back in 1945. He contacted the island game warden, Sandra Lucas, who to his surprise was already observing the netters from the Squibnocket cliffs; a caretaker working on a house overlooking the water had spotted them and called her. But even if she could catch them in the illegal act, Lucas said, she had no craft of her own to pursue them. "We'll join you right away," Anthony said.

Gude and Richie Guerin hopped into a car and raced toward Squibnocket. On the road they met up with Jesse Reinfelder, son of the late bass fisherman and conservationist Al Reinfelder. Soon they had all joined the warden and caretaker outside a hunting lodge in the cliffs, spying on the netters through binoculars and a bird-watching telescope. They counted several striped bass when, around 7 p.m., the fishermen hauled in one of the nets.

Lucas used her two-way radio to explain the situation to her superiors at Massachusetts's Department of Fish and Game in Boston and obtain permission to ask the Coast Guard for assistance. Tim Rich, the police chief in the nearby town of Chilmark, overheard the conversation and decided to lend his aid. A sports fisherman himself, the six-foot-six Rich drove to the cliffs in his squad car. He didn't recognize the boat as a local one. It was rapidly getting dark and Rich knew that if they waited much longer, the netters would escape.

"Obviously they knew they were being watched," Anthony would recount. "Jesse and I positioned ourselves right on the shoreline directly opposite their second net, where they'd have to come in very close to get it. I stood on a rock trying to read the numbers, but there weren't any on the hull. We did get the make of the boat, Chaparral, and of the biggest engine, a 200-horsepower Johnson outboard. Then they hauled the net up and got ready to take off."

By the time Anthony and Jesse reached the clifftop again, Lucas and Rich were gone. The chief had persuaded the Coast Guard to try an intercept. Guerin had stayed behind to let them know which direction the netters were heading. On board the Coast Guard cutter, Chief Rich and six crewmen were armed with .45 automatics and shotguns. As the cutter sped toward the point of Gay Head, Guerin notified them that the speedboat was approaching them from the west. The fishermen were apparently making a getaway toward the Elizabeth Islands.

The two vessels met at a spot called Devil's Bridge. "Miraculously," Lucas would recall, "when we came around the corner, they were heading right towards us with no running lights. The Coast Guard aimed the spotlight and they had no place to go. Otherwise, I'm sure with those two outboards, they could've zipped right away."

It was 9 p.m., several hours after the vigil began, when the warden and police chief boarded the boat. There on the deck, four long, lean striped bass were barely hanging onto life, flopping in the nets. Compared to the numbers taken by Rhode Island's net fishermen, this was a relatively token catch; but Massachusetts hadn't allowed *anyone* to kill stripers in nets for thirty-seven years. The fishermen, claiming they spoke no English, offered no resistance. They were handcuffed and taken aboard the cutter. From Menemsha Harbor, they were transported across the island to the Dukes County Jail. All three turned out to be

Portuguese from New Bedford, about twenty-five miles away on the mainland. "Catching them was a million-to-one shot," Warden Lucas said. "If it wasn't for the island people getting involved, none of this would've occurred."

The bust of the netters made big news. "High Speed Sea Chase Snares Fishermen with Gill Net Haul," a Page One story in the *Vineyard Gazette* was headlined. "Netting Striper Violators," the *Boston Globe* headline read. Coming when it did, in the midst of the controversy in Rhode Island, the message could not have been lost on Mendonsa and his fellow council members.

A few days later the Conflict of Interest Commission turned down an appeal by Mendonsa's lawyer and added Michael Parascondolo—another trap-netter who sat on the council—to its warning list. "This doesn't imply that they can't participate in the public hearings," said the commission's executive director, Rae Condon, "but our advice is that they should not attempt to influence the rest of the council or vote on the striped bass issue."

With the tide turning against them, Mendonsa and Parascondolo filed for a court injunction, attempting to stop the public hearings and bar the Fisheries Council from taking any action on the bass until they could appeal the conflict-of-interest ruling. "This could take months!" Jim White fumed one afternoon, as he watched the sun set over Narragansett Bay. "It's another delaying tactic so they'll still be able to kill off undersized bass during the fall migration."

While agreeing to hear the net fishermen's appeal, a Superior Court judge dismissed their effort to halt the public hearing process. The first hearing on the minimum size limit was held in Middletown, where Mendonsa resided. About a hundred fishermen showed up, far more than usually bothered to attend such gatherings. State fisheries officials from Massachusetts took the unprecedented step of coming to Rhode Island to offer testimony. They made it clear that the Bay State's own commercial bass fishery had given up at least 300,000 pounds of potential earnings because of the new size limit, but that there was no other option if the fish were to survive. I brought up the netters' arrest on Martha's Vineyard, adding, "Nobody is asking Rhode Island net fishermen to pay that kind of price." One of the fishermen nevertheless still rolled up his sleeves and wanted me to "step outside," and then nearly assaulted a TV cameraman.

The decision-making process, as Jim White had predicted, dragged on over much of the summer. On July 1, the Superior Court dismissed all claims made by Mendonsa and Parascondolo. Shortly thereafter, John Cronan's Division of Fish and Wildlife tried to avoid the issue by recommending that the minimum size limit be raised to twenty-four inches for rod-and-reel fishermen (but stay at sixteen inches for commercial netters) and postponed the vote for another month.

In Massachusetts, talk of vigilantism on the waters began to be heard. *Boston Globe* columnist Tony Chamberlain expressed the prevailing sentiment: "If one poisons the water in the river in front of his own house, he poisons his downstream neighbor's as well. If the Ocean State thinks it has heard criticism from Massachusetts thus far, it is hard to imagine what will happen when a Cape Cod fisherman realizes that he has released a juvenile fish that a Rhode Island boat fishing the same waters can catch and take home to sell."

I went on more local TV talk shows, including one with White, trying to rouse public support. Finally, after three public hearings at which more than 80 percent of those in attendance spoke in favor of a no-exception, twenty-four-inch minimum limit, on August 10 the Marine Fisheries Council voted. Mendonsa did not remain silent, as the Conflict of Interest Commission had advised, but both he and Parascondolo abstained from the voting. Nonetheless, the decision was 4-2 to support the Division of Fish and Wildlife's halfway measure raising the limit for rod-and-reelers and keeping the same sixteen-inch minimum for the netters. Public opinion didn't seem to matter, nor the long-term viability of the species.

Jim White stood up at the end and asked: "When did Rhode Island secede from the United States of America?" He was almost in tears outside. "We're not giving up," he told me. "This is only the beginning."

About the only heartening news came from the Vineyard, where the captured net fishermen each received the maximum fine of $1,000 and lost their boat permanently. New York, too, was continuing to hold out against raising its size limit above sixteen inches.

Late that summer, a photographer friend and I decided to make the drive from Boston to accompany Maryland's marine biologists on part of their annual young-of-the-year survey. It was time to visit the Chesapeake Bay, the cradle of striped bass civilization, and we arranged to rendezvous along the way with Bob Pond and Avis Boyd. Not far from Maryland's capital of Annapolis, we met up with Joe Boone. I'd read Boone's research reports on the striped bass population, but this was the first time we'd met. He was a ruddy-faced, sandy-haired man in his late forties, wearing sneakers and a pair of cutoff jeans. I liked him instantly. His bearded research assistant, Jim Uphoff, was younger, with a strong Southern accent. He, too, was a congenial sort, and obviously dedicated to his work.

Boone and I hopped into a four-wheel-drive truck towing a sixteen-foot motorboat. While our two-car caravan followed behind, we continued on past the sprawling cornfields of Kent County, headed toward two rivers that Boone estimated had traditionally supplied close to half of the bay's successful striper production. Last year, when Boone had made his seine hauls in this region, he hadn't found a single one. "These are the critical sites," he was saying. "To get *none* in rivers where we've often seen forty or even fifty in a haul, well, that's something I would have never considered possible."

Even under optimum conditions, Pond had explained, scientists estimate that less than 1 percent of the eggs and larvae will live to be adult fish. So many natural factors are in play. The less buoyant eggs will sink to the bottom and be smothered by sediment. Adequate water velocity is also important in keeping the eggs suspended in the water column. But they can easily succumb to changes in weather patterns. An abrupt cold snap, for example, or spring flood or capricious currents may kill large numbers outright.

So might other environmental conditions, such as high concentrations of chlorinated chemicals or abrupt changes in the water's acidity (more-toxic pH levels) brought on by spring freshets. The geography of the bay and its surrounding watershed makes it especially vulnerable to nutrient overloading. Its shallowness (an average depth of about twenty-one feet) is what accounts for its productivity, by allowing large amounts of sunlight to penetrate to the bottom—but these same shallow waters are not well flushed. The striped bass's vulnerability to environmental shifts is also evident in the larval stage. A sudden change in water temperature, for example, has been shown to have direct lethality on larvae.

The first site we visited was amid the willow trees on the Bohemia River, by an eroded beach known as Long Point. Ever since the survey's inception nearly thirty years earlier, this had been one of thirty locations where a pair of seine samples would be taken at three different intervals, in July, August, and September. Wading out, I helped hold one end of the 100-yard-long net as we circled the shoreline. After we brought it in, Boone pawed through the resulting piles of tiny fish on the beach. As he held up each one to scrutinize it, he looked more like an Indian shaman than a modern biologist. Along with hundreds of baitfish, and a carp and catfish, there was a single baby bass.

It was a beastly hot day, and I went swimming in each of the four rivers we visited. The next survey area required taking the boat out to a sandy beach. But the story was the same: three bass in two hauls. "This is terrible," Boone said. "This is almost always the second-best site. It's low flood tide and the conditions couldn't be more perfect, and we get almost nothing."

The day's work ended no better in the strong currents of the Sassafras River. The day's total was twelve bass in eight hauls. "Ain't gonna send many north at that rate," Boone said, shaking his head disappointedly.

Bob Pond had been trying to get federal scientists to do a study of the eggs and larvae, to no avail. "They will only look at thirty-day-old fish. They measure the food content and say, 'Well, there's not enough food around, so they're starving to death.' Then the chemical company or the agricultural company isn't responsible, because it's 'mass starvation.' But you've got to go back to the eggs and larvae stage and see, are the fish active? Can they feed if they want to? I have a strong feeling that the chemical and farm industry has simply put a 'no' on looking into this."

One could gather, even just from offhand remarks, that a major split existed between the marine biologists and the hierarchy within Maryland's Department of Natural Resources. The previous winter, the DNR *had* announced another spring closure of the spawning reaches to fishing—but not of the rivers that the bass had to swim to get there. At the same time, Tidewater Fisheries division director Pete Jensen had quietly opened tens of thousands of acres in the Aberdeen, Maryland, area to gillnetting for the first time.

As in Rhode Island, commercial fishing interests here were formidable. Larry Simns, lobbyist for the Maryland Watermen's Association, put it like this in a speech to the state legislature, which had the final say on the DNR's recommendations: "As long as the rock [bass] will support a commercial fishery, that's proof the fishery is not dying. Would we catch the last fish that swims, like a lot of people claim? Oh, probably we would, but only if it were economically feasible. Only a sport fisherman can afford to catch the last fish."

One ray of hope was formation of the Maryland Saltwater Sportfishermen's Association, which had attracted 2,000 members in its first year of existence and been instrumental in getting the DNR to close the spawning reaches to fishing. There was also Jim Price, head of the state's Charter Boat Association, who had single-handedly blocked a legislative effort, instigated by watermen, to legalize transparent monofilament nets, which would have been much harder for the bass to see than the opaque nylon kind then in use.

"How do you go up against the commercial fishing tradition?" Joe Boone asked rhetorically at a seafood restaurant that night. "You do it by organizing. So far, the people whose pocketbooks are affected are a lot more vociferous than the people who really care about the bass. That's what has to change."

We said our goodbyes, and the next day my friend and I drove north to a meeting of Save Our Stripers on Long Island. The group was feeling desperate because, despite the desire of New York fisheries officials to comply with the management plan, heavy pressure from commercial interests was being applied to the state legislature, which in this case was charged with enacting fisheries regulations. Most dismaying of all, I learned that some of the commercial fishermen in Massachusetts were now trucking their undersized striped bass directly to New York's Fulton Fish Market.

I kept on writing articles, getting word of the stripers' plight to readers of publications such as *Boston* magazine, *Yankee* magazine, *The Fisherman*, and *The Amicus Journal*. *Newsweek* responded with a short piece, "Schemes to Save the Noble 'Striper.'" Bob Pond called in late September with the news that the Chesapeake young-of-the-year survey showed improvement during its last rounds. The average count was 8.4 juvenile striped bass per seine haul, the best figure since 1978. It was encouraging enough, we both hoped, to keep the population going—and the fight alive.

Chapter Five

How the Striped Bass Stopped a Highway and Eluded the Mob

Striped bass were very much in the consciousness of New Yorkers in 1982—and not just fishermen fighting for a higher minimum size limit. They had swum right into the controversy over a proposed new 4.2-mile, six-lane highway along Manhattan's West Side. The Westway Project, as it was known, had been conceived ten years earlier, hailed by its backers as a "prime example of a new era of urban highway planning and design." It would replace the old West Side Highway and be built underground by filling part of the Hudson River shore, with the highway tunneled inside—and the flat top available for real estate development. The $2–$4 billion project qualified for 90 percent funding from the federal Department of Transportation, a windfall for the city. The project's main attraction included 242 acres of new riverfront property for luxury apartments, office buildings, and parkland. "It's a hustle," as developer Donald Trump confided to *Sports Illustrated* writer Robert H. Boyle. The Westway was David Rockefeller's brainchild, and it was strongly backed by Governor Hugh Carey and Mayor Ed Koch.

Not everyone supported it, of course. A huge increase in traffic would surely result in a decline in air quality. If they chose, New York officials could have withdrawn the highway development plan and still have been entitled to put the same amount of federal money into improving mass transit. In 1980, after two and a half years of acrimonious public hearings, the EPA finally granted an initially denied permit under the Clean Air Act. The only potential remaining obstacle to construction, it appeared, was the U.S. Army Corps of Engineers, which is required to grant dredge-and-fill permits on all navigable waters. In this case, the Westway Project heads wanted to create a massive landfill on 181 acres along the shore—requiring a mountain of dirt big enough, by one estimate, to bury all of Central Park six feet deep. This was where the future real estate would be built, replacing a series of dilapidated piers along a two-mile strip of Hudson River waterfront.

And that's where striped bass came into the picture. The story began, appropriately enough, with Robert H. Boyle.

"Westway had been going on, but it was very quiet," Boyle remembers. "Then one day, I got a call from Peter Silverstein. He'd been thirteen [years old] when he became a director of our Hudson River Fishermen's Association. Peter was then already in the Bronx High School of Science, a brilliant kid. When Con Edison would send lecturers around to talk about the need for power and not doing anything bad on the river, it got to the point where the power company would ask, 'Is that kid Silverstein gonna be in the audience?' Because he'd just tear them apart!"

In 1980, Silverstein, recently graduated from Duke University, was about to begin pursuing a master's degree on genetic variation of striped bass in the Hudson River. He'd stumbled upon a newspaper reference indicating that the National Marine Fisheries Service (NMFS) had dropped its "objections" to the Westway Project. "I asked Bob about that," Silverstein recalls, "and he said he had no idea that the Fisheries Service ever *had* objections! . . . Bob told me to get in touch with Al Butzel, an attorney who'd been involved with the Storm King case and was now representing the New York City Clean Air Campaign in the Westway case."

As it happened, around the same time, Butzel had been called into the office of Chuck Warren, the EPA's Region II administrator. "Chuck was sympathetic," Butzel would remember, "but he said something like 'Godspeed, I hope your lawsuit works, but the politics are such that we really can't veto the air permit.' Then, on my way out the door, he added, 'By the way, have you seen the National Marine Fisheries

Service's analysis of what's going on under those piers?' I said no. Chuck said, 'You'd better get hold of that.'"

Butzel knew that a study of the interpier area had been conducted by the New York State Department of Transportation and passed along to the Army Corps. "I had heard rumors they were pulling up fish left and right, but I never thought that would make any difference. But when I saw the comments that the NMFS had sent in, basically saying the Westway Project had the potential to decimate the striped bass population of the Hudson, well, suddenly a lawsuit on this seemed like an awfully good idea."

After word came back from Silverstein and Butzel, Boyle consulted the board of his Fishermen's Association. They quickly moved to join the case. The initial Environmental Impact Assessment completed by Westway's planners in 1977 had dismissed the area to be filled between the Hudson piers as "biologically impoverished," a "wasteland" containing only barnacles, tunicates, oligochaetes, mud snails, mud crabs, and blue crabs. But in 1979, Westway's proponents had reluctantly acceded to the EPA's insistence that more thorough research be undertaken. The consulting firm of Lawler, Matusky & Skelly (LMS) was hired to conduct a twelve-month biological survey, using boat trawls at eight Hudson River locations. What LMS discovered did not please its employers. There was, in fact, "an astonishing amount of fish life" in the interpier area, fifty times more fish than were found in the river channel.

The most prolific fish of all, especially in the winter months, were juvenile striped bass. Until now, it had been thought these Hudson-spawned fish migrated south only as far as Haverstraw Bay in their first year of life. Apparently, though, the majority of young stripers swam downstream into the lower Hudson. Substantial numbers found calm waters among the broken piers and littered bulkheads of the New York City shoreline and—thanks, ironically, to about 150 million gallons a day of raw sewage discharged by Manhattan into this sector of the river—also found warmer water temperatures and nutrient-rich sustenance in the form of micro-organisms. The LMS survey discovered 15 to 105 times more bass here than at any other site along the river. In fact, a considerable portion of the Hudson's striper population stayed their first couple of winters precisely where the Westway Project proposed to tear down their habitat.

These findings made the Westway planners *very* uptight, and they took action accordingly. First, with the collusion of the state Department of Transportation and the Federal Highway Administration (FHA), they held the LMS study back from the Army Corps for some months. Then the Corps's own "experts" fiddled around with the LMS data. They decided that a "mild winter" explained the bass's choice of an interpier nursery during the trawl survey. "For the striped bass population, the project area appears to represent one of many available habitats," they said. Rather than heed LMS's findings, they concurred with the conclusion of the first survey, that the highway would "cause minimal loss of estuary and productivity for species other than macro-organisms." Ignoring the contrary advice of the NMFS, the Army Corps granted the landfill permit.

In the spring of 1981, Al Butzel filed suit against the government in Federal District Court on behalf of the Hudson River Fishermen's Association, New York's Sierra Club, the Clean Air Campaign, and other groups; simultaneously, Boyle's organization put Ian Fletcher on a shuttle from Boston to New York. Fletcher had been the Fishermen's Association's expert biostatistician on striped bass in the Con Ed case, and now he was heading for Butzel's office to review the Westway files. Fletcher would recall having the day's *New York Times* spread out on his lap during the flight, featuring a front-page picture of President Reagan handing Mayor Koch a poster-sized $85 million check with which to commence construction on the highway. "I said to myself, 'We don't have a chance.'"

What Fletcher found when he began sifting through dozens of Westway-related file boxes is recounted in Robert F. Kennedy Jr. and John Cronin's 1997 book *The Riverkeepers*. Included among the documents were hundreds of pages of data on fish populations collected from thousands of spots in the Hudson River Harbor. "Using this information," Kennedy and Cronin wrote, "the New York State Department of Transportation and the Army Corps of Engineers had projected that

Westway construction would harm the habitat for less than .04 percent of the Hudson River's commercially valuable striped bass. But when Fletcher examined the same data, he concluded that 64 percent of the striped bass habitat would be destroyed. He then retraced the Corps's formulations and proved that their consultants had known the true impacts and had deliberately massaged the numbers to mislead the court."

Fletcher realized that the consultants "had created a giant imaginary grid stretching from Westway across the river to New Jersey. In order to deceive the court into believing that the fish were distributed homogeneously across the Hudson, they then averaged the number of fish in the grid area and assigned an equal share to each block in the grid, thus concealing the high concentrations in the Westway area."

Boyle remembers standing outside the courtroom as the hearing was about to begin, talking to Marcy Benstock, the spearhead of the Clean Air Campaign. "Marcy and her people had been talking about all the cars, the smoke and the pollution, that the Westway would cause. 'Really, Bob, there are striped bass here?' she asked me. I said, 'Marcy, the striped bass are going to win this case.'"

Indeed, in November 1981, Judge Thomas P. Griesa dismissed the air-quality and traffic issues, but agreed to hear arguments on a single issue—the fish. In mid-January 1982, Butzel's witnesses began unraveling the Westway Project's efforts to conceal the truth about the striped bass's piers, starting with Fletcher's damning testimony. Mayor Koch had stated publicly, "Let's hope the striped bass will have the sense to move across to the Jersey shore." A biologist for the U.S. Fish and Wildlife Service testified that this suggestion was comparable to "amputating your right hand because you have another on the left." Boyle put it succinctly: "Striped bass don't go shopping for apartments!"

The NMFS's expert maintained that if young stripers were denied the Manhattan pier area, the very survival of the Hudson population "and its substantial contribution to the Atlantic Coast fisheries would be jeopardized." Given the drastic decline of the Chesapeake's population, the Hudson might now be contributing as much as half of the migratory stock of the species. Especially because many of its other habitats had already been destroyed, this one was especially critical to the bass. Fletcher estimated that somewhere between two million and eight million young stripers enjoyed the wintertime protection afforded by the old piers.

Still, it was hard to read which way the fifty-one-year-old Judge Griesa would go. He'd been on the federal bench for ten years and was regarded less as an interventionist who tried to influence public policy from the bench than as a strict constructionist with a narrower view of the law's interpretation. The question was whether the Army Corps of Engineers' discharges of dredge and fill material would pose a significant threat of degradation and thus violate the Clean Water Act. Judge Griesa could be seen taking copious notes as the trial progressed.

On March 31, 1982, the judge handed down a meticulously written ninety-four-page decision. Having known nothing about the Westway case at the time, and with the ruling coming on the very day that Massachusetts went for the twenty-four-inch size limit, I would never forget the front-page headline in the *New York Times:* "U.S. Judge Blocks Westway Landfill as Threat to Bass." Judge Griesa had invalidated the Corps's permit and enjoined any landfill along the most expensive proposed stretch of highway in human history. If the Corps still wanted to go ahead with the permitting process, it would have to play by the rules.

Judge Griesa did not mince words about the evidence he'd seen presented. The government agencies involved had "manipulated the data in a misleading fashion," and the "obvious purpose was to detract from the startling revelations about the presence of fish in the proposed landfill area." The Army Corps "had no right to swallow up these issues in the privacy of its bosom." Added the judge: "I have sentenced people to prison in securities fraud cases where the conduct was less blatant than the drafting of [the scientific reports]. . . . I am deadly serious about this."

The trial had provided evidence, the judge continued, of "the importance and value of striped bass. . . . The striped bass fishery contributes to the economic well-being and enjoyment of literally millions of citizens. Moreover, because of the environmental problems of the present age, the health of this fishery is a matter of concern. . . . The proposed landfill would have the impact of destroying this habitat."

Governor Carey's office termed it all a "result of a procedural error by a federal agency," and Mayor Koch said that his support for Westway remained undiminished. Joe Conason, writing about the case in the *Village Voice*, said: "Ultimately, as ironic as it may be, this could be a victory provided by the striped bass for transit riders over the highway lobby."

The Army Corps said it would do the job again, and right. At the end of April, new studies were announced to assess how much environmental damage would occur from the Westway. Then, on June 30, Judge Griesa blocked all federal funds for the project. Federal agencies, he asserted, had concealed important information about the striped bass, which constituted a $100-million-a-year recreational fishery. The Federal Highway Administration and the Westway Project had, in fact, "colluded in a successful effort to persuade the Corps of Engineers to refrain from issuing an environmental impact statement . . . on the subject of fisheries." Testimony on behalf of the project and the agencies "was characterized not only by a striking lack of plausibility on critical points, but also by a remarkable amount of consistency, evasion and asserted loss of memory on matters where memory would be expected." Judge Griesa proceeded to appoint a Special Master to monitor compliance with his orders requiring new environmental reports.

That summer, on my way back from the Chesapeake, the judge granted me an interview in his chambers. He was tall and thin, an avid tennis player, a harpsichordist and pianist who often performed chamber music with his friends. He told me that he'd had his first meal of striped bass when he took the case—and loved it! Now, he added, he also wanted to go on a fishing trip.

"I think it would be a disservice to the public to let things just go back to the Corps and FHA and let them take their course," the judge said. "There is no ironclad, surefire way I can make them do their duty but, given their performance so far, I wonder if it's going to be another farce, a charade. That's why I've appointed a Special Master."

Looking back on the Westway controversy years later, Al Butzel recounted: "One of the many things that Bob Boyle did for me was to make me understand how significant striped bass are, and have been. It was exciting to challenge Con Edison, but I came to feel that was also an opportunity to really do something that would protect the fish. Then when we got involved in Westway, it was—hey, we're extending what people know about the life cycle of the striped bass! Only as a result of this effort did people begin to realize that the piers had actually become the substitute for the wetlands and the shallows that used to be along the edge of Manhattan when people first settled here. So we'd created an artificial environment which, in its own way, had replicated at least to some degree the natural environment that had allowed the bass to survive and do well.

"When we put on this case in front of Judge Griesa, it was storytelling, educating this judge who didn't know anything about striped bass—about the species, what it meant, how they moved, why they were in jeopardy, and how it made logical sense. It turned out he spent his summers in Cornwall, so he had a lot more affection for the Hudson than we'd ever known going in. But he listened! By the time we got to final arguments, I was so utterly familiar with the striped bass, I could make them come alive for the judge. While I may have been helping them, well, they were helping me."

Bob Boyle put it this way at the time, in *Sports Illustrated:* "The striped bass is a symbol of where we stand in the world. The federal government was going to give up a couple of billion dollars to New York to screw up the Hudson. . . . But Westway is flat on its back and the referee is counting nine. . . . The striped bass has been the noble creature that has led all our fights in the Hudson, and so far he remains undefeated."

Whether Boyle's optimism about the striped bass could prevail along the coast remained, at best, an open question. Commercial landings for 1982 were recorded at 2,164,000 pounds, the lowest figure since 1935 and a 90 percent drop in less than ten years. Maryland biologist Joe Boone, in his *1983 Tidewater Fishing Forecast and Homily*, noted that the "moderate" year class born the previous year would offer a temporary reprieve from full collapse, but if current fishing practices continued, the bass would be virtually wiped out by 1986. Such warnings in the spring of 1983 didn't seem to deter commercial fishermen on Cape Cod, who turned in a petition with 500 signatures asking to go back to the old sixteen-inch minimum size limit. In early May, Massachusetts's Fisheries Council voted unanimously to keep the twenty-four-inch size limit in place.

That same spring, the Maryland state legislature finally decided to act. In late February, 200 sports fishermen had marched on the state capitol demanding stronger regulations. Intensive lobbying by Charter Boat Association leader Jim Price had led to a bitter wrangle with commercial watermen, whose efforts to maintain the old twelve-inch minimum size inside the Chesapeake Bay were supported by state fisheries officials. On the last day of the legislative session, a bill was passed raising the Maryland bay limit to fourteen inches (twenty-four inches in coastal waters), as called for by the interstate management plan, an action that would protect thousands of young striped bass from Chesapeake gillnetters. Neighboring Virginia adopted the same measures and followed Maryland's lead in closing its spawning reaches to bass fishing during the spring.

Rhode Island was still recalcitrant. Their fisheries people were reportedly working to weaken the management plan, to allow states with traditional net fisheries more freedom in implementing regulations. A bill was introduced in the state legislature that would have forced all fishermen to adhere to the twenty-four-inch minimum and changed the makeup of the Fisheries Council. But when I went to testify at a legislative hearing in Providence, commercial netters packed the hall and the proposal didn't even get voted out of committee.

New York, it appeared, had become the key battleground. Now that Massachusetts had held the line on the twenty-four-inch rule, a bill to raise the size limit from sixteen to twenty-four inches was again introduced in New York. There were two major voices of opposition. And both carried a lot of weight.

Fred Schwab, a founding member in 1967 of Save Our Stripers (SOS) on Long Island, had been fighting one of those voices, that of the haul seine netters, for fifteen years. I'd met Fred, an administrator for the Department of Public Works in Nassau County, the previous summer at an SOS meeting on my way back from the Chesapeake. We now spoke regularly by phone and exchanged newspaper clippings from our respective regions by mail. The word wasn't yet in fashion, but I guess we were "networking." He told me how wasteful the haul seine fishery was. It encircled fish with a 2,400-foot net and was utterly nonselective in terms of species and size, resulting in an extremely high fishing mortality in all age groups.

Like Bob Pond, by 1970 Schwab and SOS were already starting to worry about the future of the striper population. On April 1 of that year, an Ad Hoc Committee for the Preservation of Striped Bass had taken out a full-page ad in the *New York Times*. "One man against the survival of a species," it was headlined. The ad told of a bill to make the bass a no-sale-allowed gamefish in New York waters. It had already received 54-1 approval in the state senate but had been pocketed by the speaker, Perry B. Duryea Jr. of eastern Long Island, so that it would not reach the assembly floor. A wealthy Republican and, at the time, the most influential politician in the state besides the governor, Duryea owned a wholesale fish business in Montauk and was closely aligned with the haul seiners.

The striper gamefish bill sailed twice more through the senate, only to be blocked each time by Duryea in the assembly. In the summer of 1972, SOS even launched an economic boycott of the Montauk area. Forty-two organizations representing 70,000 sports fishermen pledged to support it. Motel reservations and charter-boat trips were cancelled in droves. The boycott hurt—but it didn't stop the haul seiners.

"SOS from a Seaside Slaughter," Boyle's November 6, 1972, article in *Sports Illustrated* was headlined. It told of recreational fishermen slashing nets out on Long Island, where the seiners had left beaches cluttered with unwanted dead fish, including striped bass under sixteen inches. The same crew of haul seiners had branched out, Boyle noted, carting their double-end nets to the Outer Banks of North Carolina to scoop up tremendous hauls of roe-laden striped bass wintering-over before returning to the Chesapeake. After meeting stiff opposition from local North Carolina fishermen, the haul seiners confined their operations to Long Island. But their animosity toward those they termed "the sporties" did not abate. In Schwab's view, those "sporties" would turn out to be instrumental in defeating Duryea's bid for the governorship against Hugh Carey in 1978.

Besides his championing of Westway, Carey hadn't proven to be any friend of striped bass management either. In 1980, a twenty-four-inch limit bill had passed the legislature, only to be vetoed by Carey on grounds that other states had not enacted similar measures. By 1983, New York had just elected a new governor, Democrat Mario Cuomo, and SOS's longtime nemesis Duryea had left the assembly. So there seemed hope that a striper bill might finally sail through.

The haul seiners, however, found a new legislative champion who cited allegedly scientific calculations that raising the striper size limit to twenty-four inches would reduce the landings of the net fishermen by half. This, complained the haul seiners and their allies among the gill-netters, would drive them out of the bass business and effectively end a several-centuries-old tradition.

The state senate, having supported striper protection bills in years past only to find them bottlenecked in the assembly, insisted this time that the assembly act first. A freshman assemblyman, Patrick Halpin, took up the striped bass as his personal crusade. SOS and a nascent New York Sportfishing Federation lobbied on the bill's behalf and generated a massive letter-writing campaign. Their opponents included more than the commercial fishermen, however. A lobbyist for the Fulton Fish Market maintained that any size increase would "substantially raise prices for the already overburdened consumer."

Located next to the East River, in a shabby two-block area just north of Wall Street, the Fulton Fish Market was at the time the biggest sales outlet of its kind in the United States. Between two and three in the morning, while most New Yorkers slept, it came alive as fish retailers from across the Eastern seaboard began conducting the day's business with some sixty wholesalers. In 1981, those wholesalers had grossed about $325 million from the sale of 168 million pounds of fish. And, it had been revealed by the *Wall Street Journal* on April 14, 1982, they had shelled out $700,000 in protection money since 1975 to the Genovese organized-crime family. According to the U.S. Attorney's office, all the fish coming into Fulton were unloaded by a half-dozen companies that had been granted monopolies by the Mob. A federal investigation into Fulton's practices, begun in June 1979, had so far resulted in the conviction of eleven firms and eleven individuals. But this hadn't really cramped their style.

About 80 percent of the East Coast's commercial catch of striped bass typically was shipped into Fulton. Fishermen from Connecticut—where the sale of bass was outlawed altogether—simply traveled to neighboring New York to sell their catch. Small fillets were the most popular among Fulton's grocer and restaurant customers, but if New York's minimum size limit were raised, those lucrative smaller fish would no longer be legally available.

Rumor had it that folks from Fulton had twisted Carey's arm to block any such legislation. A New York State marine biologist told me, off the record: "The Fulton Market was contacted when these bass management proposals were first drawn up, and they weren't the slightest bit responsive. They have their own lobbyists, and we think they are a very strong force in the state legislature. They may have a lot more to do with all this than we can begin to understand."

Years later, fisheries biologist John Boreman would tell me this story: "We received a visit once, during one of our ASMFC technical committee meetings, from a fellow who was involved with the Fulton Fish Market. He came wearing a three-piece suit and addressed our whole group of scientists. Basically, he told us that his uncle had set him up in the business and would be very unhappy if we did *anything* to hurt that business. Then he went on to try to convince us that the striped bass coming into the Fulton Fish Market were a different stock than the one we were trying to manage on the coast—so we shouldn't worry about it. 'Trust me,' he said. Then he left. We all looked around at each other. Finally, one of us said, 'Well, we're still here, so. . . .'"

Such was the behind-the-scenes climate in the late spring of 1983 as the minimum size bill again went to the full New York Assembly for consideration. A four-hour floor debate on the afternoon of June 15 proved the longest and most acrimonious on *any* issue to come before the assembly in several years. Patrick Halpin even brought in a mounted fifty-six-inch striped bass and held it up to his fellow legislators, declaring that this showed just how large a bass could grow if given time to mature. The haul seiners' man shot back that a bass this big constituted clear evidence that the species was thriving nicely and needed no further help. But when the tumult settled down, the vote was 98-49 in favor of the fish. Halpin's picture with the mighty striper appeared on the front page of *Newsday*. The commercial netters hired a full-time lobbyist and turned their ire on the senate. A couple of days before its scheduled vote, Fred Schwab got word that the senate majority leader—who had the power to keep the bill from reaching the floor—was leaning against it.

"The president of SOS, Bob Buss, and I drove about 165 miles to Albany and walked into the man's office," Schwab recalled. "We told a secretary that we wanted to speak to the majority leader about the striped bass bill. She said, 'Oh, of course, somebody else was just here.' She mentioned the name. We knew the person was from the Fulton Fish Market. We thought, 'Oh God, we're dead!'"

Schwab figured the wisest course of action was to forget about seeing the senate majority leader. Making an abrupt turnabout from his office, he and Buss went instead to see the number two man, the majority whip. "We told him that our bill was in trouble, that the senator who was supposedly carrying the ball hadn't pushed it at all. 'For chrissakes,' the whip said, called up the prime sponsor of the bill, and really chewed his ass out."

That, perhaps coupled with a threat from the wealthy leader of a Staten Island environmental group to land his private plane on the senate lawn, apparently saved the bill. After an hour-long debate on a Sunday, the last day of 1983's summer legislative session, the New York Senate voted 37-13 to pass a landmark striper protection law. It not only raised the size limit to twenty-four inches but also banned any bass fishing between December 1 and April 21, and instituted a $100 licensing fee for anyone selling a bass.

Even with votes in both houses secured by handsome margins, the bill's primary sponsor, Assemblyman Halpin, delayed for a short time sending it on for consideration by the new governor, Mario Cuomo. Halpin wanted to allow time for supporters to generate letters and phone calls to Cuomo's office. For, almost immediately upon the bill's passage in the legislature, a massive counterattack had been mounted. Local newspapers and TV stations ran numerous stories about the haul seiners' time-honored fishing tradition. Pressure from the Fulton Fish Market was subtler but no less intense.

Rhode Island, not surprisingly, was waiting anxiously to see whether New York would move. A reluctant Cronan, the fisheries director, told the Marine Fisheries Council at the June meeting: "If the New York legislation passes and we retain the twenty-four [inch limit for rod-and-reel] and sixteen [inch limit for netters], what it boils down to is that, regardless of our personal feelings, we've got some real problems." Any further action was tabled pending the outcome in New York.

Meanwhile, the fish situation was worsening again in the Chesapeake. A spring trawl survey had turned up very few surviving newborn fish of *any* species in the Choptank River.

Even in purely economic terms, the facts seemed clearly on the side of more meaningful conservation. According to the latest bulletin from the federal government's Emergency Striped Bass Study, issued in July: "almost $220 million of economic activity and 7,000 jobs have been lost to the coastal areas of the ten [Atlantic] states as a result of the decline, since the early 1970s, of the striped bass resource."

Was Governor Cuomo paying heed? The assistant publisher of *Salt Water Sportsman*, Spider Andresen, who docked his boat next to ours on Martha's Vineyard, didn't think so. A source of Spider's in New York had lunch with the governor and reported that he had made up his mind to veto the bill when it crossed his desk.

Besides the pressure from the haul seiners and the Fulton Fish Market, there was the Westway Project for Cuomo to consider. "Since Cuomo's ascension to the state house," wrote Joe Conason in the *Village Voice*, "the state has done everything in its power to frustrate the intention of the judge's orders and to cover the misdoings revealed in [Thomas] Griesa's court" when the judge stopped the project the year before. Early in 1983, Judge Griesa was so incensed that he ordered a trial to determine whether the state should be held in contempt, an edict that was withdrawn only after assurances that New York and the Army Corps would proceed with drafting a new environmental analysis in good faith.

Al Butzel would remember: "It became very personal between Cuomo and us, and me. I remember once, shortly after he was elected, he invited the environmental groups to meet with him. Somebody called me and said, 'Will you join us, because we want to talk about Westway.' So I went to Albany. Cuomo walked into the room and announced, 'If Al Butzel's here, I'm not gonna stay!' Well, he didn't leave. I didn't say anything, either."

After the regional head of the Army Corps was found to have a conflict of interest because he had sought a job with a Westway contractor, Colonel Fletcher Griffis came to replace him. The colonel undertook to learn about Westway's potential impact on the striped bass and, as the court had directed, consulted formally with a large group of independent scientists. During the summer of 1983, around the same time the striper size-limit bill passed the senate and was soon to reach his desk, these scientists reported something Governor Cuomo didn't want to hear: it would be impossible to determine the project's impact on the striper fishery—let alone how to mitigate it—without at least two more years of study. The state's response was to try to overrule Colonel Griffis politically, by appealing directly to the Corps's overseer, the assistant secretary of the Army—a man regarded, according to reporter Joe Conason, "as an enemy of the environmental movement in the mold of [Reagan administration Interior Secretary] James Watt."

Why was Cuomo so adamant about the Westway Project? Some said he was beholden to the building trades and operating engineers unions. Others spoke of the influence of Chase Manhattan Bank president David Rockefeller, who was at the center of Westway's group of investors, bankers, and big businessmen. Whatever it was, the legislature's passage of a much stronger protection bill for the striped bass only accentuated the fact that the fish was in too much trouble to sacrifice its piered Manhattan habitat. If the governor acquiesced and signed the bill, it was admitting as much.

Given the circumstances, the only possible salvation for the bass seemed by way of the media. In July 1983, I got hold of someone at the *American Sportsman* series, who in turn contacted ABC News. They called about doing a possible network story the following week. Five minutes after I hung up the phone, out of the blue came a call from NBC's national news headquarters. The correspondent, Fred Briggs,

had read my article in *Yankee* magazine about the striped bass's plight
and was interested in doing a segment. Talking fast, I filled him in on
what was happening in New York.

A few days later, the *New York Times* came out with a lead editorial urging the governor to veto the striper bill. The new size limit, the
Times maintained, would "make life impossible for commercial fishermen on Long Island." Better that New York wait on any action until
Congress completed its emergency study of the fish's decline in 1985.
The editorial made no mention of the cooperative interstate management plan, nor of the adoption by nearly every other coastal state of
that plan's twenty-four-inch minimum size requirement. Indeed, the
bass bill was made to seem an arbitrary piece of legislation being
imposed on the hapless netters by New York sportsmen.

Whose lead was the *Times* following? I'd learned a good bit over the
past year about how such influence works, having found allies among
the *Boston Globe* and *Providence Journal* editorial writers. The editorial
pages of the *Times* also leaned toward the establishment's desires to
build Westway. And who could ever say about the influence of the folks
from Fulton?

The *Times*'s outdoor columnist, Nelson Bryant, told me he'd never
been consulted by the editorial staff, despite his lifelong expertise on
striped bass. Indeed, when Bryant followed up with a column refuting
much of the editorial, the *Times*' sports department killed it. It seemed
a blatant example of censorship, though Bryant was informed that what
the editorial staff wrote was not the sports page's concern. In another
development, after Assemblyman Halpin responded with an immediate
letter to the editor via express mail, the newspaper told him that his mis-
sive couldn't be located. Instead, the *Times* ran a letter from a commercial netter supporting the paper's stand. If editors at the *Times* hadn't
made up their minds long since, someone certainly appeared to be
applying effective pressure on the city's "paper of record."

Other New York media, including the *Daily News* and the local
ABC-TV station, carried considerable editorial coverage also condemn-
ing the striper bill and bemoaning the seiners' financial losses. The

Boston Globe fired back with "The Striped Bass Massacre," urging that "Massachusetts residents, beginning with Governor Dukakis, should mount pressure to bring New York and Rhode Island into line with the interstate management plan."

Governor Michael Dukakis took heed. On July 25—the same day the *New York Times* editorial appeared—he penned a strongly worded two-page letter to Governor Cuomo. "Massachusetts was the first state to implement the twenty-four-inch minimum size provision," Dukakis wrote. "Contrary to rumor, this measure was very costly to the Commonwealth in terms of lost poundage and was not popular with a sizable contingent of commercial fishermen. However, we proceeded in spite of this opposition for several reasons. First and foremost was our growing concern over the depressed state of striped bass stocks. Secondly, we hoped that timely action on our part would prompt other states to follow suit." Dukakis concluded: "In the interest of striped bass conservation and in the spirit of interstate cooperation, I urge you to sign this measure into law."

Not long after, Dukakis reportedly cornered Cuomo about the bass bill at a meeting in Maine of Northeastern governors. But New York's new governor remained noncommittal. By the time the bill reached his desk on July 27, 1983, Cuomo's legal counsel, Benjamin Wiles, had already held a two-hour-long meeting with each side. "This took by far more time than any other bill this year," Wiles later told me. "Rarely during the bill period do I meet with anybody. In this instance, I felt I had to. I don't know how many telephone conversations I had, but there were dozens."

Several thousand letters poured into Governor Cuomo's office concerning the striped bass. The day he received the bill, sports fishermen and conservation groups spent all day deluging Albany's Executive Chamber with telephone calls. They were not alone. An official with New York's Department of Environmental Conservation said only: "It's gone well beyond us. Every group opposing this bill is using old-time union tactics as far as getting to people, sending new faces into Albany every day."

On July 28, Cuomo attended the grand opening of a newly refurbished Fulton Fish Market. On July 29 he traveled to Long Island, into the haul seiners' territory. The striped bass had become the most controversial issue in New York politics. The outlook for the bill could only be deemed bleak.

It was also at the end of July that a camera crew from *NBC Nightly News* arrived, at 7 a.m., at the Martha's Vineyard airport. I'd arranged with correspondent Fred Briggs to meet them, and from there he hoped we would encounter, at one of the local beaches, some fishermen to interview about striped bass.

When the NBC team and I arrived at the wooden walkway leading down to Squibnocket Beach that morning, I pointed—"Oh look, there are some fishermen!"—and proceeded to trip and fall flat on my face. Somehow I managed to hold my rod high enough so that sand didn't get caught in the reel. I brushed myself off, adjusted my cap, and prepared for prime time.

NBC was also filming on the Hudson River and in the Chesapeake Bay—the first round of the young-of-the-year survey was over, with about one striped bass per seine haul indicating another dismal index, and Briggs said they would be paying a visit to the Fulton Fish Market. They would also spend a day on the water with George Mendonsa and would attend the Rhode Island Marine Fisheries Council meeting scheduled that night.

When I saw Briggs at the council session, he shook his head and said he didn't know how he felt about the matter anymore. Mendonsa had obviously wooed and captivated him. But with national TV cameras rolling, fisheries director Cronan stunned the crowd by not waiting on New York but suddenly recommending that the state go ahead with the twenty-four-inch limit and adopt the management plan in full. I was as surprised as anyone, but guessed that perhaps all the pressure to do something about the bass's situation was getting to him. The measure would be sent out to public hearing before a final determination was made by the council, however.

Governor Cuomo's decision now loomed more critical than ever. Monday, August 8, was the deadline for him to act on all legislation passed in the 1983 legislative session. By then, he had already followed through with approvals or vetoes on more than 700 other bills passed by the state legislature. He had only one left to consider—the striped bass bill. If the governor did not sign the striper protection bill by midnight on August 8, this would constitute a pocket veto.

Legal counsel Wiles had prepared a lengthy memo for Cuomo, and the governor had called him back with a set of additional questions. Twice more on that critical Monday, the governor telephoned Wiles for further consultation. According to sources within his office, Cuomo almost requested an additional twenty days to make up his mind—something that he legally could do, because the state senate had recessed late this year, but that no New York governor had done for *any* legislative matter for the last thirty-five years.

Rather than act during the workday, Governor Cuomo decided to wait. He wanted to see what *NBC Nightly News* would report across the country. The striped bass segment, his aides were informed, would air that very night.

On Martha's Vineyard, my friends and I waited, too. It seemed almost surreal that the network's story would appear just as Cuomo was up against the deadline on the bill. But then, that was the way of the striped bass. My nervousness only increased when I reached Wiles by phone and learned that the governor planned to watch the newscast before finally making his decision.

Late that afternoon, I managed to get in touch with correspondent Briggs. "If you do nothing else," I said, "you've got to say that Governor Cuomo is getting pressure from the Fulton Fish Market *not* to sign this bill!" Briggs was noncommittal.

The governor, I knew, was of Italian heritage. Raising the Fulton red flag would not look good to his constituents. A governor with presidential aspirations could not afford to appear beholden to the Mafia.

At 7 p.m., the face of NBC anchorman Roger Mudd flashed across the screen. This was before the days of CNN and dozens of other cable news channels. Millions of Americans basically got their day's news from one of the three networks. What NBC said tonight would put the striped bass on the national map, for better or for worse.

A segment that lasted for a full two minutes and twenty seconds began with Mudd intoning: "Anybody who knows anything about salt-water fishing, from Maine to the Carolinas, knows the striped bass is in trouble. Because of pollution, because of weather, the striped bass catch is getting smaller every year. And as Fred Briggs reports tonight, there is now a battle between the sport fishermen and the commercial fishermen over one of the country's great game and table fish."

There was a sound bite of yours truly, saying, "The bass is a special fish; it's got a magic to it," which cut to: "To commercial fishermen, the magic is the price stripers bring." Considerable footage of Mendonsa and his crew hauling their traps followed, with George claiming that we wanted to get rid of net fishermen. The "fundamental conflict," as Briggs put it, was "fishing for fun versus fishing to make a living."

It was hard to tell where the telecast was headed. More conflict: "Catching a striped bass today is like trying to catch a zebra," said one of the sports fishermen along Squibnocket Beach, with Mendonsa countering, "If the state of Rhode Island went to a twenty-four-inch measure for everybody, it would put us out of the business."

Then, zeroing in on a busy, wet street crowded with trucks and pushcarts: "The Fulton Fish Market, major buyer and distributor of fish for the East Coast—it's in New York, and New York's legislature recently passed a bill for the twenty-four-inch limit. But Fulton, and commercial fishermen on Long Island, have put pressure on Governor Mario Cuomo to veto it. The limit may be no cure-all. Chesapeake Bay pollution, even changing weather patterns, could be a part of the reason for the striped bass decline. All that is certain is that there are fewer of them every year, for sportsmen, for net fishermen. And both will be poorer without them."

Fulton pressuring the governor! They said it! Damned if they didn't say it! My friends and I stood up and cheered.

At eleven o'clock that night, I couldn't wait any longer. I dialed the home phone number of Governor Cuomo's legal counsel. "Did the governor watch the news?" I asked.

"Yes," Benjamin Wiles replied. There was a pregnant pause before he continued: "I can't tell you officially, but I don't think you have anything to worry about. I think you can sleep well tonight."

We cracked open a bottle of champagne.

Out on Long Island, Fred Schwab recalled years later: "I went fishing that night. I hadn't fished much that summer. Spent most every other day in Albany. My wife said, 'Don't you want to stay home and hear whether the governor signs it or not?' Well, I didn't think he was going to, I really didn't. So I went fishing. Didn't catch any fish.

"My reward, I guess, came later that same year. Thanksgiving morning here, 1983. I caught the biggest bass in my life. I had a bad back that night, I'll never forget that, and I was alone. She was the only fish I caught. I remember trying to lift her up to weigh her. Her tail tip was still on the ground. She bottomed the tube scale at fifty-five [pounds], but I know she was well over that, probably sixty-five. I've only seen one that was ever bigger. But I didn't want to take a lot of time, because I knew I was gonna put her back in the water. If it had been three years earlier, I would've put her on the wall. But by then, I wasn't gonna kill 'em. So . . . she went back. Wish I'd had a camera with me, though."

Two days after Cuomo signed the bill, I got a call from Rhode Island. A new group, the Saltwater Fishing Alliance, had been formed. Jim White was going to be its president. The Fisheries Council planned to vote on the twenty-four-inch limit at its next meeting, in three weeks.

And nobody could have guessed what George Mendonsa had up his sleeve.

Chapter Six

How Rhode Island
Changed the World

Once New York acted to raise its minimum size limit, the pressure intensified on Rhode Island to follow suit. "Rhode Island's turn on bass," the *Boston Globe* editorialized. Governor J. Joseph Garrahy was being deluged with letters. Fred Schwab, writing from Long Island, sounded like a delegate to the Constitutional Convention: "What one state does, or fails to do, is the proper concern of all!" Jim White castigated the governor's apparent "unawareness of this situation" and added: "What this issue is really all about [is] . . . not the survival of the Bass but personal greed."

After my article featuring White had been published in *Yankee* magazine, he'd become something of a hero along his mail route. People had even sent him several hundred dollars to help the striper cause. He'd seen a school of little bass in a Rhode Island river, all playing with their tails out of the water, the likes of which hadn't happened since White was a boy. It seemed a propitious sign. After fisheries director John Cronan in early August had endorsed the twenty-four-inch limit for everyone as his department's position, we were looking forward to the public hearing. Fred Schwab even drove all the way to Providence from Long Island to attend, accompanied by a colleague from SOS and both of their wives. Schwab spoke impressively about the successful attempt to override the opposition of net fishermen and bring New York into the cooperative conservation fold.

Citizen after citizen at the hearing called for going to twenty-four inches. Only one fisherman spoke for maintaining the status quo. There were two other proposals besides Cronan's up for consideration. One called for a compromise, raising the limit to somewhere between sixteen and twenty-four inches. Another, which none of us took seriously, called for shutting down the striped bass fishery completely, for commercial as well as sports fishermen. I probably should have paid heed to the fact that a few of George Mendonsa's trap-netters spoke for just such a solution—a moratorium on *anyone* fishing for striped bass. Big George himself, still facing potential conflict-of-interest charges, stayed silent and stony-faced throughout the hearing.

Less than a week later, on September 6, 1983, the Marine Fisheries Council gathered in Wakefield to render its decision. It was a crisp night, a good night for fishing, but the room was full. Mendonsa sat next to another trap-net company owner, Francis Manchester, an elderly gentleman fond of quoting figures about striped bass abundance from a 1941 newspaper clipping.

Years after, Mendonsa would recall: "The other commercial guy and I, we had it rigged. We knew we was gonna lose, we couldn't fight the sports people. So I'd said to Francis, 'We're goin' down the drain on this, we're gonna lose the striped bass fishery. Well, let's take 'em down with us!'"

Manchester opened the night's proceedings with a long justification of the current sixteen-inch minimum size limit, manipulating figures as he went along to prove his point. Then, defying his own logic, Manchester said: "To put everyone on an even basis, I move to prohibit the taking and possession of striped bass in Rhode Island . . . to become effective no earlier than December 1, 1983, and expire one year after."

There was what could only be called a deafening silence in the room. Mendonsa broke it. "I will be affected by any rule adopted," he said, "but to no greater extent than any other commercial striped bass fisherman. As will be seen by my *vote*"—he all but trumpeted the word—"I intend to vote in such a manner which places the greatest restrictions on the catch of striped bass in Rhode Island. I do this to show my commitment to the conservation of marine fisheries, specifically striped bass—and as a challenge to those who would use the Marine Fisheries Council as a tool to advance their own selfish ends!"

My mind was a swirl of thoughts. Obviously, if Mendonsa and his bass-trapping brethren voted against a twenty-four-inch limit, they faced more serious accusations of conflict-of-interest problems. If they abstained from such a vote, the twenty-four-inch law would almost surely pass. But if they voted to ban bass fishing altogether, for themselves and everyone else, there was no clear conflict of interest. The ban wouldn't take effect until *after* the 1983 fall run of small migrating bass, and, after biting the bullet for a year, the commercial interests could move to go right back to the old regulation. Perhaps they reasoned that, by then, the rod-and-reelers who were denied the right to sell their bass or bring one home for dinner would be ready to play ball.

I counted heads at the conference table. Saul Saila, the marine biologist who generally offered a voice of reason, was absent. So was a boatbuilder who'd begged off from attending at the last minute, saying he had to rise early to go to New York. This left the three trap-netters and a commercial dragger fisherman, plus two representatives of the sportfishing community. One of the latter, Louis Othote, responded by making a motion to adopt the twenty-four-inch size limit. "We're very close to coastwide success with putting this plan into effect, and to go in this other direction is as bad as not doing anything," he said. Then the dragger fisherman retorted, "In my opinion, Lou is completely wrong"—and the die was cast.

Robert Bendick, the new young head of the Rhode Island Department of Environmental Management (DEM), was as taken aback by the startling turn of events as anyone. "I think if the council is to maintain its credibility, it has to vote based on the best facts available to it at the time," he said, not based on "anger and disappointment at the way something has evolved. . . . I think this will lead to an endless war between the commercial and recreational fishermen, which is the last thing we need."

Loud applause greeted Bendick's tough stand. Manchester then turned on Cronan, demanding to know whether the fisheries division's change in position was based on new biological reasons or social considerations. The ever-equivocating Cronan conceded that the latter were key to his supporting the size limit increase.

Bendick was getting hot. "Yes, biologically nothing has changed since last year, which is part of the point. We keep looking for an upturn in the bass population, and it hasn't come." Manchester interrupted to read aloud from a 1982 National Marine Fisheries Service press release, stating that there were more striped bass larvae in the Chesapeake than ever before.

"Yeah, and in the next paragraph it says they all died!" Bendick shouted.

"Well, if they're all gonna die, why are we letting the damn fish go?" cried Manchester. "Why don't we take 'em while they're here?!"

A chorus of boos came from the audience. State troopers shuffled their feet nervously by the doorway. Manchester pressed for a vote on which motion to consider first and prevailed with his own, 4-3. "I think we're playing games, fellas," said Bendick, "and it's not gonna end well."

The vote on the one-year total fishing ban proceeded. Mendonsa, Manchester, trapper Michael Parascondolo, and dragger Bob Smith voted aye. Othote, Bendick, and sportsman Joe Dawson voted nay. The twenty-four-inch limit never came up. The room erupted in fury.

"For three years we tried to do something, and I never saw you open your mouth!" one sports fisherman shouted at Mendonsa.

"You tried to give us the shaft for three years!" Mendonsa fired back.

A local bait-and-tackle shop owner named Joe Mollica stood up, shaking with rage. "I live in front of your net and people watch you through binoculars in disgust! You kill every goddamn thing that swims up and down that area. That area is vacant!"

"I invite you to come out on my boat, spend a year with me!" Mendonsa yelled back, his huge hands clenched in the air.

"I wouldn't spit on the same ocean with you!" Mollica shouted.

Years later, Mendonsa reflected: "Well, the sportsmen in that room were jumpin' up and down like yo-yos! I was riding with Francis. I said, 'Francis, your car's out there in the parking lot. You see the attitude of them guys? We better get the hell outta here before they do something to your car!'"

In unison, Mendonsa and Manchester stood up. Bendick hastily adjourned the meeting as everyone headed for the exits. Outside, I stood next to Jim White, Will Barbeau, and others who had carried the fight. Nobody could say a word.

Overnight, as it was in New York, the striped bass became the hottest political issue in Rhode Island. Governor Garrahy stepped in, his top aides meeting for several hours with the commercial industry. "The state told me right to my face, 'This is a disgrace that Rhode Island is now known as a non-sportfishing state, and we don't want that reputation,'" Mendonsa would recall. "I said, 'Well, I don't want it either, but don't try to give me the shaft while the other guys gain by the regulations. Make it even.'"

Reaction to the "even" idea was swift in both national and regional media. A short article in *Sports Illustrated* termed the vote "a perverse boost" for the imperiled striper. The *Boston Globe*, in a Sunday lead editorial called "Captains Outrageous," compared the council members to "petulant children who flounce off with the marbles when the game isn't going their way," while the *Providence Journal*'s editorial called for the vote to be quickly rescinded.

Mendonsa and Manchester were laying low and refusing any comment to reporters. Governor Garrahy wrote a letter to the fisheries division asking for a compromise. It was rumored that the

DEM's Bendick had asked the council to meet in emergency session, but had been rebuffed. Meanwhile, Rhode Island's sports fishermen were busy organizing their counterattack. Membership soared in the newly formed Saltwater Fishing Alliance, presided over by its founder, Jim White, with tackle shop owner Mollica signing up eighty members by himself. A "gathering of forces," as Will Barbeau put it, took place at White's home. Barbeau had drawn up a proposal to restructure the Fisheries Council. He even enlisted the assistance of the public-interest organization Common Cause. And complaints against the net fishermen were resubmitted to the Conflict of Interest Commission.

In the waters off Newport, someone cut the buoys attached to Mendonsa's nets. He spent the next few nights sitting on a promontory overlooking the sea, rifle in hand.

By the time the council reconvened in early October, this season's young-of-the-year survey results would be due in from the Chesapeake. So far, it looked to be another dismal year for survival of newborn bass. I made arrangements to meet up again with Joe Boone and Jim Uphoff, and to attend a meeting of federal scientists in Washington discussing their latest findings on the bass. Anthony Gude, my younger fisherman friend who'd been instrumental in the fight, had never been to the region. We left Martha's Vineyard together on September 21; Bob Pond would meet us in Woods Hole after the forty-five-minute ferry ride. The three of us headed south in his station wagon.

Pond was simultaneously distressed and pleased. He had just received a letter from John P. Wourms, a noted developmental biologist at Clemson University. Having examined the materials and documents Pond had sent him, Wourms wrote: "Even on the basis of preliminary information, the evidence is striking. There is something radically wrong with the early embryonic development of the

Chesapeake Bay population of striped bass." Pond's fear was that a genetic change had taken place in the bass born over the past decade or so. The eyes, nerves, and immune system, he observed, were not functioning as well. "That's why, as they grow older, they don't go after lures the way they used to," he said. "They just can't see them."

Anthony, who was meeting Pond for the first time, nodded in agreement. "That's what I've been noticing," he said.

"Everything is so compartmentalized in research," Pond continued, shaking his head. "The fishery biologists can only work from the surface up, not down. In other words, they take a fish that's swimming around in the environment and that's where they work. Below that belongs to the developmental embryologist, who starts at the first cell. The trouble is, once people get money in their mitts, they aren't going to pass it to somebody else. We testified at the Chafee [Senate] hearings that we need a multidisciplinary approach and suggested a committee of scientists—biologists, biochemists, all the way down the line."

In a heavy rainstorm, we arrived in Delaware by early evening, then rose before dawn to drive into Washington for the Emergency Striped Bass Study conclave. Phil Goodyear of the U.S. Fish and Wildlife Service presented the latest results of toxicity research. One chemical "cocktail"—a mixture of contaminants developed at the Columbia (Mo.) National Fisheries Lab—increased the rate of striper mortality four days after hatching. A Fordham University project had turned up a fascinating indication that some heavy metals, primarily copper, actually provided protection to striped bass against disease. Others, such as arsenic, enhanced their susceptibility.

Sulfur and nitrogen oxides, emanating from coal-burning power plants and other sources of combustion, can travel long distances in the atmosphere and create acid rain. In quite a number of Chesapeake tributaries, pollutants seemed to be more toxic to the bass when acid rain problems were acute. This was a most important finding. Acidity is measured on a pH scale, ranging from acidic at the low end to alkaline at the high. The neutral point is 7, with values below that being increasingly acidic. The precipitation that customarily fell on

Maryland's Eastern Shore was among the most acidic detected any-where, ranging from 4.45 to 3.5. The flat, coastal-plain soils here were extremely vulnerable to runoff from heavy rains, leading to sudden acid pulses in streams and rivers.

This past spring of 1983 had seen some deluges along the Chesapeake during striped bass spawning periods, and the survival rate of larvae into juvenile had once again been poor. By contrast, the Hudson River had shown a nice surviving hatch again this spring. While containing many of the same chemicals as the Chesapeake sys-tem, the Hudson displayed one big difference—its waters were far more alkaline. So was acid rain the main factor resulting in the bay's fish problems? As far as striped bass, Bob Pond was not convinced.

After the meeting, Anthony and I walked over to the Capitol and stood inside before the statue of one of his ancestors, Missouri senator Thomas Hart Benton. Then we passed along the mall to visit the Washington Monument and the Lincoln Memorial. A sentence from the Gettysburg Address was especially haunting. "It is for us, the living, rather, to be dedicated here to the unfinished work which they who fought here have thus far so nobly advanced."

The annual survey of Chesapeake spawning—132 hauls at 22 sites in four major areas of the bay system—had ended. Joe Boone wasn't at liberty yet to give the precise figures on the striper index, but he pre-dicted it was going to be terrible. The day that Anthony and I caught up with Boone and Uphoff, they were doing some additional seining on the Choptank River. About 100,000 hatchery-raised hybrid fish, all marked for easy identification, had been raised on two acres of ponds in southern Maryland and released into the Chesapeake in the spring. These were male white bass from Tennessee, crossed with female striped bass from Maryland, and they might eventually grow as big as ten or twenty pounds. They weren't counted in the young-of-the-year survey, but "if we find more hybrids than wild striped

bass, it's another indication of how poor the striper hatch was," Boone said. "And we're probably getting ten to twenty hybrids for every striper."

Anthony and Boone struck up an immediate rapport. Holding up a wriggling fish a few inches long, Anthony asked the biologist to show him the mark. Boone placed a finger on a near-invisible dark speck. There were no wild stripers in the net.

"Maybe we'll hit them next site," Uphoff said.

"Gotta keep trying," Boone said.

A twenty-foot Mako pulled in not far away and tied up along the riverbank. It was Jim Price, the charter-boat captain who was leading the charge for the bass in Maryland. I'd met him for the first time at the meeting in Washington. The group of us sat down together in the weeds. Boone chewed on a stalk, looked around, and asked what we thought could be affecting the entire ecosystem. We conjectured about genetic changes, climate, acid rain. "Baffling, isn't it?" said Boone. "We can't find a common thread."

Price observed how fast some of the 1982 year class appeared to be growing. Yes, Boone agreed, they seemed to have a wide range of sizes—from four inches all the way to twelve or thirteen. An accelerated growth rate, of course, made them vulnerable to the bay's watermen that much sooner. "By 1984–85, they could be up to 60 percent of the commercial take, maybe even higher, at age three," Price predicted.

"We have essentially one year class left in the world," Boone said, "the '82. How are we going to manage them? That's the big question." Boone used the analogy of pigweed in his cornfield. Over a few years, the pigweed had completely overcome an herbicide being used to control it. "Look, a lot of organisms can gradually adjust to less than fatal conditions," he went on. "Every year in the Chesapeake, we see surviving healthy new striped bass, in very small numbers. These could be the few that are slightly different genetically from all their brothers and sisters, and so can survive whatever pollutants killed the others. They may in time be able to transmit this immunity to their offspring. So by leaving these fish alone, we may be able to build up a stronger population of stripers which can tolerate the degradation of the present environment. Sure, we've got to clean up the bay. But in the meantime we should protect completely the bass we do have left and hope for the best. If I were in charge, we'd impose significant, if not drastic, limitations on harvest —and do it immediately."

Suddenly, the twenty-four-inch minimum size limit that we'd fought so hard for in Massachusetts, in Rhode Island, in New York, along the entire Atlantic coast—suddenly, it didn't look like it was going to be enough. Boone wondered what the sports fishermen would say.

At the end of the afternoon, Anthony and I joined Price in his boat. We headed off in some choppy water to try to catch a few weakfish to bring home. Price told us, in confidence, that he was probably going to file for threatened species status for the bass in Maryland. If he was successful, that would close down the fishery.

Anthony looked intently at me. "You know," he said, "maybe what Rhode Island just did, for all the wrong reasons, is the right thing to do after all." It seemed a crowning irony that we might now come to adopt the platform of our biggest opponent.

On October 4, the same day as the next scheduled Fisheries Council meeting in Rhode Island, Maryland announced the dire results of the 1983 young-of-the-year survey. The average number of juvenile striped bass found in seine sampling of the Chesapeake Bay's rivers was 1.4—the second-lowest figure in the survey's thirty-year history. Maryland Governor Harry Hughes responded immediately, calling it "imperative that additional conservation measures" be adopted and ordering his Department of Natural Resources to draw up a new management plan for the entire coast.

That night, a standing-room-only crowd spilled out into the hallway in Wakefield, Rhode Island. There was a single item on the council agenda. Reporters scribbled furiously, and TV cameras rolled. The DEM's chief, Robert Bendick, began by reading aloud a letter from Governor Garrahy. Calling the striper fishing ban "an exceptionally far-reaching restriction," the governor urged the council to "undertake careful reconsideration of its action."

Louis Othote, one of the two appointed sportsfishing representatives, suggested a compromise: establish the twenty-four-inch minimum size limit, but allow any fisherman to keep one or two larger bass per day. While effectively eliminating commercial fishing, Othote said, this would still give locals and tourists an opportunity to keep a bass for dinner or as a trophy. Not surprisingly, this did not sit well with the net fishermen on the council and in the audience. Mendonsa scoffed at Othote's rationale, saying, "The man who gets into bass, catches one fish, and quits for the night, I'd like to meet and shake his hand."

Othote responded, "The recreational fishermen, if we allow any fish to be kept, would have to police their own ranks." Mendonsa retorted, "The number of fish being sold can *never* be enforced. A fellow can catch a dozen fish and spread 'em out through the markets."

Mendonsa must know whereof he speaks, I thought. He claimed he had sold 130,000 pounds of striped bass in 1981, most supposedly going to the Fulton Fish Market. But a sports fisherman who'd checked Fulton's list of incoming fish from Rhode Island that fall "didn't see anything *close* to that figure." It was a puzzlement.

Othote made his motion: the twenty-four-inch measure, a single fish per day allowed over that size, a regulation that would be self-renewing until scientific research showed a change in the striper situation. Othote and his ally Joe Dawson raised their hands in favor of the compromise. So did dragger fisherman Bob Smith, who had voted for the moratorium in September. On the other side of the fence, as expected, Mendonsa, Manchester, and Parascondolo voted no. This left the swing vote to Saul Saila, the generally reticent fisheries biologist from the university.

Earlier that evening, Saila had read a prepared statement. He had missed the previous two council meetings, he mentioned, because he had been in Malaysia and the People's Republic of China "as a scientific adviser and lecturer on fisheries, at the invitation of those governments." He then described his recent lengthy telephone discussions with John Boreman. The latest evidence revealed that practically *no* striped bass between three and fourteen years old had shown up on the Chesapeake spawning grounds in the spring, a big change from only a year earlier. New studies also indicated that 80 percent of the eggs are produced by seven- to nine-year-old bass who fall in the thirty- to thirty-five-inch size range; a twenty-four-inch minimum size limit wasn't going to do as much as the managers had hoped. "At this time," Saila said, "the consensus of the experts is that the greater the restrictions, the greater the benefit to the striped bass."

Saila had yet to mention where he personally stood on the question of the night. Now, in a move that stunned many in the room, he raised his hand in opposition to Othote's compromise. The new motion failed, 4-3, and the total ban remained in place.

Then Joe Dawson raised the stakes. "I want to make another motion," he announced. "What we have now as a regulation allows no taking of striped bass for one year. I move that this apply for *three* years!"

"Parliamentary roulette," someone in the audience muttered. A murmur of surprise swept the crowd, but there was a curious sense of respect as Dawson changed the rules of the netters' game. He'd taken what had been a seeming mockery and made things real. Othote seconded the motion. Saila spoke again. "It *would* be on the order of three years before we begin to see any substantial improvement in the striped bass fishery. That is a biological fact."

Another vote was taken. Othote, Dawson, Smith, and Saila voted in favor of extending the striped bass fishing ban to three years. Mendonsa, Parascondolo, and Manchester all abstained. The vote was 4-0, effective January 1, 1984.

"I am honestly convinced that this action clearly indicates we are concerned about the fish," Saila said afterward. Dawson said, "If we're gonna do this, let's make a real commitment." After their initial shock, many sports fishermen swung around to concur. Will Barbeau didn't feel angry this time. "If we've got to give up fishing for bass to keep these netters from slaughtering them, if that's what it takes, then that's what we've got to do."

Support was far from unanimous, though. Jim White, as president of the new Saltwater Fishing Alliance, was especially on the hot seat. "Personally, I think this is good," he told me over the phone two days later. "Know what our problem is? The commercial guys are speaking with one voice, but when you get to us, everybody has an opinion. There's no general consensus. My phone has been ringing off the hook—fishermen asking what are we gonna do, how are we gonna stop this? I'm trying to be a politician, and I'm not a very good one."

Bait-and-tackle shop owner Joe Mollica, a longtime friend of White's, was particularly irate. "I feel for Joe; he's ready to sell his shop," White said. "But I told him, there's no way we're going to get this changed. Let's take it for what it is, come out in favor of it. He says if that's the way I feel, he wants nothing to do with me. And he's responsible for 50 percent of our membership. He must be reckoned with."

White had also heard of vigilantism on the waters. "One big trap has been cut, lost $15,000 worth of gear, I think it was Manchester's. An Asian guy, a gillnetter, got beaten up by somebody. It's getting to be a scary situation."

What soon thereafter transpired among Maryland officials didn't make it any less frightening—when it came to the striped bass's future. At a meeting of the Atlantic States Marine Fisheries Commission in Rockport, Maine, on October 25, Maryland unveiled new proposals for revising the interstate management plan. It called for somewhat tougher protective measures in the other states, but Maryland's provisions for the critical Chesapeake Bay itself fell far short of what seemed needed.

On the eve of the Maryland announcement, there had been a private meeting between the state's fisheries people and the commercial watermen, I learned. Department biologists came to call it Black Thursday—because of the way their bosses caved in to the money interests. Tougher restrictions on commercial netting were deleted and new rules put in place. Anchor gillnets, which accounted for about half of the striped bass landings in bay waters, were prohibited in the spawning reaches between March 15 and May 31. But as Jim Price pointed out, "We estimate that 70 percent of their stripers are caught *before* March 15. So this action is really not going to save many fish. Besides, the watermen will simply put three stakes on their anchor nets around the shore and call them stake nets—which are legal."

A Maryland senior biologist, whose identity I decided to keep anonymous in an article I wrote for the weekly *Long Island Fisherman*, said: "What we considered minimal restrictions have now been gutted. This is just window dressing, and a lot of people feel they've been knifed in the back. It doesn't make sense. People who are concerned about the bass are getting to the point where they don't know where to turn or how to fight this."

I placed a desperate phone call to Alan Peterson, the Northeast fisheries director at Woods Hole, who had butted heads with Rhode Island officials a year earlier. I'd always found Peterson a straight shooter. I told him I had no idea anymore what to do.

There was a long pause. Then Peterson said simply, "Why don't you get everybody together and talk about it?"

"You mean a national conference of some kind?" I asked.

"What have you got to lose?" he responded.

Chapter Seven

Showdown
at Friendship Airport

The ensuing months were filled with meetings in many locations, with more unexpected twists, clashes, and reconciliations surrounding the grave situation facing striped bass. First, there turned out to be widespread interest in a citizens' gathering to discuss what might be done, as Alan Peterson had suggested. The National Parks and Conservation Association in Washington, D.C., agreed to donate their conference room. Numerous media expressed a desire to provide coverage. I needed, though, the imprimatur of a government scientist willing to lay on the line the striper's plight. In early December 1983, I called John Boreman.

He'd spent his boyhood summers in the Catskills, "a diehard brook trout fisherman." Those summers aroused a keen interest in studying streams and forests, and like Bob Pond, Boreman ended up at the Syracuse College of Forestry. From there he moved on to Cornell to do graduate work, specifically on rainbow trout in New York's Finger Lakes region. "I was a population modeler back in the days when that was something new," he recalled. "When all the power plant issues

129

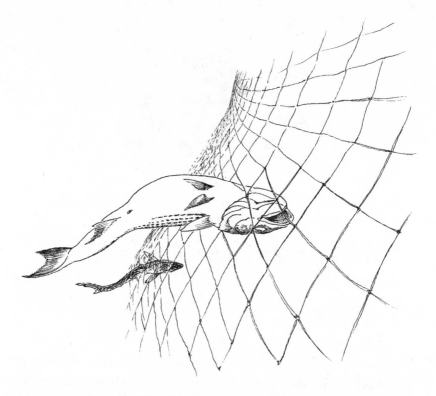

came up in the early '70s, the Department of the Interior was looking for such types. The poster child species was striped bass, and I kind of backed into the job."

Assigned to the Storm King Mountain task force in 1974, Boreman remembered "walking onto the job and meeting Phil Goodyear, who'd been responsible for writing the impact statement on the Indian Point nuclear facility. I said, 'How do I catch up on the science?' Phil pointed to an entire *wall* of documents on striped bass. He said, 'Start at that end, read, and when you get to the other end, then come talk.'"

During the waning years of the power plant controversy, Boreman was still at work on his doctorate from Cornell while both he and Goodyear were simultaneously consultants for the EPA. Both were later also named co-managers of the Emergency Striped Bass Study, albeit for different federal agencies—Boreman having moved to Woods

Hole in September 1980, to work for the National Marine Fisheries Service (NMFS).

We were about the same age, though John could then have passed for someone in his early twenties. He was soft-spoken and seemed painfully shy, certainly not a fellow to leap full-bore into a tempest like this one. Still, things had changed since our first encounter in Buzzards Bay almost two years ago. Boreman and Goodyear had both worked diligently to get a standardized data collection program off the ground—landings, fishing effort, spawning stock size—and Boreman knew that the Chesapeake's reproductive surveys correlated with precision to the subsequent coastal harvest figures.

"After a year or so of just doing the modeling, it still wasn't obvious what was causing the decline, even though things were pointing fairly heavily towards overfishing," Boreman would later recall. "We couldn't really say that was the smoking gun, but the bottom line was, striped bass spend so many years in the fishery. They are at risk of being caught for fifteen or twenty years. You compound their sensitivity to fishing mortality out over that time frame, it makes quite a difference in their spawning stock size. It was really very simple math—and made it obvious what you had to do."

As Boreman had phrased it when I called him that December day: "No fishing mortality is the best you can ask for the bass right now." Every year, he said, about half of the available fish out there were getting caught. For the current reproductive trend to start reversing, egg deposition in the Chesapeake would need to be three times its current rate. A coastwide moratorium, Boreman estimated, ought to increase egg production sixfold. As things stood, the bass "can't compensate anymore for increased mortality; they have nothing to fight back with."

Would he be willing to come speak at the conference I was setting up? Yes, Boreman said, he would.

I needed someone from the Maryland scientific staff, too, and I found him in Jim Uphoff. Tension was rising in the state. Jim Price had petitioned the Department of Natural Resources to declare striped bass a threatened species in the Chesapeake and had hired a lawyer. The commercial watermen, working with seafood packers and shippers, had countered by enlisting a former state biologist who maintained that annual changes in the young-of-the-year survey didn't mean that much. And Governor Harry Hughes, at a meeting with his bay counterparts from Virginia and Delaware, had just introduced a new supposed panacea—substantial investment in a striped bass hatchery program.

Uphoff scoffed at that notion when I contacted him. The total cost—including obtaining land for 200-plus acres of ponds, building the hatcheries, and putting them into production—was estimated at some $50 million. Maryland simply couldn't afford this outlay. The survival record for hatchery-raised bass wasn't all that good either. "Hatcheries are great for research," Uphoff added, "but it's unlimited faith in technology that got us into this mess in the first place." Far better to remove fishing pressure and slowly rebuild the wild population.

Opposition to any substantive goals coming out of the conference arose from an unexpected quarter—John Cronin, head of the Hudson Riverkeeper program. A ban on sale of Hudson stripers remained in effect, due to detected levels of PCB contamination in the fish. Nonetheless, Cronin said that any talk of a coastwide ban on commercial fishing or a moratorium raised "a significant problem" for New York fishermen, who had already acceded to "severe limitations." He didn't want to see any federal intervention either. Still, Cronin said he'd like to attend the meeting. So I issued a press release listing Hudson Riverkeeper as a participant in the Atlantic States Emergency Conference on the Future of the Striped Bass, as I called the meeting I was organizing. Cronin then called to lambaste me for including Riverkeeper among the participants, going so far as to threaten a lawsuit. I tried to assure him that I wasn't trying to box Riverkeeper into supporting something it didn't want to support, but he was unmollified.

Reactions like this, my friend George Peper said, were simply indication that we were at the center of something bigger than we'd ever anticipated. I shouldn't panic, but had to accept that attacks could come from anywhere. I'd implored Peper to fly in from L.A. and help see me through this. He'd been my most astute adviser from the beginning. The morning before the conference, we packed up the car and embarked on the eleven-hour drive from Boston to the nation's capital.

Our event on December 12, 1983, was preceded by a morning meeting with Jim Range, a top aide for Senate Majority Leader Howard Baker. Range was himself a striper fisherman. He had helped draft the Endangered Species Act as well as the legislation creating the federal emergency bass study, and he had consulted on the interstate bass management plan. With about twenty of us settled into chairs in his office adjacent to the Capitol Rotunda—the assembled including staff from several other Senate offices—Range gave us a remarkable two hours for discussion. I felt as if I finally understood what power was all about. The man exuded it.

There were problems with the federal government stepping directly into the striped bass management process, Range told us. Jurisdiction inside the three-mile coastal limit customarily belonged to the individual states. True, it was a federal/state striper management plan—"the best politically available" when drawn up—but its governing Atlantic States Marine Fisheries Commission was basically a compact among the various states. It had no real teeth. Despite ASMFC's lack of regulatory authority to force reluctant states into line with its management plans, this remained "tender territory" for federal encroachment. The legislation Gerry Studds initiated in the House, threatening a cutoff of federal fisheries funds by a certain date for noncompliant states, might be the best we could hope for. More direct involvement, Range thought, wasn't going to fly in Congress. This was disappointing, but still, Range was clearly sympathetic.

Several miles from the Capitol, at the National Parks and Conservation Association, the folding chairs we set up were barely enough to hold an overflow crowd of about sixty people. Dick Reston—editor of the *Vineyard Gazette*—had given me the home phone for his father, renowned *New York Times* columnist James Reston, whose ear I'd bent for five minutes the previous morning about the conference. Sure enough, the *Times*—which had yet to cover the striped bass story beyond its outdoor pages—had a national reporter on hand.

In my opening presentation, I described the bass in history and its human significance before turning to the management plan, which was considered the answer when drawn up four years ago but now seemed "a day late and a dollar short," as the NMFS's Alan Peterson had put it. "The only immediate step that can be taken is to limit fishing pressure to whatever extreme is necessary," I said.

John Boreman followed. He was not here, he said, to recommend a moratorium or gamefish status; that was up to the managers. He was here to lay out the scientific findings. And he did so, cogently, brilliantly, undeniably, from the probable contaminant problems "including hydrocarbons and heavy metals . . . measured in Chesapeake waters" to the conclusion that the fish needed "a significant reduction" in fishing pressure in order to survive.

Challenged during the question-and-answer period by Arnold Leo, the New York haul seiners' representative, Boreman became even stronger. Leo, seeking to shift blame onto the recreational fishermen, taunted the scientist: "Your job is to identify where the pressure is coming from!"

"Right now the striped bass cannot withstand *any* fishing pressure!" Boreman fired back. Well, the Chesapeake might be in bad shape, but the Hudson bass population was doing fine, Leo retorted. Boreman was unfazed. "Urgency means yesterday, as far as I'm concerned," he said.

Uphoff spoke next. The 1982 index of juveniles might seem to indicate a fairly successful hatch, but the numbers at the head of the Chesapeake Bay—the largest spawning area in terms of acreage and pounds of fish harvested—had been below average. "This may make the '82 year class, the last we've got, lower than it appeared on the surface."

The old guard of striped bass conservationists—Hal Lyman of *Salt Water Sportsman* and Chris Weld of the National Coalition for Marine Conservation—talked about the morning meeting with Jim Range and what "will not fly" in Washington. George Peper brought the discussion back to Boreman's analysis of what dire straits the bass were in. "A lot of people told the Wright Brothers that their plane couldn't fly, and it did. I think we need to look rather at what *has* to be done, or this species will cease to exist."

Spokespeople from the Sport Fishing Institute, and then the American Fisheries Society, added their voices in favor of a complete fishing moratorium. Hudson Riverkeeper John Cronin responded that such a radical move definitely wasn't on his agenda. Arnold Leo's constant interruptions marred the proceedings, and I came away from the long day wondering how effective this exercise had really been. There was no strong consensus, as I'd hoped for. At our wine-and-cheese reception, though, there was palpable excitement—not only among fishing enthusiasts but also among groups such as Greenpeace and Friends of the Earth, which expressed interest in greater involvement. At that point, none of the big environmental organizations had ever taken on a fisheries-related issue. Whales and ocean pollution, yes, but fish weren't considered to have a large enough constituency.

Seeing the next day's press coverage cinched it—the conference had made some waves, all right. It was the lead national news story in the *New York Times:* "Marine Experts Say Urgent Steps Are Required to Aid Striped Bass." The *Baltimore Sun's* coverage began: "Maryland and other coastal states aren't protecting rockfish enough to stop their decline, biologists told a conference here yesterday." Interviewed by the Associated Press, Jim Range said: "In the opinion of a lot of folks, we need a federal catalyst to save the species."

In two days, the ASMFC's Striped Bass Board would be convening at Friendship Airport in Baltimore. Let's see what the directors of coastal fisheries have to say now, I thought.

December 15: All afternoon, I'd listened to Lee Zeni—the former Navy commander now in charge of the Maryland Tidewater Administration—go on about how much his state was doing to protect

striped bass. He seemed wary when he felt out John Boreman on what steps the scientist hoped the federal government would take and defensive when Massachusetts's Phil Coates suggested that the 1982 year class fish were already being "hit" by Maryland fishermen. I remembered what Joe Boone had told me about Zeni: "He's got no understanding of biology. But he doesn't think any of us are worth anything, doesn't listen to us and never has."

Pretty much all the state directors agreed that they now needed to achieve at least a 50 percent reduction of what their striper landings would have been with the existing management scheme in place. Zeni kept proclaiming that his state was way out in front on implementing this, but I knew his facts and figures were bogus. Finally I couldn't take it any longer and asked for permission to speak. My request was granted.

After a few remarks, I read aloud from the article I'd recently published, quoting an unnamed Maryland biologist as saying that the state's ballyhooed plan might serve to protect only 15 to 20 percent more fish at best.

"You're calling me a liar!" Zeni shouted. His countenance was beet-red.

"I didn't use those words, Mr. Zeni," I replied evenly.

"Well, you tell me the name of that biologist and I'll fire him in the morning!" the ex-commander cried.

"I'm under no obligation to do that, Mr. Zeni."

When the day's proceedings shortly adjourned, he confronted me in the hallway. "You had no right to say that to me!"

"I had every right!"

"Unless you're irresponsible!" Zeni said, walking on with his back toward me.

"*I* am responsible, Mr. Zeni!" I called after him.

I walked over to a pay phone and dialed Anthony Gude. "Anthony, something's just happened," I said. "I'm not sure just what it means, but this fellow Zeni really lost it with me." I told him the story. Anthony listened and chuckled. It must have felt to him a bit like the clandestine chase when they'd busted the gillnetters. Neither of us, though, had any idea of what this little moment would ultimately generate.

The next day, the Atlantic states voted for an immediate 55 percent reduction in striped bass landings. It would be up to each state to generate proposals on how to achieve this goal, by the time they reconvened in April. With the meeting drawing to a close, Zeni brought up my insubordinate behavior. "For someone to insult a state based on an unnamed biologist—who I suspect I know the name of, and who knows nothing of statistical analysis—seems inappropriate. Is the public free to say anything it wants, or must it be as responsible for its statements as we are?" A biologist representing Rhode Island and Massachusetts's Coates responded with mild reprimands of my behavior.

After the *Baltimore Sun* quickly followed with an editorial calling for a moratorium on all striper catches in its waters, six days after my confrontation with Zeni a memorandum was issued to all Maryland Fisheries Division staff. "Conflicts Between Personal Attitudes and Departmental Position," the December 22, 1983, edict was titled. Recently at a formal session of the ASMFC, "a very embarrassing and intolerable incident occurred as a result of indiscretion on the part of one or more Fisheries Division staff . . . [that] may cause a tremendous negative impact on our efforts to have the coastal states take conservation actions to protect striped bass." My comments, the memo continued, had "openly accused the department of misrepresenting the level of conservation" to be "achieved by the Maryland initiatives." The statistical database had been called into question, embarrassing the governors of Maryland and Virginia and many others. "Such actions on the part of department biologists are inexcusable." It was "assumed" that these intolerable conversations had taken place during a meeting in Washington on December 12. "We cannot accept the existence of such a divergent set of viewpoints. . . . I am not suggesting that you compromise any biological data that you possess, just that you accurately represent it."

They should have checked the date on my article, I thought. It had been published prior to December 12, when Jim Uphoff had been the only department biologist present at my conference. In fact, it wasn't

Uphoff who had been my Deep Throat. Mr. Zeni was jumping to some rather unfortunate conclusions. He also might have guessed that what essentially amounted to a gag order on his field people would rather quickly be leaked straight to the "environment writer" of "zealous nature" who'd created all the stir.

It was. In turn, I promptly leaked the memorandum to somebody else—Tom Horton, the environmental reporter for the *Baltimore Sun*. In later years, Horton's outstanding writing on the Chesapeake Bay—for *National Geographic* and in several books—would receive many accolades. Back then, he was already *the* local expert on rockfish and all matters aquatic. We'd actually first run into each other years earlier, in 1971. I'd been traveling through Ethiopia with a backpack and a portable typewriter, and he was editing the Army base newspaper at remote Kagnew Station. When I first came across his byline on an article about striped bass, I couldn't imagine it was the same Tom Horton. Now I usually camped out on his floor during my sojourns to the Chesapeake, and we kept each other regularly apprised of new developments.

When he saw the Fisheries Division memo, Horton set up an interview with Zeni. As Horton later recounted, he asked whether the Tidewater director didn't think this could be considered a form of intimidation. Zeni hemmed and hawed before saying, "I guess so." But his biologists didn't tell him the real story, he added. "It's that Uphoff fella, isn't it," Zeni asked, "who talked to Russell?" Horton said, truthfully, that he had no idea.

"That guy Russell's got a hidden agenda," Zeni said confidingly.

"He does?" Horton asked.

"Some kinda deal to make him a lotta money if the commercial guys can't catch bass anymore—a hidden agenda."

Apparently, the Maryland Tidewater people had made some inquiries among Massachusetts officials, trying to find out what they could about me. They'd gotten wind that, at one public hearing, I'd mentioned that sometimes my friends and I would send a bass out to others in our circle who lived on a farm in Kansas. In return, they'd occasionally ship us back a box of grass-fed, homegrown beef. Somehow this translated into an effort to corner the striped bass market in a barter arrangement for corn and turkeys. Apparently, it was inconceivable to Zeni that someone would confront him so strongly for the sake of the fish itself.

After their formal interview ended, Zeni admitted to Horton privately that Maryland's regulations as currently proposed would *not* reduce their striper harvest by 55 percent—but they'd get it right by spring. "Tom, I tell you, a moratorium would be the easy way out, and I'm not a man who likes to take the easy way," Zeni said, showing Horton out the door.

"State's Rockfish Conservation Claims Called Inaccurate" read the headline on Horton's lengthy Sunday news piece. Maryland's new regulations were "a move in the right direction, but no more than a first step," John Boreman was quoted as saying. The article noted that Zeni had shuffled three biologists in as many years as head of his administration's rockfish section and recited the history of the state's waffling under the watermen's pressure, leading up to the confrontation at Friendship Airport and the gag order on the biologists. Losing credibility over the bass might place the governor's entire Chesapeake Bay Program in jeopardy, the article suggested.

The next day, a small plane crashed into Jim Uphoff's backyard. "I guess when it rains, it pours," the biologist told me. Besides the memo designed to muzzle him and his peers, Uphoff had received a separate "confidential" memo accusing him—falsely—of being my source. "If Zeni could, he'd fire somebody. He's on a witch hunt," Uphoff added. "Now I'm getting blamed for everything under the sun."

Other newspapers began to pick up on the story. In the *Boston Globe*, another "nameless" Maryland biologist—"not the same person to whom Russell spoke"—blasted his superiors, saying they "aren't about to let facts slow them down." Whatever new plan they were proposing to save the bass was unclear. "That's because," said the biologist, "they change it every day. . . . They have stacked the information against the moratorium."

Zeni's heavy-handed approach was backfiring. His field staff was going into revolt.

The pace of events only quickened as 1984 dawned. I incorporated a new nonprofit organization, the Striped Bass Emergency Council, under the umbrella of the National Coalition for Marine Conservation. Bob Pond filed a petition with the federal government asking that all striped bass over fifty pounds be declared an endangered species and that all Chesapeake bass be designated as threatened species. I learned that a staff meeting to discuss Maryland's latest proposals to cut commercial landings had ended in a shouting match. "I said my piece," one of the biologists said afterward. "They are certain now what they're faced with." When I prepared to drive south again in a snowstorm, Joe Boone cautioned: "Wear your flak jacket."

At a meeting of environmentalists in Washington that I attended, the Maryland Wildlife Federation said it was going on record in favor of a coastwide moratorium for a minimum of three years. More media attention was needed to build a consensus, it was felt, even though the press would tend to zero in on the sports-commercial conflict rather than the fish. Our point needed constant repetition: "What if there are no critters to reinhabit an improved [Chesapeake] Bay?" a gentleman from the National Wildlife Federation emphasized. A spokesperson from the Audubon Society, however, predicted that federal legislation was unlikely, given that this was an election year.

Still, a number of congressional staff attended a meeting of the Emergency Striped Bass Study group the next day, where Boreman said that interim measures should "not exclude consideration of a total moratorium." The latest findings were to be presented in February at a full briefing of House and Senate members.

It was on to Annapolis, Maryland, where a state senator had introduced a bill calling for a moratorium. Dr. Torrey Brown, head of the Department of Natural Resources (DNR), defended his fisheries agency's "comprehensive approach" to protecting the bass. Expectedly, the watermen sided with Brown. Unfortunately, so did the leader of the 3,300-member Maryland Saltwater Sportfishermen's Association, who testified that he'd supported the moratorium two weeks ago but changed his mind after a private meeting with Brown. The Maryland Charter Boat Association had divided, too; Jim Price's effort to get support for the moratorium had been blocked. "Those guys are trying to get a special permit for themselves," Uphoff told me. "The DNR loves that—divide and conquer." With these forces against it, the Maryland moratorium bill was resoundingly defeated.

In Rhode Island, the Fisheries Council held the line on the fishing ban at its next meeting, although George Mendonsa wore a big smile and hinted that he was ready for compromise. Appearing before the Conflict of Interest Commission, he admitted he didn't honestly support the moratorium. Meanwhile, Jim White's Saltwater Fishing Alliance bitterly split apart over the moratorium. Sports fishermen who opposed the new regulation came together with the commercial netters in a secret meeting. Joe Mollica said his group would support a twenty-inch minimum size limit. George Mendonsa wanted eighteen inches. They weren't quite in accord, but clearly a setup was on the horizon.

In fact, White would recall, "*We* put a sting on *them*. The Conflict of Interest Commission wound up there." Right afterward, on January 10, the commission ordered Mendonsa to "cease and desist" from any deliberations on striped bass and fined him $5,000 for having defied their warnings and "using one's public office for private financial gain." It was the highest penalty ever levied by the commission.

For his stand in support of the moratorium, White had been receiving threatening phone calls in the middle of the night. "You'll never fish in Narragansett again," one said. "If you go to Charlestown, that'll be the last time anybody'll see you," another said. He was all but certain these were sports fishermen—his own people. Finally, after

White and his remaining allies made it known that they would arm themselves if necessary, the phone calls ceased. But White had decided to step down as president of the Saltwater Fishing Alliance. "We've lost 50 to 65 percent of our membership" since the moratorium, he said. "The organization will probably go the way of the dodo bird."

He wrote a long letter to Senator Chafee: "The Doomsday Clock for the Striper is at one minute to twelve." He talked of how John Cole, Chafee's old college roommate whose book had inspired the senator to initiate the emergency study, had called him two weeks before to ask how the fight was going. "I could only give him bad news." The Rhode Island moratorium "was all a bluff to get the sportsmen to back off on their demands for a twenty-four-inch limit."

White continued, "In the time we have worked to save this fish, we have always worked within the American system. I was raised to believe that in America when things are wrong you can work to change them. . . . All it takes is hard work and dedication and the system will work. Our children are still taught this myth in school today.

"Well Senator, what nobody told us or our children is that there's one exception, 'THE ALMIGHTY BUCK.' No one on earth can compete [against] or beat such a foe. . . . I do not think that God intended the Striped Bass to end in such a way, to die at the hands of man's own greed. For if it does . . . then maybe we as human beings are the endangered species."

White had another political ace up his sleeve. Claudine Schneider, a Republican and one of Rhode Island's two representatives in Congress, happened to have an office in Coventry on his mail route. She also happened to sit on the House Fisheries Subcommittee. "It was almost like divine intervention," White would remember. Every day, without fail, he would time his deliveries to spend some extra minutes at the congresswoman's local headquarters on Rolfe Street. If she was in Washington, he'd lobby her staff assistants. If she was in Rhode Island, he'd corner her personally. "I'd go upstairs to her office in my uniform, sit down with my pack of mail, and say, 'We've gotta talk about striped bass, Congresswoman.'" He always carried copies in his mailbag of the latest clippings I'd sent along—a kind of striper news service. (Between the December bass conference in Washington and the end of January, more than fifty articles appeared.) The postman, as they say, always rings twice.

Through January, Representative Schneider held a series of meetings. Her first appeal came from White, Will Barbeau, and a couple of other recreational fishermen who supported the moratorium. The story of the secret Mendonsa-Mollica meeting, and the uninvited appearance of the Conflict of Interest Commission, "led Claudine to really sit up and pay attention to what we were saying," White would recall. Her staff followed up by consulting with Fisheries Council members Mendonsa, Manchester, and Smith. To the staff's surprise, the commercial fishermen indicated that they could continue to live with a ban on striper fishing if it applied equally along the coast.

Early in February, someone from Representative Schneider's Rhode Island office called. Could I come down there to meet the congresswoman? I did. She appeared, if not passionate, certainly sincere in her desire to protect the fish on behalf of her constituents.

The next day, I flew to Washington to attend the briefing that Boreman and Goodyear were giving to more than thirty legislative staff. "Most of the female stripers left in the spawning population are from the 1970 and older year classes," Boreman grimly informed them. Goodyear's summary described "numerous organic and inorganic contaminants that have been detected in striped bass eggs, young of the year, and adults." The fifty-page report of the Emergency Striped Bass Study concluded: "This indicates that striped bass may be one of the first species to succumb to these toxicants in estuarine waters."

The debate over the striper's fate had gone beyond overfishing. The next day, Jim White received a phone call from Schneider's office. His presence was requested at 10 a.m. on February 16, presuming he could find a replacement for his morning's round of mail. A number of his friends and associates would be there. As Jim recounted the conversation long after:

"We've also invited some gentlemen from the commercial sector," the staff assistant said.

"You did?" Jim responded.

"Yes, you guys are going to talk."

"You kidding me? Are you gonna call the state police?"

"Oh no, we've talked to the commercial fishermen, and they're in agreement with you."

"In agreement?!"

Someone had placed a mount of a large striped bass on the wall behind her as Representative Schneider began the press conference. "Right now we are witnessing the demise of one of the most prized gamefish in the world," she began. "It has been said that pretty soon we're going to be able to name the ones that are left." Citing a 90 percent decline in landings over the past decade, she continued: "We have to recognize that this is an East Coast problem which must be addressed from a national perspective. I believe this kind of emergency measure is very necessary and the most equitable solution that will bring about the quickest results. I don't introduce bills that are easy. They are always controversial, and part of my way of proceeding is to bring both sides of an issue together and develop compromises to see what is in the best interests of everyone. But in this particular case, it's obvious that we have to put social and political special interests aside."

Schneider was going to introduce federal legislation calling for a three-year moratorium on all fishing for striped bass, to be rescinded only when the Chesapeake's annual reproductive surveys showed clear signs of improvement. The Atlantic Striped Bass Restoration Act of 1984, as it was to be called, would give regulatory authority to the administrator of the National Oceanic and Atmospheric Administration. Any harvesting of striped bass north of South Carolina, and any interstate commerce in the fish, would be illegal and subject to stiff civil and criminal penalties.

In her conclusion, Schneider held up a copy of my *Yankee* magazine article featuring Jim White, headlined "Who Will Save the Striped Bass?" The congresswoman said, "I saw that question and figured we'd better come up with an answer."

Asked about Rhode Island's talk of modifying its existing statewide ban, Schneider quickly responded, "Now they won't be able to." After a moment's thought, she added, "Part of what made me get this moving was because some of the fishermen were concerned that the state wasn't serious about enforcement of the ban. So a federal requirement can only help."

It was a precedent-shattering move, the first time anyone in the federal government had sought to regulate fishing in state territorial waters. What the experts had said would "never fly" was in fact airborne only two months later. There would be acrimonious legislative hearings ahead, certain opposition from crucial officials in Maryland, probably resentment from Schneider's colleagues feeling that she'd jumped the gun. There was, for sure, a long way to go, with no certainty of success.

We'd come a long way, though. In an exchange that one outdoor writer figured probably hadn't happened in 150 years between a sports fisherman and a commercial rival, photographers duly took note as two men of Portuguese heritage, Jim White and George Mendonsa, shook hands.

I went over to Mendonsa afterward to express my sympathy about his $5,000 fine. "Yeah, I see the tears in your eyes," he said. "When they put me in jail, send me some cake, will ya?"

Chapter Eight

Revolt of the Biologists

Momentum in the media built quickly on the heels of Representative Schneider's dramatic announcement. An editorial in the *Providence Journal* urged Congress to pass her moratorium bill. The *Boston Globe* came out similarly, as did the *Baltimore Sun*.

At a Rhode Island Fisheries Council meeting on February 28, state fisheries director John Cronan opined that "there isn't a very good chance of the legislation going anywhere," based on Congress's traditional reluctance on questions of states' rights. There was wisdom in what he said, but at home, his efforts to introduce a compromise measure in lieu of his state's existing moratorium were going nowhere too. Marine scientist Saul Saila, his voice quavering, came out for a coastwide ban. Mendonsa and Parascondolo remained mute observers through a long evening. There wasn't enough support even to merit a vote, so Rhode Island's three-year moratorium stayed in place.

Representative Schneider called me in Boston to voice concern about the stand of Massachusetts's Gerry Studds. He was the key voice among Democrats on the fisheries subcommittee, having championed the federal Emergency Striped Bass Study through the House in 1979. But now he was taking a wait-and-see attitude, wanting to know what the various states would come up with on their own to meet the newly recommended 55 percent reduction in fishing landings before signing on as a cosponsor for her bill.

Invited to testify at the first subcommittee session in Washington on March 20, I decided I could size up the situation then. And I liked what I saw of both Schneider and Studds in action. Two federal officials, the Reagan administration's top appointees from the National Oceanic and Atmospheric Administration (NOAA) and the U.S. Fish and Wildlife Service, not only came out against the moratorium but even opposed continued funding of the emergency study. Asked by Schneider how important the striped bass's survival was to them on a scale of one to ten, both men said ten. Asked by Studds how much peril they thought the bass was in, one of them again said ten.

"Well," Studds responded, "I'm sure if the gentle lady from Rhode Island still had the floor, she would question either your passion or your analysis. If indeed you are committed to the species, and if indeed it is near its maximum peril, then surely something needs to be done and fast. . . . When you say that *most* states are responding, I can only infer that some of them are not. That could be a fatal omission, could it not?"

Yes, both officials replied sheepishly.

What about a federal fishing ban if the states failed to act within a particular time frame? NOAA's John Byrne conceded that the threat of a moratorium on all striped bass fishing "would go a long way toward getting the states to work together in their own best interests."

"I think that is about as close to an endorsement of this proposal by the administration as we can reasonably anticipate," Studds said, and asked for the next witness.

The reports from the emergency study's co-directors only buttressed the case for a moratorium. John Boreman testified: "Reported commercial landings of striped bass from Maine to North Carolina for 1983 (1.6 million pounds) represent a drop of over 25 percent from the

1982 level (2.2 million pounds) and are equal to the lowest reported catch on a record that extends back to 1929." Phil Goodyear said that even under the 55 percent reduction plan, "it would take several decades to restore the stock to its former level of abundance." However, "the elimination of fishing as a source of mortality has the potential to cause the population to increase ten- to twentyfold from generation to generation. Thus, it is theoretically possible to restore the stock to the level of abundance of the 1960s in little more than a decade."

Among those testifying, it was a 5-5 split on the moratorium question. John Breaux of Louisiana, the Republican chairman of the subcommittee, appeared strongly opposed to federal intervention based on states' rights. I tried to defuse his argument by showing how ineffective it was to leave states to their own devices in this kind of situation. Fully a year and a half after such action was called for by the ASMFC, for example, only three of the dozen coastal states had enacted the earlier management plan, which called for a twenty-four-inch minimum size limit by 1983. Maryland biologist Joe Boone had made the point with his usual eloquence in his 1984 *Fishing Forecast:* "Though not the resource of a particular state, each state establishes harvest regulations within its boundaries. Consequently, each state warily eyes the catch elsewhere, intent on getting a 'fair share' of the booty. Naturally, self-interest and politics permeate every phase of harvest deliberations. Management becomes cumbersome, chaotic, and exasperatingly slow, even in times of perceived emergency. What is every state's responsibility is no one's responsibility."

Within a week after the hearing, Schneider's number of cosponsors had risen to fifty, including ten of the twenty-eight subcommittee members. Meanwhile, each coastal state was to present proposals on how to achieve 55 percent cutbacks in their striper harvests at an early April meeting in Woods Hole. The meeting was closed to all outsiders. When the proposals were announced two weeks later, they were not encouraging. New York's proposed reduction in harvest was a mere 5 percent, Delaware had nothing new to suggest, Virginia was at 35 percent, Maryland at 27 percent. It was just as Joe Boone had written.

Shortly thereafter, Studds introduced his own bill in the House. While it didn't go as far as Schneider's, it called for a federal moratorium in waters of any state that hadn't met the 55 percent target by December 1. When subcommittee chairman Breaux decided to consider Studds's bill—and not Schneider's—she managed to halt all discussion because a quorum was not present. Studds was furious. "A total moratorium is not politically feasible," he said, "and she's deprived us of the opportunity to do something about the bass's problem. It makes no sense. Chairman Breaux had told me that my bill was a 'golden opportunity' for action."

The bass had become snared in a political net. Things didn't shape up much better on the Senate side where, three days after the Studds-Schneider imbroglio, Rhode Island's Chafee introduced the Coastal Migratory Fish Conservation Act—which would impose a temporary moratorium *until* the states got their act together—and Massachusetts's Edward Kennedy responded with a less stringent, Studds-type bill of his own. It was looking like another Massachusetts versus Rhode Island battle over striped bass regulations. Except this time, Li'l Rhody was taking the lead.

While Congress was bogged down in procedure, the news out of Maryland was grim. Several commercial netters had obtained a court injunction blocking new regulations that would have shut down striper fishing during the Chesapeake spawning season. Meanwhile, a paucity of adult male fish, and probably the lowest number of breeding females in memory, was showing up in the rivers. The only positive sign was that rainfall had been slight through the second week of May, Jim Uphoff reported, and "so far we have not experienced a die-off of striped bass larvae."

Uphoff had his work featured in a *Sports Illustrated* article by Robert H. Boyle that focused on acid rain. Within the Maryland Tidewater Administration, though, the bosses were still after Uphoff.

Two months after the incident at Friendship Airport, he was confronted with a phone bill showing that he'd called my number on December
2, 1983—the "smoking gun" that Lee Zeni and company thought confirmed their suspicions about who my source was. (My *Fisherman* article, however, had gone to press four days before that.) Zeni took
Uphoff aside. "He tried to scare me into confessing," Uphoff later told
me. "He said, 'There are procedures to dismiss you, but we're letting
you go this time.' I told Zeni, 'It wasn't me. You have problems with
your entire staff.'"

After further attempts to force him out, Uphoff, fed up, contacted
a local attorney and filed a grievance complaint charging harassment by
the Maryland Tidewater Administration. I wrote a letter on his behalf.
"They've completely fabricated a case against me," Uphoff told reporter
Tom Horton, "trying to silence disagreement."

In addition to its news coverage of the story, the *Baltimore Sun*
blasted Zeni and the Tidewater Administration in an editorial,
"Fingering the Whistleblower": "Those who muzzle their staff to hide
the truth are themselves subverting the public cause. It is they who
bring shame to the age-old and honorable role of civil servant."

A few days after that, the Maryland DNR announced that a formal
apology was being made to Uphoff. "I do think there's an understanding now that there's a right to speak up without being persecuted," he
told the *Sun*. Exactly how strongly Uphoff and his colleagues were soon
to speak up, nobody could have predicted.

With so much resting on what the various states would do, the federal
legislative process seemed to move at its often-customary glacial pace.
In the House, the subcommittee chairman agreed to bring up the bills
by Schneider and Studds for consideration in June, but only after he
had a full report on the ASMFC's annual meeting. Schneider's aide
Charles Swezey attended it and wrote: "My overall impression is that
there are now three things threatening the bass: pollution, overfishing,

and the Atlantic States Marine Fisheries Commission. . . . The state fishery managers seem only concerned with maximizing harvest."

The National Marine Fisheries Service, the agency that would be charged with overseeing any federal legislation, sent a letter to Schneider on May 30. "Unfortunately, despite repeated prodding by concerned citizens and members of my staff since 1978, the states have not been able to act in a timely way," NMFS director William Gordon wrote. "We stand ready to implement a coastwide moratorium and monitoring program." The next day, the congresswoman took this new ammunition onto the House floor. "Pity the plight of the striped bass," she said. "It is time for the federal government to act."

If there wasn't much movement at the level of state government, among the citizenry, there were encouraging signs. On Martha's Vineyard, after my friends and I made a few rounds of local eateries, eight restaurants catering to the tourist trade decided to remove striped bass from their menus. The movement spread to Cape Cod, and soon to four major seafood dealers in the Boston area. Legal Seafoods, the city's largest retail and restaurant fish market and purchaser of up to 50,000 pounds of bass in 1983, along with three other dealers, announced that they would no longer buy stripers and joined the call for the moratorium.

In response to a mass mailing by Spider Andresen at *Salt Water Sportsman*, a number of leading boat and tackle manufacturers also wrote letters to Congress expressing support for the moratorium bill. The Fisherman Tournament stopped recognizing stripers for prizes, instead offering an award for catch-and-release. The Cape May, New Jersey, tournament ended its forty-year tradition of awards for the biggest stripers, and the Virginia Saltwater Fishing Tournament and Nantucket Anglers' Club took similar action.

Still, it often seemed a case of one step forward and two steps back. In Rhode Island, a bill in the state legislature to exempt Fisheries Council members from the conflict-of-interest laws passed the House, though it failed to get out of the Senate Finance Committee. Mendonsa and Parascondolo didn't even bother to show up for the next council meeting on June 27. But they didn't have to. Scientist Saila was absent as well, on a six-month leave to Japan. Trapper Francis Manchester had stepped down, replaced by a member of the sportfishing community, but ironically, the three recreational representatives seemed the ones most eager for compromise. Tackle shop owner Joe Mollica packed the room with rod-and-reelers opposing the moratorium.

The council did an about-face. It voted 3-1 to rescind the moratorium. It approved instead, effective in July, a package that gave lip service to meeting the 55 percent reduction goal in commercial landings— but that allowed a minimum size limit for everyone of only eighteen inches! The trap-net companies would be subject to a seasonal quota and gillnetters limited to keeping two bass a day. But how these limits would be enforced was anybody's guess. Rhode Island was back to square one.

The morning after Rhode Island's turnabout, the Fisheries Subcommittee met again in Washington to take up the Schneider and Studds proposals. A bitter dispute erupted over economic self-interest, conservation, and states rights. Virginia's representative labeled Schneider's approach "draconian," while a Maryland congressman sought to water down the Studds bill, pushing the date back to December 1985 for states to meet the 55 percent target. Studds compromised at June 1985, and it was his bill, not Schneider's, that was voted out for consideration by the full Merchant Marine and Fisheries Committee. Breaux, I was told, had demanded the compromise, because several states had said they couldn't achieve a 55 percent decrease by the end of 1984.

One of those states was New York, where the battle over Westway continued to rage. A report commissioned by Governor Cuomo had recommended abandoning the highway and trading in the money for funds to benefit mass transit. Cuomo rejected that idea, then tried to circumvent Judge Griesa's orders by seeking an exemption from Congress.

When that failed, Cuomo next turned to the Army Corps's chief of engineers, who called off the scheduled two-year study on the bass. Sufficient information already existed, Lieutenant General J. K. Bratton said, for the Army Corps to proceed with crafting a new Environmental Impact Statement. Cuomo hailed this as "a very significant, an extremely positive step." Attorney Al Butzel called the move "illegal."

Fisheries scientist John Boreman was enraged. "They are using very poor data to judge the effect on the bass," he told me. "It's nowhere near good enough to proceed. Thirty striped bass experts have said that a two-year study is needed before *anything* should be considered. With the Chesapeake gone to hell in a handbasket, this could cause the demise of the Hudson fish, too." He was planning to organize the marine scientific community to fight it.

Late in the spring, the Army Corps finally conceded that Westway could pose a "significant adverse impact" on juvenile stripers, but added that there was a "reasonable chance" this impact could be lessened. Westway's sponsors were arguing that striper habitat could be built into a new seawall. The riprap design of the wall, combined with posts and rocks placed in the water, would supposedly provide a good replacement for the piers. When I asked Hudson River activist Bob Boyle about this, he sneered. "With the existing piers, the bass are in shelter, the backwash. They like the discontinuity in the bottom. Maybe this *could* all be duplicated in the river by erecting something like steel bars. But first you'd have to do extensive studies of certain depth, certain light or lack of light that the bass need."

In the *New York Times*, one op-ed columnist revealed new corruption around Westway that was linked directly to Governor Cuomo, while another, John B. Oakes, deplored "Wasteway" as "a fraud on New York's taxpaying and voting public." Yet on May 18, 1984, the paper took an editorial stand headed: "Host, Not Hostage, to the Bass." Noting an Army Corps report estimating that only 20 percent of the Hudson's striper population might be displaced, the *Times* concluded: "The Engineers' supplemental impact statement offers a sound basis for believing that construction can finally proceed."

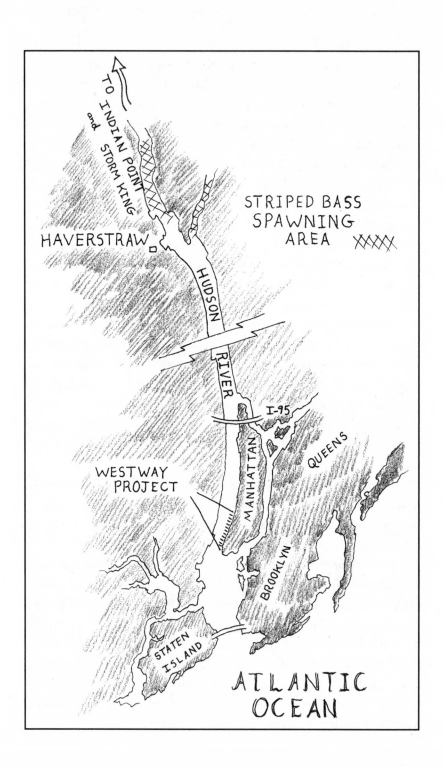

The Corps held public hearings in late June on whether to grant the dredge-and-fill permit. The location was Madison Square Garden's Felt Forum, the customary locale for heavyweight title fights, with eighty-nine witnesses testifying over two marathon days. Boyle, with his usual élan, said: "Allowing Westway to be built would be to fire a torpedo at Noah's Ark." Others described the striped bass as a "political football" and a "red herring." I was quoted in the *New York Times* for rising "to the rhetorical demands of the occasion" by hailing the bass as "a grand creature whose existence is intrinsically linked to the spirit that forged this nation." But the *Times* gave the last word to a Westway supporter who stated: "Bass eat sewage and garbage; I wouldn't think of eating bass. A study should be made of people who eat bass."

As the summer wore on, it remained impossible to predict the fate of the most expensive highway ever contemplated. The hard-fought victory for the twenty-four-inch minimum size limit in New York was suddenly up in the air, too. Long Island's haul seiners, led by Arnold Leo, had persuaded their assemblyman to introduce a bill allowing them to keep a daily 100 pounds of striped bass under twenty-four inches. My friend with Save Our Stripers, Fred Schwab, though, had something different in mind to turn the tide.

It had to do with the levels of polychlorinated biphenyls (PCBs) in striped bass. It was unclear how severely PCBs might affect the fish themselves, although one scientific study indicated that PCBs weakened their bones. Humans, however, were a different matter. For eight years, ever since tests showed that many of the Hudson River striped bass had PCBs in excess of five parts per million in their flesh (considered at the time above the safe threshold for human consumption), a ban on any commercial fishing of stripers in the river had been in effect. In 1983, the latest analysis revealed that the majority of Hudson stripers were now below that threshold, at 4.9 parts per million. Compared to the Chesapeake region, the Hudson seemed to be thriving—its latest young-of-the-year count showed the highest index of juvenile stripers since the survey began in 1969—and there was now talk of reopening the commercial fishery.

Fred Schwab was aware that the Food and Drug Administration (FDA) had proposed, a full seven years earlier, that the PCB tolerance level for fish consumption be lowered to two parts per million because the compounds were cumulative and so potentially dangerous. The agency hadn't yet acted, in part because the National Marine Fisheries Service said they had to consider the economic impact of such a decision on commercial fishermen. Fred thought a letter-writing campaign to the FDA might be in order. I wrote one, urging the FDA's commissioner to consider that the Hudson River's striped bass now comprised a larger portion of the Atlantic coast's migratory population. "To adequately safeguard public health," it seemed time for the FDA to move forward on its earlier proposal.

"Where I worked, we got the *Federal Register*," Schwab would recall. "I think there'd been something in it several months earlier, which had convinced me that maybe I should take a look at it every once in a while. One day in May, I'm thumbing through the pages and there it is—I guess our letters got the FDA off its butt—they're going to finally make a ruling and drop the PCB tolerance level from five to two parts per million, effective in August."

Schwab and an SOS colleague were already planning a trip to Albany to lobby against the haul seiners' bycatch proposal. After Fred saw the *Federal Register* notice, they departed immediately. They got an appointment to see Pat Halpin, the young assemblyman who had shepherded the twenty-four-inch bill through the legislature in 1983.

"Pat was worried," Schwab remembered. "He said, 'Gee, there's a lot of pressure to pass this bill giving the haul seiners a break.' I said, 'Pat, look at what the FDA just did.' Well, he called the FDA and had the whole decision faxed over from Washington. That afternoon, he put copies out to all the assemblymen and senators. Within a few days, that nonsense from the seiners was labeled the 'Poison Fish Bill.' Down the tubes it went!"

As the word spread from there, the FDA's ruling had import for stripers well beyond New York. Not only would the Hudson commercial fishery stay closed, but other states where substantial numbers of

Hudson stripers migrated—like New Jersey, Rhode Island, or even Massachusetts—might have to put no-sale bans in place. (Connecticut had already declared striped bass a gamefish, with no commercial trafficking allowed within its borders.) It was more fuel for Representative Schneider, who had this to say on the House floor: "Some scientists have speculated that you cannot find one striped bass on the entire East Coast that can presently meet the new PCB standard. If this is the case, a commercial moratorium will have to be enforced anyway."

It paid to read the *Federal Register*'s fine print.

Maryland was still the crucible for saving the bass. Jim Uphoff had been monitoring the Choptank River all through the stripers' spring run. After it ended, he had more bad news. "The week before last," Uphoff told me early in June, "we found 1,000 or better larvae of striped bass and white perch. By last week, we could find only two dozen! There's apparently been a big die-off. . . . The past couple of years, when it's been bad like this in the Choptank, it's bad everywhere else. And the very low numbers of brood stock spawning this spring just aggravate the hell out of the situation."

The hatchery efforts, once touted as a possible panacea for the decline in striper numbers, had gone poorly as well. Of 84 million eggs processed in Cedarville, Maryland, the rate of fertilization was only 10 percent. Of 2.1 million striper fry planted in ponds, only 2.5 percent survived, and even half of those few survivors failed to inflate their swim bladders, meaning most of them would soon die, too.

Hatchery-raised fish almost skewed the latest young-of-the-year count. Boone had been surprised to find a substantial number of fingerling bass in an area of the upper bay that had offered poor results for several years. It looked as though the spawning survey might show enough improvement to let Maryland officials off the hook. Then it was discovered that about 50,000 hatchery-raised bass had been released shortly before the survey started in July, only a few miles away. Apparently these fish had congregated in the area where Boone and Uphoff were sampling. Subtract those numbers, and the reproductive success index looked likely to fall to a critical level for the fourth time in five years.

But the NMFS had turned down Bob Pond's petition to have striped bass designated a threatened species under the Endangered Species Act—which would have guaranteed a fishing moratorium—and he now planned to challenge the ruling in the courts.

In August, I drove to Annapolis to talk directly to Maryland's fisheries biologists about where matters stood inside the Tidewater Administration. Commenting on Maryland's management of the bass to the *Boston Globe*, one had said recently—anonymously—"I'm not sure whether it's graft or ignorance. I hope it's graft, though, because I would hate to believe that they could be so stupid." I made arrangements to meet several of the scientific staff clandestinely, one on one, under some trees in the parking lot across from the Department of Natural Resources (DNR). They were ready to talk—about a lot of things. It was to prove an eye-opening day.

The schism between the biologists and their superiors actually dated back to 1978, when Lee Zeni had been brought in to run the Tidewater ship. A civil engineer who had headed the DNR's power-plant siting program, Zeni had a reputation as a fine administrator when it came to carrying out contracts and keeping people on deadlines. Above all, he was a firm believer in systems. Fish biology, Zeni reasoned, simply needed better computer-type analysis. Staffers like Joe Boone and Ben Florence, however, knew that the imponderables of the water did not work so logically. Especially in the troubled Chesapeake region.

Mutual distrust increased after the state called in Pete Jensen, on loan from the federal government, to head up Tidal Fisheries under Zeni. "The first time I ever met Pete Jensen, we fell out right then," Boone would recall. "Ben Florence and I were in his office. There was a new regulation restricting the mesh size on gillnets. Pete said, 'We've gotta figure out a way to circumvent this.' I don't remember just what I said, but he never came to me after that."

The way Jensen himself saw things, "Really you're managing people, not the resource. You can be biologically correct, but if it doesn't fit the circumstances of people, it doesn't make any difference." In Maryland, the people Jensen had in mind were basically the commercial watermen, 1,500 of whom paid an annual $50 from their incomes to a Watermen's Association headed by a legislative lobbyist. Jensen and Zeni insisted that the bass's decline was not at a crisis stage. Inside Tidewater, however, some of the staff suspected that the Watermen's Association was exerting undue influence on decisions. Shortly after assuming his new post in January 1979, Jensen had attended the annual Watermen's Association trade show, I was told. His wife happened to win the grand prize in the watermen's raffle—a new pickup truck. Shortly thereafter, Jensen and Zeni lifted a spring spawning area closure that Florence had put in place the year before and opened thousands of new acres to commercial netting at the Aberdeen Proving Grounds.

In 1980, Florence was quoted in the press as saying that the latest young-of-the-year index was the second lowest in the survey's history—which it was. "Jensen came in and jumped all over Ben about this," Uphoff recalled. "In the official department release, the figure became 'relatively low reproduction.'" Florence found himself shifted into a new post, keeping tabs on the state's recreational fishing industry. According to Uphoff, "They started editing Joe Boone's statements, too. That's when Joe began circumventing the process, doing these yearly 'Forecasts and Homilies' on his own and sending the real story out to people."

In this year, 1984, Boone was even forced by his superiors to allow commercial watermen to run their own companion young-of-the-year surveys. The netters had long disparaged his sampling methods as inaccurate, even accusing him of falsifying his data. So Boone was delighted when the watermen were unable to prove their contention that he simply wasn't sampling the sites where the baby stripers were. All of their efforts actually turned up fewer fish than Boone's.

Following the brouhaha over Uphoff, the gag order, the grievance complaint, and the DNR's apology, the Tidewater tycoons seemed to be running scared. The biological staff was frozen out of their discussions, while consulting statisticians were brought in from outside, seeking ways to computerize and justify the managers' bass protection plans. Another employee was appointed to oversee the estuary program, replacing David Wharton, who was asked to become a "hearing officer" in charge of public meetings. Florence, told that he still needed an "attitude adjustment," was downgraded again to "acting hatchery director." Boone, who would be up for retirement in 1985 after thirty years at Tidewater, was offered a new job running a trout program. He turned it down.

When Jensen returned, temporarily, to the federal NMFS, the Maryland DNR replaced him with George Krantz. In one meeting with the field staff, Krantz stated his belief that in the future all such personnel should hold doctoral degrees. "They would then be contractual and appointed, thus subject to political whims," said Florence. "If you don't accept what the boss says, they cancel your ticket. It's not the merit system anymore."

There were skeletons in some of those managerial closets, though. And the biologists, their own backs to the wall, were no longer holding back on discussing them. There had been a payoff to the watermen, in the form of a lucrative research contract. There had been a deal cut with a local outdoor writer, in exchange for muting his criticism of the protection plan.

Later, I called Pete Jensen about that grand prize his wife allegedly garnered in the watermen's raffle. "Yes, she won a pickup truck they were raffling off. We chose not to take it," he replied. Why was that? I asked. "We didn't want it." So the Jensens didn't receive anything? "Oh yeah, we received something, but we chose not to take the truck."

I was able to confirm, through other sources, all the tales I'd been told that long afternoon under the trees. I planned a magazine exposé. I made sure that Tom Horton at the *Baltimore Sun* got the scoop, too. One story didn't preclude another. It was just a matter of time—and timing. Suddenly, there was more than one Deep Throat. I wondered whether the Maryland Tidewater hierarchy realized it.

On my home turf, by a slim margin the Martha's Vineyard Striped Bass and Bluefish Derby committee had voted to keep the bass in its thirty-ninth annual fall competition. My friends and I mounted a petition drive that, by the end of August, had gathered more than 1,200 signatures of island residents asking the derby committee to change its mind. The petition, the Chamber of Commerce responded, was too late; commitments had already been made with sponsors. Besides, why should they be made scapegoats for the striper's decline? I responded that it was a moral question: how could you continue to put a price on the head of a vanishing species?

The *Vineyard Gazette* took a tough editorial position: "The Hunt Must End." Editor Dick Reston, who privately called the derby's stance "gillnetting of tourists," said he feared soon being run off the island on a rail. The *Boston Globe's* editorial page backed him up, noting that though entry forms had been printed, "that is a poor excuse not to do the right thing." The *Gazette* published a letter from Peter B. Greenough, the husband of opera star Beverly Sills and a longtime derby participant, who tore into the committee and announced that he was boycotting the derby that year.

As the furor mounted, the derby committee sought to portray this as a personal vendetta by me against them. After a bluefishing trip one afternoon out to No Man's Island, my friend George Peper came up with an idea to push the envelope in a different direction. Why not take out a full-page ad in the *Vineyard Gazette* promoting catch-and-release fishing and asking people to forsake their prizes?

We went to work. More than fifty fishermen and conservationists from all along the Atlantic coast offered their support and sponsorship of the ad. They included broadcaster Curt Gowdy, writers John Cole and George Reiger, Representative Schneider, and seven former derby grand-prize winners. It was a powerful statement that included a section on "How to Release Your Bass."

I was in Boston, making a few last-minute calls on the ad, when the phone rang in my home office. It was the midafternoon of September 11, 1984. It was Will Baker in Annapolis, the executive director of the Chesapeake Bay Foundation, a fellow I knew only slightly. He seemed out of breath.

"Are you sitting down?" he asked.

Earlier that afternoon, behind closed doors, the staff of the Maryland Tidewater Administration sat around a conference table waiting for the arrival of George Krantz, their fisheries chief. He was coming to announce the state's final decision on new regulations to protect striped bass. Joe Boone reclined resignedly in a corner chair, contemplating early retirement. Nearby were David Wharton and Ben Florence, who had both been shifted into different programs when their views on striper protection had clashed with those of the powers that be. Jim Uphoff had not come, "for fear I'd shoot off my mouth and get in even more trouble," he said later.

The burly, bespectacled Krantz entered, leafed quietly through a sheaf of papers, and cleared his throat. He did not sit down. "Today," he said finally, "we are announcing that the striped bass in Maryland has been designated a threatened species. A moratorium will take effect on January 1, 1985."

Boone would never forget the stunned silence that settled over the room. "We just sat there with our mouths hanging open."

Dr. Torrey Brown, director of the DNR, told the media that the ban "will be in effect until the fish return to a population size that we can once again allow people to go fishing for them"—an expected minimum, he added, of at least four years. This was Page One news in the *Washington Post* and *Baltimore Sun*, the latter running a banner headline as if it were the end of a war: "State Bans Harvest of Rockfish in 1985."

Governor Hughes had accompanied Boone and Uphoff on the young-of-the-year survey the year before, "wearing rubber boots and asking lots of questions," as Boone remembered. The governor's high-profile, $40 million program to clean up the Chesapeake Bay was being placed in jeopardy by the controversy over the fish that graced Maryland's state seal, and he wanted to resolve the issue quickly. Hughes formed a special committee to compensate commercial watermen for their financial losses. He said he was making a personal appeal to every other governor along the Atlantic coast, urging the rest of the states to follow Maryland's lead. Years later, Governor Hughes would remember also "going to see Senator Chafee from Rhode Island; as a matter of fact, I took him a framed picture of a striped bass. He was willing to do legislation. He was very interested in that fish."

Lee Zeni, when asked to pinpoint the date of the decision, answered with another question: "When do you begin to suspect your cure is not working?" I flashed back to those few minutes in time at Baltimore's Friendship Airport, but reserved comment.

In fact, the decision may have been spawned at what Joe Boone says came to be "jokingly referred to as 'the potato conference.'" One sultry day in August, Boone remembers, "I was helping my father-in-law dig potatoes when my wife called and said there was someone here to see me. A good friend of mine who worked for the Port Authority was outside with a gentleman I'd never seen before. His name was Ron Moser. He was the director of the state's Department of Transportation and had just been appointed to Maryland's striped bass citizens' advisory committee. He wanted firsthand knowledge of everything that was going on. He stayed for an hour and left after he was satisfied there was a real problem. I was told that Verna Harrison, the assistant director of DNR, once worked for Moser. Some think the moratorium happened because he convinced her of how serious this was."

Harrison, who today runs the Keith Campbell Foundation for the Environment, doesn't recall Moser (now deceased) approaching her about the moratorium. She does remember coming to the conclusion, along with DNR director Brown, that it was necessary. The real heroes, in her view, were the staff biologists. There were, she says, some "very tough meetings with the Eastern Shore [legislative] delegation, and very rewarding meetings with Governor Hughes, who was a trouper."

For his part, Governor Hughes would recall DNR director Torrey Brown coming into his office with Zeni and perhaps others. "Torrey basically said, 'We think the best thing we can do is [declare] a moratorium.' Realizing that there would be those who wouldn't like that idea very much—particularly the commercial watermen—I said, 'Well, if that's what we've gotta do, go ahead and do it.' It was really that simple."

The day after the announcement, Joe Boone drove out to a river where Jim Uphoff was doing some seining and beckoned the younger biologist to his side. "The war is over," Boone said. "We can stop fighting 'em now. They finally admitted the emperor had no clothes." That evening, a small celebration took place in one of the little towns that surround the Chesapeake Bay. The men who had long known the tragic reality of the bay's fish management, and fought against the odds to change it, could finally lift their glasses—and hold their heads—high.

We celebrated, too. On the Vineyard, George Peper went out and scooped up some blue crabs. With Anthony Gude and others, we cracked open a bottle of champagne. Our full-page ad was running the next day—"The Striped Bass: A Plea for Release."

Chapter Nine

Striper Magic

How would the rest of the Atlantic coastal states react to the emergency moratorium declared in Maryland, the striper's main producing area? When the ASMFC held its annual meeting the first week of October, the fisheries directors, at Maryland's urging, did vote to reconsider whether the proposed 55 percent commercial harvest cutbacks would be enough. They also voted on Rhode Island's new "flexibility" approach and shot it down. The state would have to implement a twenty-four-inch minimum size limit for all fishermen, plus additional restrictions, or face a potential federally imposed moratorium.

Despite such well-intentioned efforts, however, Representative Schneider lamented that the sum total of the state-by-state plans submitted so far amounted to only a 25 to 35 percent reduction in landings. Senator Chafee was also upset with what he saw as ASMFC foot-dragging, telling his staff he now favored a blanket one-year moratorium. A conference ensued between House and Senate staff, seeking to reconcile any differences in the striper protection bills before they came to a vote. When the Studds bill came up for consideration by the full House, an acrimonious floor debate followed. Attempts to weaken the language or kill the legislation altogether by Virginia and Maryland failed, with Studds castigating both representatives. On a voice vote, the bill passed overwhelmingly and went to the Senate for action.

Many senators no doubt read a syndicated column by well-known conservative George Will on October 7, 1984, saying that, despite the federal government's time-honored reluctance "to intrude into coastal fishing regulation, traditionally a matter of states' rights . . . [t]he stakes are great and the threatened asset is a national asset, so Maryland's moratorium should be national policy." (I'd had no idea that Will was a striper fisherman, until he called to interview me beforehand.) Nonetheless, rumors circulated that Virginia Senator Paul Trible would try to prevent the bill from reaching the Senate floor before adjournment. Pressure from Chafee and Ted Kennedy persuaded him to withdraw his opposition, however.

Trouble then surfaced from an unanticipated quarter. Senator Robert Byrd objected to a section specifying that the federal striper study continue to look deeper into acid rain as a cause of the fish's decline. Byrd's primary constituents were the big West Virginia coal companies, and he refused to admit that their sulfur dioxide emissions were partly responsible for a growing national problem. Under congressional rules for eleventh-hour action on bills, unanimous consent was required to bring the striped bass matter to a vote. When Byrd said no, any action was postponed to the final day of the session.

The striped bass were about to teach another lesson—in parliamentary procedure. Phone calls seeking to sway Byrd poured into his office. Chafee tried to reason with him. Even House Speaker Tip O'Neill of Massachusetts sought to use his clout to turn the West Virginian around. Byrd remained stubbornly opposed. At high noon on October 11, concerned senators agreed to eliminate the brief sentence about acid rain. It didn't much matter, since the striper study would still be looking into *all* causes of contamination in the Chesapeake. Byrd withdrew his objection, and the bill passed by a voice vote. Since the Senate had changed the language, though, it now had to go back to the full House for another round. The same rule applied—one member objecting to a phrase at this point could kill it. And there were only three hours left before adjournment.

Chafee's staff rushed the bill over to Studds's office. Studds tried to keep the pending vote as quiet as possible, since certain congressmen who'd fought it might take one last stab. Maybe, just maybe they wouldn't realize it was coming up again. One hour to go. Speaker Tip O'Neill called for a voice vote, no objections were raised, and on October 31, 1984, the Atlantic Striped Bass Conservation Act was on its way to becoming the law of the land.

For the first time ever, Congress had stepped into a fisheries management crisis in traditional state waters. By June 1985, every state needed to have at least the 55 percent reduction approved and in place, or whatever stronger measure the ASMFC might decide upon. Otherwise, a moratorium would be imposed. The federal fisheries agencies would review the ASMFC's management strategy and report back to Congress on whether the plans were adequate. Two additional years of funding for the emergency federal study were included, to focus on chemical pollution. President Reagan, who'd already committed $10 million to help clean up the Chesapeake, was expected to sign the bill. And he did.

The striper's political migration now meandered back to New York, where the Westway Project was still very much in play. In late November 1984, the Army Corps of Engineers' new 217-page environmental review concluded that building the highway would have a "perceptible" but "minor" impact on the Hudson's bass population, downgraded from the "significant" impact the Corps had deemed previously.

At the end of January 1985, Colonel Fletcher Griffis, the Corps's district engineer, gave his assent for a landfill permit. "All great projects need a period of gestation. Westway has certainly gestated," the colonel said. Governor Cuomo was "delighted." Mayor Koch was "happy for the people of New York, happy for the striped bass, happy for America." Very happy campers, the politicians. But not for long.

By the summer of 1985, it all started to unravel. The EPA's regional administrator for New York, Christopher Daggett, turned out to have prepared a position paper opposing Westway on environmental grounds and emphasizing the importance of the striped bass. A few days before the EPA was to announce its final position, financier David Rockefeller and New York Senator Alfonse D'Amato met with a ranking EPA official in Washington, and soon thereafter Daggett was notified that his paper was being rewritten. An EPA lawyer recorded in her office diary that peremptory reversal of Daggett by higher officials entailed the risk of scandal. When Daggett persisted, another EPA official wrote in a memorandum: "It doesn't appear that Daggett has gotten the message; nothing will kill this project faster than this letter." The letter disappeared, at least temporarily.

At the end of June 1985, the letter, the diary, the memo, and other EPA documents were among hundreds of government records introduced as evidence by Westway's opponents during a new trial before Judge Griesa, who was to decide whether the Army Corps had followed his orders and the law in issuing its second permit. Included was a "Dear Bill" letter that David Rockefeller had written to EPA Administrator William Ruckelshaus, with a copy going to Michael Deaver, a principal adviser to President Reagan. The letter derided the regional EPA representative as well as fisheries agency scientists.

The latest trial brought out some other weaknesses in the proponents' case. The Army Corps had kept no records of what happened at its internal meetings, despite Judge Griesa's previous order to do so. Ecologist Michael Ludwig of the NMFS testified that his agency viewed the Army Corps's new environmental analysis of Westway as "fallacious" and without basis in legitimate science. Finally, there was the testimony of a marine biologist named William Dovel.

Dovel had been around for a long time, having catalogued some five million striped bass eggs for pioneering Maryland researcher

Romeo Mansuetti back in the 1950s and then doing some research of his own on the Hudson. Concerning Westway, Dovel had a novel theory. Striped bass, he speculated, only passed through the area on the Hudson that the Westway development would fill in; it wasn't a crucial habitat where the fish lingered through the winter and grew toward maturity. But as the tale unfolded before Judge Griesa, it turned out that Dovel had failed to mention this theory to anyone when he'd been the Army Corps's field supervisor taking fish samples early in 1984.

The first time Dovel's theory surfaced was late that year. It wasn't certain who approached whom first, but Dovel had somehow hooked up with George Lamb. They'd known each other for years, and Lamb happened to be a senior operating executive of a Rockfeller foundation as well as a board member of a lobbying group that was pushing for Westway's approval. According to Lamb's testimony, Dovel told him he "had a handle on the population dynamics and movement of the striped bass." Of course, he'd need some funding to prove it. Through a foundation Lamb worked for (headed by Laurance Rockefeller), arrangements were made to have Dovel paid $9,000 to put in three more months on his study for the Army Corps. When Dovel delivered his draft to Rockfeller Center and collected his fee, the environmental report contained Dovel's theory that the Westway area wasn't really an important habitat. This became the basis, in November 1984, for the Army Corps's changing Westway's impact from "significant" to "minor"—and for the Corps then granting approval to the project. Now, at the new trial six months later, Judge Griesa called Dovel's report "an astonishing document . . . just one hypothesis after another." On the witness stand for three days, the Corps's consultant changed his testimony several times.

John Boreman, co–scientific director of the government's Emergency Striped Bass Study, had been brought into the case as well. He analyzed some of the modeling data that had earlier concluded that perhaps 20 percent of the Hudson's bass population would be affected by removing the piers. Boreman recalled: "I realized that the consultant working for the developer had counted the number of months wrong between September and January. So I reran the model with the right number of months, and it turned out to be a 30 to 40 percent difference in impact on the striped bass. All of a sudden it became known as the Boreman Model. I kept saying, no no, it's just a *correction* to the old model."

All of the finagling had Judge Griesa fuming again. "You have an obligation . . . to want the truth," he told the Army Corps's attorneys. "You are not doing that." In August 1985, Griesa handed down a 131-page decision that excoriated the Army Corps for fudging its figures in order to downplay any problems for the bass. "Two failures to justify the Westway landfill and federal funding for Westway under applicable legal standards should bring the matter to an end," the judge said in denying the landfill permit. Governor Cuomo announced that the state would appeal. *Newsweek* wondered breathlessly: "The stakes are enormous and New York is agog: will it be the most expensive highway in history, or a feeding ground for the fish?" Others, like columnist Murray Kempton, felt that Griesa "may finally have sent Westway floating belly-up down the Hudson."

The U.S. Court of Appeals handed down a quick decision. It sustained the conclusion that there had been deceit, though it set aside the idea of enjoining Westway forever, as Griesa had ruled. On September 19, 1985, in Washington—twelve days before the federal deadline for deciding whether Westway would proceed as part of the interstate highway system—New York's top elected officials issued a joint statement. Cuomo, Koch, and the state's two U.S. senators blamed court challenges, defeats in Congress, and above all mistakes by the Army Corps for putting them at an "impasse." The decade-long effort to build the highway would end. "Westway sleeps with the fishes," the *Village Voice* headlined. Instead, $1.3 billion of available highway funds was used for improving the city's public transit system and $500 million was set aside for a replacement roadway.

Over the years since, portions of the old West Side Highway have undergone reconstruction, and a Hudson River Park Act was approved by the state legislature. This enabled the creation of some 550 acres of a park-on-piers unlike any other on the planet: thirteen piers extending out into the Hudson River as far as 800 feet along the West Side waterfront from 59th Street to Battery Park. Four hundred of those acres were designated an estuarine sanctuary, in recognition of their importance as a fish habitat, but were also available for recreational and commercial maritime use. The first sector of the park opened to the public in Greenwich Village in May 2003.

The man now in charge of a new fund-raising organization, Friends of Hudson River Park, is none other than now-retired attorney Al Butzel—one of the legal point men on both the Storm King Mountain and Westway cases. "There is no landfilling," Butzel says, "and all piers, if replaced, are required to be within the footprint of what was done there before. So that's going to keep the slow currents, which the bass like."

The state and city committed $200 million to the park project, a figure Butzel is looking to match with private-sector funding, seeking to emulate the Central Park Conservancy's partnership between property owners, businesses, and citizens. His group also works with an organization called Clearwater, taking people out on the river to teach them about the striped bass and how important the fish really were in the creation of Hudson River Park.

"When I go down and look at the one section that's so far been built [in Greenwich Village]," Butzel says, "I can't help thinking this will be one of the fabulous waterfront parks in the world. And down below there, all those striped bass are still swimming around."

Simultaneous with the shenanigans still surrounding Westway early in 1985, the issue of PCB levels in New York's striped bass arose again as well. A study by the state's Division of Fish and Wildlife had found that, besides forty-four "hot spots" previously identified in the Hudson River, both in New York Harbor and off western and eastern Long Island, the bass's flesh "consistently and significantly" contained PCB concentrations higher than the Food and Drug Administration's new standard of two parts per million. The Cuomo administration refused to release the study. Nor did it make public a letter from the state's health commissioner to the DEC commissioner stating that "in the interests of public health," these fish should not be consumed. Assemblyman Pat Halpin, who'd led the legislative effort for tougher striper regulations, later obtained the report under a Freedom of Information Act request.

At the end of March, Governor Cuomo banned commercial fishing for stripers in New York Harbor and western Long Island—but *not* in the eastern Long Island waters where Arnold Leo and his haul seine brethren operated and where 90 percent of the state's commercial catch occurred. This was in direct conflict with the recommendation by the health commissioner, but further sampling was said to be needed. Assemblyman Halpin said: "It's a dark day for the DEC and the whole state when people's health is not as important as the economics of the fishing lobby. They've given a new definition to PCB: the Politics-of-Cancer-Burlesque."

In other sectors of the coast, protection of the striped bass had become paramount. In one of the largest undercover investigations ever conducted by the U.S. Fish and Wildlife Service, even before the Maryland moratorium took effect, federal agents posing as fish dealers purchased some 6,700 pounds of striped bass—large fish beyond the maximum allowable length in Maryland and Virginia. The agents then sold the illegal fish to dealers in several other states, a violation of the Lacey Act on interstate commerce. The culmination of the "sting" operation took place on January 16, 1985, with 130 people being arrested in predawn raids on their homes.

In the spring, an advertisement for Sure deodorant that ran in newspapers across the United States caused quite a stink. Two fishermen were pictured holding up their catch, with the pronouncement "Raise your hand if you're Sure." One of the fish was a small striped bass. Alert readers started phoning the papers, complaining that this was in poor taste. I called for immediate removal of the offensive ad, in a prepared statement that contained this rather tongue-in-cheek message: "There's absolutely nothing Sure about the future of the striped bass. Personally, I think every sportsman ought to switch to something more appropriate: Ban." Deodorant maker Procter & Gamble defended itself by saying, "We weren't making an ad about fishing, but for another product." Everything rolled on, so to speak.

Then there was the Martha's Vineyard Striped Bass and Bluefish Derby. After the derby committee decided in March 1985 to once again keep the bass in the upcoming fortieth competition, the first longtime sponsor to withdraw support was *Salt Water Sportsman*. As associate publisher Spider Andresen put it in his letter, the 114 adult females killed in the tournament the previous year were very important fish, if the species was to have a future. When the *Vineyard Gazette* once more editorialized against keeping the bass in the contest, the island's Chamber of Commerce fired back, calling the newspaper's coverage "blatantly inflammatory and irresponsible" and adding, irresponsibly, "we can only state that the striped bass are in a low cycle."

The domino effect, however, was under way. The island town of Chilmark formally declared it could not support the derby. Three leading fishing gear manufacturers—all generous patrons of the contest in the past and including the world's largest rod-and-reel maker, Zebco—pulled out. On May 23, the committee reversed itself. Striped bass would be dropped from the tournament, not to be included again until the fishery recovered. It would be bluefish only. With that shift, all the sponsors moved to renew their support.

As for Rhode Island, let George Mendonsa tell the story. My onetime archenemy had mellowed a bit—but only a bit—by the time I sat down to reminisce with him, at his home in Middletown, twenty years later. Mendonsa recalled: "After all this baloney with the conflict of interest, I got a telephone call one day around midnight. This guy says, 'I'm an aide to the governor. I want to inform you that you are being released from the Fisheries Council.' I said, 'Oh, that's nice. Why the hell didn't you wait a couple more hours, till two or three o'clock in the morning?' A few hours afterwards, I left the house to go fishing. I stopped in the restaurant to get my morning coffee, like five o'clock. I pick up the *Providence Journal*, and there it is: Fisheries Council member George Mendonsa fined and kicked off the council. It was already printed in the paper before they even called me!

"So then we go back to Governor Garrahy, me and Francis [Manchester]. I says to the governor, 'You're the sonovabitch that *appointed* me to that fisheries council, and I got the shaft for $14,000 [in fines and legal bills]!' So he put a bill into the House to exempt any commercial fisherman on the council from the conflict laws. Well, it come out of the House in my favor. But when it went to this special legislative committee in the Senate, this fella from Newport never even put it on the agenda. We went back to the governor: 'What the hell is this?' Governor said, 'Well, that was a lousy way he handled that bill. But because he was chairman, it was legal what he did.'"

In the autumn of 1986, Mendonsa was placated when the U.S. Fish and Wildlife Service came around and said they'd pay him $25,000 to help out with a tagging study. He and his crew would bring in stripers in one net, keep them alive in a tank, make a buttonhole-sized incision on their bellies, insert a streamer tag, and let them go. According to Mendonsa, there were plenty—"one day alone we had 15,000 striped bass in the count, all small fish!" Later, he learned, the biggest proportion of those tagged fish turned up again in the Hudson River. Well, except for one.

"In the report they sent me afterward, they had one fish—somebody up in Toronto sent in the tag. Must've went up the St. Lawrence River, they figured. Never happened before in history." Mendonsa couldn't conceal a smile. "Yeah, well, what we did—wise guys—we took one striped bass and we shipped it to Toronto by truck. And somebody up there turned the tag back to Uncle Sam." Mendonsa was laughing now. "But the biologists got that fish swimming down the St. Lawrence River!" he guffawed.

By the time Mendonsa got enlisted by the government to tag and return striped bass, he wouldn't have been able to sell one anyway. On April 7, 1986, Rhode Island public health officials banned all sale of the fish and advised sports fishermen not to eat their catches. Forty-four percent of all striped bass tested from the Ocean State's waters showed PCB accumulations above two parts per million. Later that month, after further sampling in New York found that virtually *all* the 444 fish they tested had PCB levels higher than the allowable standard, there was no longer any choice but for Governor Cuomo to outlaw all commercial sale of striped bass taken from his state's waters. As the *Boston Globe* editorialized: "It may be that the prospect of a health threat to humans will save the striper. Fear of PCBs . . . seems a paltry excuse for doing the right thing, but any port in a storm."

This back-door method of conservation wasn't coming a moment too soon. The ASMFC, which continued to hold its technical committee meetings behind closed doors, was claiming that all member states had achieved the 55 percent reduction in landings. But as Senator Chafee put it at another congressional hearing, "You boys let yourselves off easily. I think some states are thumbing their noses at the whole business." Joe Boone called the states' efforts "a farce." New York's management plan, for example, had closed the commercial fishing season only from December to May—when the bass were scarcely in their waters anyway. The full PCB closure that followed, of course, made that a moot point.

Down in Maryland, the young-of-the-year index for 1985 had found a mere 2.9 fingerling stripers per seine haul, making it another dismal year for the Chesapeake. Any positive effects from the Maryland moratorium, of course, wouldn't show up for a few years, and nobody knew how hard the last semisuccessful spawning—the Class of '82—had been fished in the Chesapeake before the moratorium went into effect.

A fascinating tale about that particular year class surfaced in a January 1986 Tom Horton article. A scientist from Johns Hopkins,

Dr. Robert W. Chapman, had been using gene-splicing technology to trace the lineage of striped bass. Having examined more than 200 females, he'd been able to place these fish into fourteen different families—not subspecies, but "rather in the way humans have different surnames." In the course of constructing his "family Bible" for the bass, he'd come across something unusual in some of those born in 1982. It was a DNA molecule different from those in any family he'd identified elsewhere in the bay—and it appeared only in fish taken from a quiet bend of the Potomac River. This led Chapman to conclude that the 1982 spawn seemed to have depended, to an extraordinary degree, upon a relative handful of females—as few as 50, no more than 500—that spawned in just the right place at precisely the right time to permit an astronomically high survival of eggs. It also meant, as Chapman said, "that one gillnetter in one place, if he got those few fish who were going to carry the population, could have wiped out a lot of the bay's year class in a single day."

So was nature vulnerable to the hubris of man. That spring of 1982—the spring our fight began—thankfully for the striped bass, man had missed the boat.

Slowly, the bass started to respond. By 1987, the Chesapeake's marine scientists, able to begin assessing the dynamics on an "almost virgin stock" in the absence of a fishery, found their knowledge base expanded. Never before had they seen females with ripening eggs winter inside the bay; now some four- and five-year-olds were doing just that. It had also been thought that the males were not migratory, but now it appeared that at least some from the 1982 year class had left the bay.

The 1986 Maryland young-of-the-year index went up to 4.1 newborn bass per seine haul, and by 1987 it reached 4.8. Still well below par—the average over the twenty-eight years since the survey began was 8—but improving.

The various state plans were all aimed at allowing 95 percent of the small fish to spawn at least once, before they could be subject to any fishing pressure. The minimum size limit, as a consequence, kept being raised to protect, in particular, that Class of '82. It had risen to thirty-three inches by 1987, at which point only one state still allowed commercial sale of striped bass. That, ironically, was Massachusetts—the state that had formerly been in the vanguard of regulating the fishery.

It seemed fitting, too, that in October 1987, federal prosecutors filed a lawsuit looking to seize regulatory control of the Fulton Fish Market, that onetime bane of the small striped bass. This marked the first time, said Rudy Giuliani, then the United States Attorney in Manhattan, that the government had sought racketeering charges to put "an entire commercial center under court supervision." Mayor Koch, never one to miss having the last word, said he'd begun the investigation three years before, when a friend who operated a fish restaurant told him of being required to make payoffs at the market. Surprise, surprise.

That December, the governors of Maryland, Virginia, and Pennsylvania signed an agreement on a regional effort to clean up the Chesapeake, one that had taken almost a decade to achieve because of all the competing interests. They committed their states to reducing the amount of nitrogen and phosphorus entering the bay by at least 40 percent by the end of the century. Modernization of sewage systems, soil conservation, and reduction of fertilizer use were viewed as the key goals.

In 1988, the striper juvenile index dropped again, to a poor 2.7. So the announcement that Maryland's new governor, William Schaefer, made in August 1989 astonished almost everyone. The 1989 rate of surviving bass looked to be so high, the governor said, that limited sport and commercial fishing would be permitted to resume in Maryland as of fall 1990. The five-year-long moratorium could end.

The moratorium wasn't supposed to be lifted until the Maryland young-of-the-year survey achieved a three-year running average of 8 juvenile striped bass per haul. The latest survey, Maryland officials soon claimed, came in at 25.2 juveniles—already topping that "running average," the second-highest count in thirty-six years of record-keeping, and enough to justify the sudden change. Joe Boone had been retired for several years; John Boreman had left the federal bass study after 1984 to pursue new directions in his career. So who was minding the store?

After the governor's proclamation, Bob Pond and Jim Price accompanied a team of Maryland biologists to several sampling sites in September, including a place at the mouth of the Choptank River called Hambrooks Bar. A month earlier, in two seine hauls, a staggering total of 1,162 juvenile stripers had been counted there. That single site produced 45 percent of all the bass found in the Chesapeake in 1989. But, as Price observed, Hambrooks Bar wasn't the representative area it used to be. Not only had the shoreline eroded to such an extreme that the biologists were basically sampling an island a quarter-mile offshore, but the configuration of the remaining sand prevented normal seining procedures and was altered enough "to create an unnatural accumulation of fish."

Boone stepped back onstage to say: "Almost since the beginning of the survey, we discussed this problem of hot spots. Anybody who does biological sampling knows that occasionally you'll get something that's not really in line with the reality. You've got to do more than just take raw numbers, and apply a little common sense to what you've collected. If the Choptank has such a huge number of striped bass, why isn't population pressure pushing them out all through the river? Why are they concentrating? Meanwhile, the Potomac and the Nanticoke are still in poor shape. The least we could do is wait and see if this is an aberration or actually the beginning of some improvement in recruitment. But the powers-that-be have decided with the first ray of hope, let's jump in and start fishing again. Something smells there."

Bob Boyle sprang into action, too, writing a piece for *Sports Illustrated* titled "A Hasty Call on a Fast Count." Boyle and I sat next to each other at a congressional hearing called by Studds—now the fisheries subcommittee chairman—early in 1990. The ASMFC, after raising the minimum size limit again, this time to thirty-six inches, had voted the previous autumn to approve a "transition plan" for resuming bass fishing on a larger scale. Representative Schneider said she was struck by sports fishermen's strong opposition to relaxing the regulations. "They are begging for Congress to require them to make sacrifices in order to assure that striped bass stocks are truly healthy," she said. "That is, to say the least, very unusual."

After testimony by Massachusetts's fisheries director Phil Coates that extolled the ASMFC's "long, orderly, very flexible" process toward achieving the restoration of striped bass, Studds responded wryly: "Are you available for consultation on the federal budget?" After the hearing, Studds told reporters he wished the ASMFC would wait a year before making any changes in its management plan. Boyle, meanwhile, challenged Coates directly. "What's the rush?" Boyle asked. Coates insisted that "there is no rush; we've been working on a new plan for three and a half years." Boyle also wondered why the ASMFC's fateful decision to notch up the fishery was made far from public scrutiny, "up with the moose in the woods" of Dixville Notch, New Hampshire. As for the organization relying on survey data from Hambrooks Bar as indicative of a striper comeback, Boyle told Coates: "If that happened in a poker game, you'd be reaching for your pistol."

When the managers decided to lower the coastal minimum size limit from thirty-six inches back to twenty-eight inches, however, fishermen in Massachusetts and elsewhere said they didn't *want* to retreat from the current regime. So the limits generally went no lower than thirty-four inches, until the mid-1990s. A new conservation ethic had taken hold among fishermen. Catch-and-release fishing on striped bass was, among sportsmen, now the norm rather than the exception. This represented an attitudinal sea change. In the course of the striper wars, the language had changed, too. No more were macho fishermen afraid to call a striped bass what it is: a magnificent fish, even a magical creature.

Nobody had come to understand striper magic more than postman Jim White. We were sitting in his living room in Coventry, Rhode Island, early in 2004. I realized it had been going on twenty years since we'd seen each other. White still sported his dark mustache, but his hair, like mine, had begun to gray. The little boy he'd written a poem for—a poem that had deeply touched Representative Schneider, about the boy someday knowing the magic of bass fishing—that little boy was grown up now, preparing to teach Jim's grandson how to "wet a line."

After the whirlwind we'd both found ourselves caught up in, at my suggestion White had taken up writing for the publication *The Fisherman*. Damned if he wasn't about to have his second book published on saltwater fishing! Now retired from the post office, he gave slide presentations at numerous sportsmen's events and, once summer came, you still couldn't keep him off the water. He'd become a charter-boat captain of quite some renown in Rhode Island.

Some years ago, he'd related part of a striper story in an article titled "More Than Just a Lucky Rod." But it wasn't the whole story. No, that was almost too far-out for anybody to believe. He wanted me to hear it, though. Because it was true, and I'd understand why. Because it spoke worlds about the fish we'd fought to save, and about the invisible world it comes from.

It began in the summer of 1986. He was fishing off southern Rhode Island's Point Judith one afternoon, and the fish were there, when his reel broke. The only place nearby he could go to fix it was Top of the Dock, the bait-and-tackle shop owned by Joe Mollica. Longtime friends, cofounders of the Rhode Island Saltwater Fishing Alliance, they had become bitter enemies after White decided to support the Rhode Island moratorium. They hadn't spoken for more than two years. Mollica was the last person Jim wanted to see, but he didn't really have much choice.

So White walked in and Mollica said sure, he'd fix the reel. White looked up. A fishing rod hanging from the ceiling caught his eye. "Something about it, for whatever reason I couldn't get it out of my head," he remembered. "I kept getting drawn back to his shop, even when I didn't have to go there." He would just stare at the rod and finally asked Mollica whose it was. "Oh, it's mine," Mollica said.

White continued: "I said to myself, 'I can't have him build me a rod, I just can't—what would everybody think?'—we hated each other!

Finally my obsession overcame any kind of good judgment I had. I went in and said, 'Can you build me one of those?' Joe said sure, he could."

A week or two later, White's phone rang. It was done, an eleven-foot Lamiglas surf rod. Inexplicably, Jim was trembling when he arrived to pick it up. Mollica suggested they walk down to the pier together, where White could try it out. "I put on a two-ounce Pencil Popper [lure] and threw it out there. Man, that thing just sailed! Oh boy, this casts beautifully! I start bringing the Popper in—bang, I get a fish!"

It was a seven-pound bluefish. "That's going to be a lucky rod for you," Mollica said as they parted company.

White got back in his pickup and started to drive along Ocean Road in the direction of Point Judith, looking out at the water. A storm was brewing. By the time he reached the lighthouse not far away, it was pouring rain. He pulled into the parking lot anyway. Nobody was there, except two people inside their cars watching the white water roll in. As the rain let up a bit, White readied his tackle and climbed down some rocks toward the pounding surf. Mollica had said something about this rod being great in the wind, and this one was blowing about 20 knots, so White would soon find out.

Looking around, he realized he was completely alone. His mind flashed back to a different time, a time long past. He was standing next to his father throwing Atom swimmers—Bob Pond's lures—to feeding bass. Strangely, this many years later, White could almost hear his father's voice instructing him in what to do. First cast. Suddenly now, his rod bent, and he sensed something trying to drag him seaward. Out of instinct he set the hook. Instantly the line began to peel off the spinning reel. The power he felt was awesome and completely unexpected. The next few minutes were give-and-take, until the fish came close enough to see and grab. A bass! Maybe twenty-five or thirty pounds. He felt his heart pounding, as if it was trying to leave his chest. He couldn't believe it. After all, it was only eleven o'clock in the morning, not a time such things ordinarily happened. He hadn't even seen a bass this big in several years.

Putting a tag in her, White gently released the fish back into her domain. He looked up to see the water alive with fish. Bass and blues mixed together, almost at his feet. Stripers maybe forty pounds or more were devouring mullet in no more than two feet of water. He cast out again—another thirty-pound fish! And again—an eighteen-pound blue! A fourth time—a forty-pound bass! As he let each one go, White found he was talking to himself. Looking skyward, he shouted as loud as he could above the crashing waves: "Lord, if you're going to take me, then please take me now!" He was sure that if there was a fisherman's heaven, he was in it.

It was then that he saw a silver-and-black image erupt under his plug inside an incoming comber. "Oh my God!" The rod bent in two. Line screamed from the reel as if he'd hooked into the back of a truck. The rough sea conditions gave her even more strength, as she masterfully used the waves and current to her advantage. White concentrated with everything he had, as Mollica's words resounded in his head: "This is a lucky rod."

Finally White turned the fish and began looking for a place to bring her ashore, among boulders that now seemed like giant mountains lurking all around him. The waves were getting bigger. He was drenched. How could he land her in this surf with all these rocks and not hurt or kill her? Fifteen minutes later, she was three or four feet out in front and he got his first good look at her. His body shook. Fifty, maybe fifty-five, he thought. Then she turned and took thirty more yards of line off the spool. Carefully White worked his way into the water to get a better shot at grabbing her gill. As he did so, the bass came rolling toward his feet. He reached down and touched her tail. With his hand now on her, he could feel the very essence of her being in every part of his body. For that moment they were one, the hunter and the hunted. This is what I live for, he thought. The sea, the fish, the battle. But somehow, this one felt different from all the rest. This one had touched his soul. He felt a respect and reverence he'd never known before.

As he looked to see where the plug had her hooked, he caught the fish's eye. She was looking right at him, right through him, looking to see what he would do next. The treble hooks barely had the corner of her bony jaw. Just as he sensed the contest was over, with one last, quick turn she set herself free. Slowly she swam off through the heavy surf that continued, incessantly, roaring in.

White glanced at his watch. He had to get home; it was time to pick up his son from school. On the way back he raced into Mollica's shop, dripping wet, covered in scales and bluefish blood. "What did you do to that rod?" he cried. "Did you put a spell on it? Say some kind of magic words? Did you make it that I *wasn't* gonna catch fish and I did? We're enemies, Joe!"

That day was the beginning. But only the beginning. Between the end of August and when he quit fishing at the end of December, White caught and released scores of bass using this rod. Nobody else on the beach could get a strike, and he'd cast out and hook a bass. He'd try a different rod, and nothing would happen; he'd return to his truck for Mollica's rod, and immediately a fish would hit. Never once, in all those months, did the rod fail to land at least one striped bass.

It got to the point where his fishing buddies were convinced that the rod was haunted. White tried raising this matter again with Mollica, who scoffed and said, "No, you're just lucky." Then, as abruptly as it all began, suddenly the charm stopped. The eightieth time White used the rod, he failed to catch a fish. He knew it had been that many fishing trips, because he kept a numbered logbook.

White didn't record only his fishing. In a separate little diary, he kept track of all the meetings he'd attended—the private gatherings, the public hearings, the Fisheries Council sessions—all devoted, over the years, to the striped bass. He and Joe Mollica, without fail, had been together at all those meetings. On the same side, and later on opposite sides.

The entry for the last meeting was numbered "79." There hadn't been an 80.

"I still have the rod," White said. "It's hanging down in the cellar. I haven't used it in I don't know how long. Maybe my kids will someday, or my grandson. Maybe he'll ask me about what that inscription means: 'Custom Built by Joe Mollica for Jim White.'"

The two men had renewed their friendship, and they remain friends to this day.

For them, and for thousands of others, the resurgence of the striped bass has been no less miraculous. They have returned in such numbers along the Eastern seaboard that many newcomers to fishing have no idea their abundance was ever in jeopardy or that battles large and small were fought over their fate. Indeed, stripers are now so commonplace in contemporary fishing havens that imagining a season of scarcity was—until very recently—unfathomable.

Chapter Ten

The Double-Edged Sword of "Full Recovery"

In 2003, not long after his promotion to director of the Northeast Fisheries Science Center, John Boreman was on a conference call to National Marine Fisheries Service headquarters. Just beyond the picture window of his spacious new office in Woods Hole, Massachusetts, Boreman's gaze fell upon two fishermen standing on a jetty. One of them had just reeled in a striped bass about thirty inches long and was holding it high for his friend to take a picture before he returned the fish to the sea. Boreman couldn't help but describe what was unfolding in front of him. "You know, folks, this is why we're doing our jobs," he said. "If you could see this guy's face, it's what makes worthwhile everything we try to do."

The next night in New York, a dinner was held to commemorate Boreman's ten years on the board of the Hudson River Foundation. The scientist was presented with a large watercolor print of a striped bass that now hangs on the wall behind his desk. Boreman found it uncanny that the striper was about the same size as the fish he'd witnessed being caught and released the day before.

Boreman, today a youthful-looking fifty-five, became almost boyish as he told me the story. We were having lunch together at the Captain Kidd, a block from his office. It had been almost twenty years since we'd seen each other. Not long after Maryland declared its striper fishing moratorium in 1984, Boreman had left as co-director of the federal government's Emergency Striped Bass Study.

With the Maryland moratorium in place, coupled with regulations nearly as stringent in the other coastal states, Boreman expected it would be only a matter of time before the striped bass would proliferate again. Despite poor year classes in the Chesapeake continuing into the early 1990s, Maryland's 1993 young-of-the-year survey set a new record of 39.8 juvenile stripers per seine haul. In 1995 the ASMFC declared the fishery "fully recovered" and thus ready to be opened to higher levels of harvest. Minimum size limits had gradually been dropped from thirty-six inches down to twenty-eight inches along the coast. Annual commercial quotas for the various states were also gradually increased, under a series of amendments to the ASMFC's management plan. At the end of 2002, a 43 percent rise in commercial landings was allowed in the coastal ocean fisheries. (Five eastern seaboard states—Connecticut, Maine, New Hampshire, New Jersey, and Pennsylvania—still allow no sale of stripers taken in their waters.)

During this period of the fish's recovery, Boreman had been a professor of fisheries at the University of Massachusetts as well as chief of research coordination, later deputy director, and ultimately director of the Northeast Fisheries Science Center in Woods Hole. Today his primary responsibilities focus on cod, haddock, and even endangered right whales that become entangled in fishing gear.

I'd moved on, too, since the early 1980s, becoming part of successful fishermen's campaigns to outlaw commercial sale of marlin and to stop ocean dumping off New York. Over the years I'd written dozens of articles about other environmental concerns and a book about protecting the gray whales. Fighting to save the striped bass had been my awakening. Now I wondered how the bass battle had changed Boreman's life. What lessons had he learned that might be applied to other, still-troubled fisheries?

"Well, my attitude is, you just tell it like you see it and let the chips fall where they may," he replied. "Sometimes it may not be politically wise, but with striped bass it was so blatantly obvious. I attribute the bass's comeback to the fact that industry trusted the science—enough to police themselves into complying with the regulations. Someday maybe we'll get back to that. That's my job now, to rekindle and rebuild that kind of trust. Right now when I go to meetings with fishermen, I usually have an armed escort."

Boreman laughed, but ruefully. He was referring to the extremely controversial task of assessing the Northeast's groundfishery, where things aren't nearly so clear-cut as they were with striped bass. "There are twenty different stocks, each with its own little story," as Boreman says. Besides which, the cod, haddock, yellowtail flounder, and other offshore species are pursued by large, expensive vessels and are often the lifeblood of entire communities such as Gloucester, Massachusetts (site of *The Perfect Storm*). Trust is often hard to come by, as fishermen constantly question the accuracy of scientists' data, sometimes with cause.

Still, there were important instances in which much could be learned about the science and management of fisheries from the striper saga. In the case of striped bass, having exactly the same regulation for each state wouldn't have worked well, Boreman reflected, "because each state had its own type of fishery—trap nets in Rhode Island, pound nets in Virginia, gillnets in Maryland, and so on. So we suggested what's called a 'conservation equivalency.'. . . We didn't try a one-size-fits-all regulation. The states would come in with their data and build a convincing scientific argument that whatever they were doing, it would reduce the fishing mortality rate by 50 to 55 percent. It makes things more difficult to evaluate, but in the long run keeps the historical fisheries alive—recognizing that there are different cultures out there, and we need to respect that."

Conservation equivalency, Boreman commented, is now a common practice, but back then it was pretty novel. It has since been used in regulating the take of other species, such as lobsters. With striped bass, Boreman added, "the disheartening part was that a lot of these same individuals from the states came back to the next meeting looking for loopholes in their own plans." Besides the Maryland moratorium, what changed the equation was Congress's passage first of the Studds amendment to the Anadromous Fish Conservation Act, which withdrew federal funding from states that did not comply with the ASMFC plan, and then of the Atlantic Striped Bass Conservation Act, which mandated federally imposed moratoriums for recalcitrant states.

The director of the National Marine Fisheries Service in the early '90s, Bill Fox, termed the act "a brilliant stroke of policy [that] established a model for state-federal fisheries management cooperation." He recalled sending threatening letters out on two occasions. "Largely the state fisheries directors were pleased to receive these, because they were having some problem with the legislature or area of their state—and this allowed them to follow through."

In 1993, the Atlantic Coastal Act was passed by Congress. In large part modeled on the Atlantic Striped Bass Conservation Act, it has also provided a stable federal funding base for cooperative management programs ($5 million in 2001). Just as had happened with striped bass, this new law charged all the ASMFC's member states with implementing fishery plans to safeguard the future of other species. And, as with stripers, it held out the threat of a federal moratorium for states that didn't comply with the plans. It's made a huge difference with summer flounder, a once-depleted population considered largely rebuilt by 2003, and to a lesser degree with weakfish, also known as gray trout. When younger weakfish were being decimated off North Carolina, the individual states proved reluctant to take action. Once the ASMFC got enforcement power, thanks to the threat of a federal closure, the weakfish had a chance.

The relief provided to striped bass by the near-total closure of the fishery has since proven necessary—and effective—with other species.

During the mid-1980s, Florida imposed a three-year moratorium on red drum that paid amazing dividends. After the Gulf of Mexico's red drum fishery was curtailed with a ban on sale and a one-fish limit for sports anglers, a good recovery also resulted. And once the 600 square miles of New England's Jeffries Ledge were closed to commercial fishing for five years, the 2003 year class of haddock was the largest ever recorded.

Boreman has found, though, that he can't always apply approaches gleaned from the striper recovery. He was certain he could prove that if you cut back on fishing for winter flounder, they would bounce back as the bass did. This turned out to be a different story because, Boreman came to realize, winter flounder are more vulnerable to ecological changes in their estuary habitat, where they spend a lot of time, than to fishing pressure.

Above all, Boreman saw, you had to consider the different life histories of the various fishes. "Sturgeon are incredibly sensitive to fishing mortality, because they live for thirty to fifty years. But bluefish are even less sensitive than winter flounder to changes in fishing mortality, because they are short-lived [their maximum life span is but six or seven years] and are only in the fishery for a few years."

The same paradigm carried over to sea turtles. Boreman pointed out that in the early '80s, the idea was that their depleted populations would revive if only their nesting beaches were increased. However, their being inadvertently caught as bycatch in shrimp trawls proved to be so detrimental that even doubling the nesting beaches wouldn't make much of a difference. "You need to assess where the risk is highest over the longest period of time," Boreman says.

Sometimes when you make a change for one species, you can have a positive impact on another. Take the Northeast's scallop population. In the mid-1990s, one-third of Georges Bank was closed to mobile fishing gear in order to allow the groundfish to come back. Scallops were an unintended beneficiary. Several years ago, word came that large scallops were showing up out on Georges Bank. "Over the four to five years that area was closed to fishing, those scallops just grew tremendously," Boreman recalls. "It opened up a whole new market—for these ten-count hockey pucks! Now people can buy one scallop and feed a family of four."

The key to successful fisheries management, as striped bass have instructed, is to apply tough restrictions on catch levels before it's too late—and then keep them in place long enough to make a difference. As Bill Fox, present director of NMFS's Office of Science and Technology, puts it: "Our view evolved to realize that, even with good year classes, [it's always a mistake] to relax your regulations right at that time. With striped bass, we didn't make the mistake of opening up the fishery too soon as it began to rebuild, although there were pressures to do that."

Indeed, the bass were in more of an eleventh-hour situation than any of us had realized at the time. David Secor, of the Chesapeake Biological Laboratory, did some analysis of ear bones (known as otoliths) taken during the 1992 season from the heads of stripers over thirty-six inches long. All fish absorb calcium carbonate, which forms layered stones in their inner ears. These rings can be counted, in much the same way as trees are dated, to provide an accurate picture of the fish's age and environmental history, including its birth site, the waters it has traveled, even which days it was feeding. Secor's research failed to turn up a single striped bass spawned between 1972 and 1981. Not one. This indicated, quite conclusively, that intense fishing had been the leading cause of the decline. The species could apparently adapt to acid rain and even to the genetic abnormalities observed by Bob Pond—but not to nets and baited hooks.

Striper biologist John Waldman observes: "Between striped bass being saved in the nick of time in the right way and responding, and waiting too long with cod because of political pressures and then the fish not responding, it shows that there may be these make-or-break points. So you really have to listen to the scientists."

In many ways, of course, technological advances have made today's scientific work far more sophisticated than when Boreman was doing his striped bass modeling. "We were more seat-of-the-pants," as he says. "We had computers, but they couldn't handle the data back then. Today everything can be standardized, and a lot of biologists now realize the value of adding age-structure data to their landings data."

There is a potential downside to this approach in fisheries science, however, Boreman believes. "The biggest danger for [fish] population modelers is when you start to believe your own models. And not take them for what they really represent, which is just *a* characterization, your hypothesis about what's happening out there reduced to certain key elements. But you can get so wrapped up in the model or stock assessment that you start to believe it's truth, no matter what, instead of stepping back and questioning whether the data and assumptions hold up."

This is exemplified by the ASMFC having deemed the striper population "fully recovered" in 1995 by utilizing a new assessment model known as spawning stock biomass. The SSB is supposed to calculate the total estimated weight of sexually mature females in the population at any given time. The notion is that a striped bass population that's composed of many small females with a million eggs or so, or fewer heavier "cows" with many more eggs, would be approximately the same as far as spawning potential is concerned. "Fully recovered" meant that the total weight of mature females spawning in the Chesapeake would be the same as it was in the early 1970s, before the population crashed. A SSB model is supposed to include migration rates, growth rates, fishing mortality, natural mortality, regulations, and a young-of-the-year index.

But as noted outdoor columnist Nick Karas has written: "The young-of-the-year index . . . is only referred to as a guideline. They effectively threw out thirty-five years of striped bass spawning figures. Because this still was not enough to make a bold statement of recovery, it [the ASMFC] then insisted that the last thirty-five years of spawning indices were not a reliable way to judge the population and switched to the biomass approach. This is an estimated weight of the entire population over a fixed period of time." And little more, Karas believes, than propaganda to push a bigger fishing agenda.

Also adopted in recent years is something called Virtual Population Analysis (VPA). This incorporates data (landing records and scientific surveys) from all producing areas—the Hudson, the Delaware, even North Carolina—rather than relying solely on the juvenile index from the Chesapeake Bay. This VPA is then used to project Total Allowable Catch estimates in both coastal and estuarine jurisdictions. There are Target Mortality Rates and Resource Allocation Strategies, a whole new lexicon.

All of which add up, in Boreman's view, to a greater separation from the natural world. Boreman speaks of having gotten to know many down-to-earth people who understood the resource firsthand, the Robert Boyles and Jim Prices and Bob Ponds and even the George Mendonsas. "Fishermen don't simply look for numbers," he says, "they get a sense of what's happening in the water. We're losing that aspect of our science. We're not doing enough natural history work. I think that's because of the training coming out of school—get those numbers and those mortality estimates into assessment, crank the models through and get the output."

Biologist Joe Boone, whose young-of-the-year index in Maryland had always correlated so simply yet so perfectly with commercial landings, strongly concurs with Boreman. "You can't manage fish through mathematicians," he says, "because you've got to know a whole lot more than just how to plug numbers into a formula. A manager needs to have a lot of biological and environmental knowledge. If you can't prove your results from more than one viewpoint, how do you know they're correct? That VPA technique has often proven to be highly inaccurate, usually overstating the population."

Eric May, who's been studying striped bass for twenty-two years and is today a Distinguished Research Scientist with NOAA's Living Marine Resources Cooperative Science Center in Princess Anne, Maryland, points out another problem that may arise in model analysis. "A lot of the models we work with originally evolved in the late '60s and '70s," he says. "Yet all of the conditions which those models originally rested on have changed. Did anybody test these assumptions under conditions of the Chesapeake Bay now? No. For example, global warming increases metabolic rates in fish. Which means a greater bioenergetic demand in striped bass."

Indeed, there are signs that all may not be well with striped bass, not just the modeling of their population. As the bass came surging back, simultaneously populations of bluefish, tautog, winter flounder, scup, sea bass, sharks, and tunas were bottoming out. Both commercial and recreational pressure shifted to stripers because they were available. In 1999, an estimated 13.2 million pounds were taken by sports fishermen—more than any other marine fish. Fishing trip expenditures for striped bass soared from $85 million in 1981 to $560 million in 1996 (adjusted for inflation), an annual growth of 35 percent.

At the same time, the mortality target—a measure of the numbers of fish that can be taken without harming the stock, also known as "F-max" to the modelers—appears to have been set too high. The age structure of striped bass is not filling out properly. The older fish are disappearing. The renowned conservation writer Ted Williams poses a "disturbing question: Can a population with a grossly skewed individual size-and-age structure—in which virtually all individuals are removed before they have attained more than 50 percent of their age and growth potential—be said to be 'recovered'? I submit that the answer is no."

In December 2003, after rushing earlier in the year to increase the commercial catch by 43 percent, the ASMFC released updated calculations indicating that striped bass fishing mortality had been well in excess of the "F" target every year since 1997. In fact, it was very near the overfishing threshold again.

When you factor in the modeling that's being done on Atlantic menhaden—the striped bass's food of choice—recent interpretations of striper well-being become even more suspect. And more potentially disastrous. Menhaden, a vital source of protein for many larger fish, were long ubiquitous in the Chesapeake Bay and on the Eastern seaboard. In the ASMFC's studies of the oily little baitfish, a new forward projection model has recently replaced virtual population analysis; now the data are said to reveal about 200 times more "age-zero" (under one year old) menhaden than modelers previously thought. "If there are really that many," as Jim Price says wryly, "I'm surprised our striped bass aren't finding a few more to eat."

In 1990, when the efforts to restore striped bass first began showing obvious positive results, the mean weight of a twelve-year-old striper was about thirty-three pounds. Yet by 1996, that same age fish weighed, on average, twenty-five pounds. The decline in weight in relation to age was clear to fishermen all along the coast. Personally, I didn't have a clue as to why until a letter arrived from Joe Boone early in 1998. Since his retirement I'd maintained sporadic contact with him, but this time his letter had a sense of urgency.

Boone wrote: "Stripers, once again, are signaling that serious ecological problems exist in the Chesapeake Bay. Some persons whisper of ecological catastrophe. I'll wait and watch before pushing the panic button. The visual manifestation is skinny stripers . . . well below normal weight, generally devoid of body fat, and may exhibit abnormalities of internal organs such as spleen and liver. Some fish disease experts speculate that immunosuppression from starvation has allowed normally innocuous bacteria to invade the weakened fish."

Boone continued: "Incidentally, Bob Pond was the first person to call attention to starving stripers, and he was pooh-poohed by everyone until now. Once again Bob was years ahead of the learned ones. A quick check of prey population data revealed disturbing information. Striped bass . . . in the Chesapeake are known to feed primarily on anchovy and particularly young-of-the-year menhaden. The sein-ing survey shows that populations of both species have collapsed in recent years. . . . Yet purse seiners, with the blessing of NMFS, are operating as though the menhaden population is stable. Sounds familiar, doesn't it?"

While other scientists questioned a downward abundance trend for anchovy and menhaden, it took a few years of hammering on them by Jim Price for the managers to even begin to pay attention. "In Chesapeake Bay, predator demand is reaching unprecedented highs while available prey is at an all-time low," Ken Hinman of the National Coalition for Marine Conservation (NCMC) warned at the ASMFC's annual meeting in December 2003. "A growing number of conserva-tionists and biologists believe the continued high level of menhaden harvest in the bay, if not curtailed, could jeopardize the hard-earned recovery of striped bass and other species. . . . The situation practically cries out for an ecosystem-based approach to management. . . ." Hinman concluded, "Although well-intentioned moves are being made in this direction, the system moves without urgency while we continue to manage without caution."

Ecosystem management. As John Boreman says, "Everybody's talk-ing about it, but still trying to figure out how to do it. . . . It seems like the right way to go—but there's no clear path laid out. Right now we have problems even defining what a marine ecosystem is. So how do you manage something you don't even know how to define?"

The phrase "ecosystem approach" was first coined in the early 1980s and found formal acceptance at the Earth Summit in Rio in 1992, where it became one of the underpinnings of the Convention on Biological Diversity. It's been described as "a strategy for the integrated management of land, water, and living resources that promotes conser-vation and sustainable use in an equitable way." In a fisheries context, the goal is basically to ensure that an ecosystem's function and produc-tivity are not adversely affected by management decisions. It seeks to move out of the box of assessing fish populations and setting catch lim-its on a species-by-species basis. It looks to move from a narrow approach to a holistic one, considering the interactions between predators and prey, the effect that fishing one species has on others, and environmental changes, both human-caused and natural. It synchronizes conservation efforts.

But that's easier said than done. The Sustainable Fisheries Act, passed by Congress in 1996, introduced the ecosystem concept into the federal fisheries management process. Boreman's Science Center has since initiated pilot ecosystem projects in the Northeast. Finally, in 2001, the ASMFC began developing a way to integrate traditional single-species management into a forward-looking, multispecies framework. Its fishery management plan for menhaden was amended, with a stated objective to ensure that enough menhaden would be available to species that depend upon it for food, foremost being striped bass. However, there was no "plan" outlined to accomplish this. In this it wasn't alone. According to the NCMC, although more than a hundred federal and interstate fishery management plans were in place in the United States, "precious few are truly *plans* in the sense that they anticipate future events and prepare accordingly."

In November 2003, I attended a panel discussion on ecosystem management at a Washington, D.C., conference called "Managing Our Nation's Fisheries." Bonnie McKay, a social scientist from Rutgers University, made a fascinating distinction between traditional and ecosystem management. In traditional fishery resource management, she argued, the philosophical basis was utilitarian values, a commodity product (the idea of maximum sustainable yield), single-species models and management, and a top-down, expert-knows-best approach. By contrast, ecosystem management brings in a plurality of values—the "land ethic" of Aldo Leopold—and a science that recognizes complexities, discontinuities, surprises in both natural and human systems, and ultimately the inadequacy of single-stock and deterministic scientific models. There is much we do not know, much that is uncertain. This attitude leaves more room for other forms and sources of knowledge, those who have "traditional ecological knowledge" that is "experience-based."

There are some, like John Boreman, who believe it's vital to move toward this more open-ended direction. During our luncheon talk, Boreman added: "One of my philosophies now is, our scientists need to talk a little less and listen more to the industry. Learn how to take information from the fishermen who are out there, and put it into some kind of robust, quantifiable form. They have a lot to share with us. In the past we've had a tendency to say, 'Well, that's just one observation, it's not replicated, not testing the hypothesis.' But it *is* an empirical observation, and we've got to figure out ways to incorporate those. Otherwise it's a black hole, where all the information goes in and then gets handed back in edicts."

Boreman has gone so far as to invite fishermen to give seminars to his scientists so as to "humanize" the process. His Science Center also offers an extensive course for fishermen, bringing in twenty at a time to learn what's involved in management, policy, and legislation; to learn, for example, what a scientific assessment is, what happens to the catch data they are required to report.

Still, Boreman—who earned his fins studying striped bass—remains a relative anomaly. This is such a radical shift in perspective that, as Maryland biologist Jim Uphoff says: "The push to do ecosystem and multispecies modeling is really what needs to be done, but I get a sense that a lot of managers will be very resistant. . . . It's hard enough to give these guys sound advice that they understand on a single-species basis, let alone when you start talking about multiple interactions that reverberate throughout the ecosystem.

"With striped bass, sure the survival rate has gone up for the eggs and larvae. That's a great thing, but this is like building the Maginot Line. We geared up all our management to take care of overfishing, and now we're not looking at what the real danger or management problem is. During the recovery, no one paused to ask the question: what in the hell are we going to feed these fish? Then, when the warning signs started to develop, they just turned a blind eye and kept saying everything's fine."

One reason for the blind eye has been the continued spawning success for stripers in the Chesapeake. This wasn't consistently true—a mediocre or even relatively poor year class often followed a banner one—but it happened often enough for the managers to believe that the spawning stock biomass was still A-OK. As new research is indicating, though, a big part of the reason for these occasionally booming year classes is weather-related.

Ed Houde, of the Chesapeake Biological Laboratory at the University of Maryland, explained it to me at the fisheries management conference like this: "What we know now is, wet and cool years seem to go together. We've had several of these during spring months where big runoffs occurred: 1996, 1998, early in 2000. And we're seeing bigger-than-average recruitments [the number of fish surviving through their first year] associated with those years for striped bass. Also for white perch and all the anadromous fishes. Strong regulations had a huge effect in bringing back striped bass, but we needed that environmental push to make it happen as fast as it did. We had some really favorable conditions."

The weather pattern Houde referred to is called an Ohio River Valley low pressure system. NOAA climatologist Bob Wood has determined that such prevailing conditions assist in both the abundance and the geographical spread of zooplankton in the bay. This means more food available in more locations for newborn stripers, plankton feeders before they start eating menhaden at age one and older. The converse is known as an Azores-Bermuda high pressure system, bringing warmer and drier weather. In years when that predominates in the spring, striper spawning success is not as good.

But these weather patterns seem to create precisely the opposite effect for coastal spawning fish such as spot, summer flounder, and menhaden. Warm, dry spring weather brings on an earlier summer. It causes more saline waters to extend higher up into the estuary. That sends summer zooplankton into the range (and mouths) of newborn coastal spawners as they migrate from the ocean into the bay. And by far the most prolific of these coastal spawners are menhaden. Consequently, menhaden appear to have better spawning success under a weather regime that results in poorer recruitment for striped bass, and vice versa.

Which brings us back to looking at the overall Chesapeake Bay. If we want to preserve striped bass, we must attend to the whole ecosystem in which they swim and how it is changing. Movement toward this kind of ecosystem-based approach was recommended in 2003 by both the Pew Oceans Commission and the U.S. Commission on Ocean Policy. Some elements are already being implemented in California, Alaska, and Australia. In July 2004, seventeen of the world's foremost marine scientists detailed this "revolutionary new paradigm" in the prestigious journal *Science*. Ellen Pikitch, executive director of the Pew Institute for Ocean Science, commented: "We've been putting blinders on, but it is now clear that single-species management is inadequate, and in many cases, destructive. Ecosystem-based fishery management can be implemented right now, even in cases where very little information is available. Because of the complexity and uncertainty about marine ecosystems, this approach will inevitably require erring on the side of caution." The scientists wrote that, ideally, an ecosystem approach "would shift the burden of proof so that fishing would not take place unless it could be shown not to harm key components of the ecosystem."

Perhaps no contemporary issue better illustrates the need for this than the striped bass–menhaden story. As Les Kaufman, a prominent biologist with the Boston University marine research program, says, "The problem is, people think nature gets broken, you fix it, and the story's over. It's never over."

Chapter Eleven

The Myco Mystery

It's a steamy morning in early August 2004. I'm riding in now-retired postman Jim White's Isuzu alongside a half-dozen fishing rods. The Greenwich Cove Marina, where he keeps his charter boat, is about six miles from his home in Coventry, Rhode Island. In recent years, White has become a well-known skipper. People from thirty-nine states and eleven foreign countries have enlisted Jim's expertise to pursue stripers and blues around Narragansett Bay.

He's excited, telling me how his three-year-old grandson, Devon James, has already caught a "schoolie" striped bass. "Grampy," White says, referring to himself, "bought D.J. a rod before he was even born." The lad, a precocious member of the White clan's fifth generation of fishermen, did his first real fishing at twenty-one months and already "can identify more fish than people ten times his age," Grampy says proudly.

We pull in at a bait-and-tackle shop, where White buys fifteen dollars worth of sea worms that have been shipped here from Maine. By the shop's front door is a newsletter for the Rhode Island Saltwater

Anglers Association. An offshoot of the organization White started more than twenty years ago, the group today has 4,000 members.

At the dock, his twenty-three-foot Triton boat, *White Ghost II*, is tied up right across from *White Ghost I*, a slightly smaller Ranger model that his son, Justin, runs. In the spring, they've been known to do as many as three trips a day in tandem when the fishing's good. I ask Jim how he came up with the "White Ghost" label. When he had started doing a lot of fly-fishing charters, he says, he'd researched the literature. He'd found "ghost" flies of black, blue, green, and silver in the annals of trout and salmon fishing—but never white ones. He'd always been secretive about his own fishing. "A lot of guys used to say I fished like a ghost—'You go out, disappear, then reappear.'"

From below a canopied center console, White starts the single-engine outboard while I free the dock ropes of his boat. He eases slowly out of the marina toward the open waters of the bay. At fifty-five, two years younger than me, he sports a mustache and goatee that have turned a salt-and-pepper shade. Our camaraderie feels not unlike that of soldiers who've been through a war together—which, in many ways, we have. Yet this is the first time we've been on the water in each other's company. In the old days, the days of battle, there simply wasn't time.

"You know," Jim says, lighting a cigar, "we pulled off a miracle." He doesn't have to say he's talking about our effort to save the bass. "You know why?" he continues. "Because we didn't play by the rules." White laughs. "We never had the numbers of people that Congresswoman Schneider believed we did! I had her convinced that there were thousands."

"Well," I respond, "maybe you were speaking for the way thousands of people felt."

"Maybe so," White says. The most gratifying thing, he continues, was all the guys who came up to him starting around 1989, when the striped bass's recovery became evident, to say, "You were right."

"I started chartering in '93 with a tin boat," White says, "and never left this harbor. Fished the high tide in, then out, and back in again. That was at the height of the [bass] rebuilding. We could catch 150 fish in a day. It was amazing. Who could've dreamed we'd see these kind of numbers?"

It was all catch-and-release. White can't recall the last time he brought a bass home to eat. "Then, '97 was the first indication of—here we go again. Things went immediately down to about 50 fish [a day], and then dropped even more in '98. I started going outside and to the islands more. Went to a Florida-style flats boat in '98, but it couldn't run the whole bay all the time, so that's when I got the center-console Ranger. When I had to go out for two weeks to do what I did in three days, I knew something was wrong. A lot of people don't want to admit it again, but there is a big hole in the fishery."

It's a calm day, wind out of the southwest. White starts our drift in pretty much the exact center of Narragansett Bay, along a reef called Ohio Ledge. He puts out line from two small trolling rods, each baited with a sea worm at the end of a two-foot-long tube that resembles a slithering eel. About twelve feet down, our bait moves through the water at a slow, but hopefully enticing, pace. Before long, a pair of bluefish strike almost simultaneously. As we reel them in, both shake loose the hooks with a leap and a splash.

Perhaps a half hour passes. It's always hard to say in the time zone of the sea. Then, again within moments of each other, both rods bend hard with something writhing at the other end. We lift them out of the rod-holders and begin the familiar, rhythmic ritual: reel down, lift slowly high, reel down. . . . This doesn't feel like a blue.

Each of us brings in a striped bass. "How do you like that?" Jim cries. "Middle of the morning in August! That's not supposed to happen!" He gently lifts each fish onto the boat, carefully removes the hook from the mouth, and sets the striper down on a measuring board along the bow-rail before returning it to the sea. Each bass is exactly twenty-three inches long.

On one of the fish, prior to releasing it, White points to an area just below the mouth. Starting to become visible are a series of reddish-colored lesions. He shakes his head. After putting our lines out again, he says finally: "For a while, we were seeing this white fungus. Every year for the last four, it's gone down. But what's replaced it are the red spots and sores. When I went back to do my slide shows for this year, I couldn't believe how many fish I saw with big red sores on them. At first people would pass it off that the redness came from their grubbing for crabs on the bottom or chasing lobsters and scraping their belly. Made sense. Some fish do that, like the large-mouth bass; their tail gets all red. Turns out, though, that all along it's been disease."

A palpable silence descends. White resumes: "Last year I cleaned a fish for a customer. After I filleted it, I went to throw it away, but I said, 'Let's see what she's eaten.' I cut the stomach open some more. The spleen fell out. It was the most ungodly thing I'd ever seen in my entire life. Red, green, black, all kinds of sores. I said, 'I wouldn't eat this fish if I were you.' But the outside was absolutely perfect."

We stay out for another hour or so. I bring in a healthy-looking fish, at twenty-seven inches just under the "keeper" minimum size. White goes on to tell me some great stories about hosting fly-fishing legend Lefty Kreh and other luminaries.

Yet what lingers are these words, spoken as we near the dock once more. "I think it's gonna crash again," he is saying. "Then the question becomes, will we get lucky like last time? How many times can you roll the dice?"

Jim White is not the only fisherman observing signs of disease in striped bass along the Atlantic seaboard. "Oh yeah, been seeing lots of 'em," Charley Soares, a veteran angler from Cape Cod, commented. "In fact, some were so bad we didn't even take 'em out of the water. I just took pliers and released 'em."

The skin lesions were first observed by pound-net fishermen in the Chesapeake Bay in 1994. It wasn't until the summer of 1997, however, that the disease was first isolated, after Maryland charter-boat captain Jim Price sounded the alarm. Two of the six stripers he caught that year on August 18 had red lesions on their sides, "one of them really bad." When the same thing happened the following day, Price took all of the lesioned fish to the federal-state Cooperative Oxford Laboratory near his home. Later, Price told Robert Boyle: "I've seen millions of pounds of striped bass, but I'd never seen any diseased fish like this." Boyle says of Price: "He's Paul Revere."

That summer of 1997 was the same summer as the *Pfiesteria* scare. At the time, a microbe called *Pfiesteria piscicida*—the so-called cell from hell—was thought to have been the killer of tens of thousands of menhaden in the Chesapeake, as well as the cause of memory loss and other health problems in exposed watermen. One of the nation's leading fish pathologists, Wolfgang Vogelbein of the Virginia Institute of Marine Science, recalls: "I started getting a lot of striped bass coming into my diagnostic lab with very ugly skin sores, as suspect *Pfiesteria* cases. But we quickly determined that it was something else—a bacterial infection—and we've been working on it ever since."

Vogelbein believes that what's happening to striped bass "is much more of a concern than *Pfiesteria*. You've got a major recreational and commercial species, and a high percentage of these animals are now infected with these bacterial organisms, and we don't know what the impact on the stock is. There is a double concern with what's happening to striped bass, because this is an infection that can be transmitted to humans."

It turns out that the skin lesions are symptomatic of mycobacteriosis, an infectious disease whose effects in fish include stunted growth, emaciation, and internal lumpy gray nodules called granulomas. "Doing routine histopathological analyses," Vogelbein recalls, "one of the things we found was that the spleen was the primary target organ of the infection, followed closely by the head kidney. As the infection progresses over time, it impacts other organs—the mesenteries [in the abdomen] and liver, ultimately the gills and heart." These granulomas appeared in 53 percent of striped bass tested along Maryland's Eastern Shore in 1998–99. They are undoubtedly what Jim White observed when cleaning his customer's fish. Once these infections establish themselves, experiments have shown that, eventually, the fish will die.

There are dozens of types of mycobacteria, many of which occur naturally in the environment and one of which causes tuberculosis in humans. In fish, until recent years the onset of mycobacteriosis was associated primarily with aquariums and to a lesser degree with aquaculture—that is to say, with fish in confined spaces and close quarters. In 1983, there was a short-lived outbreak in adult striped bass from the Sacramento and San Joaquin rivers in California. Until 1997, it had never been detected in a wild species of fish on the East Coast. Then and since, striped bass are the only fish to exhibit the disease in the Chesapeake Bay.

"Why is this happening? We don't know," says Vogelbein. "We speculate it's environmentally modulated, because in general, infectious diseases in cold-blooded animals like fishes are greatly influenced by the environment. These organisms are always present out in the water. Fish always seem to be exposed to them, but become diseased only when they are stressed. The best example is aquaculture: you crowd large numbers of fish into small amounts of water, allow the water quality to deteriorate, and this stresses the fish, which break out with some sort of infectious disease. But it is extremely difficult to get at the ultimate causes of a disease in a wild population. In the case of the striped bass, there are some hypotheses, but none have been proven."

Nor is it yet known how or when the bass become infected. Young-of-the-year stripers have revealed no evidence of the disease, but that doesn't necessarily mean it's not present. Research with other species suggests that mycobacteriosis can be passed on to new generations through the eggs, but most of the literature describes a different means of transmission, according to Vogelbein: feeding on contaminated material. That was the case with an outbreak of mycobacteriosis among salmon in the Pacific Northwest during the late 1950s and early 1960s. There are other indications that mycobacteriosis, when present in large quantities, can spread through the water—through "shedding," or the excretion of mucus or feces from nearby fish. Since striped bass often travel in large schools, this certainly can't be ruled out.

Jim Price has seen the lesions on stripers as small as six inches. Of more than 400 fish examined by Chesapeake researchers in 2003, only 30 percent of two-year-old striped bass were infected, but the rate climbed with each year class until 70 percent of all five-year-olds had it. The disease was far more prevalent in males—which tend to live in the bay longer, generally not migrating at all—than in females.

Surveys in four Chesapeake tributaries in the autumn of 2003 also revealed a higher infection rate in each river compared to the previous year. Mycobacteriosis was found in fully 81 percent of the Potomac River stripers, 67 percent in the York River, 61 percent in the Rappahannock, and 54 percent in the Nanticoke.

Lesions don't have to be present for fish to be sick; indeed, they show up in as few as 20 percent of the striped bass surveyed. But fishermen who come into contact with a lesioned striped bass risk contracting what's come to be called "fish handler's disease." As Jim White described it to me: "I'd have a day of catching plenty of fish, get some cuts and scratches, and be itching them—wondering, what have I got, poison ivy? You never think, did I get this from the fish I was handling today? Never occurred to me." Now he won't even touch a bass without wearing gloves.

If the lesions on a bass come into contact with a cut or abrasion, a fisherman can develop a sore that won't heal or painful swelling of joints. The condition can be difficult to get rid of, particularly for individuals with compromised immune systems. Says Vogelbein of the infections, "Very often they're misdiagnosed because physicians don't recognize them, but they can be treated with a very effective specialized group of antibiotics." Without proper treatment, "fish handler's disease" can lead to bursitis, arthritis, or osteomyelitis and can require surgery to remove infected tissue.

The apparently exponential increase in prevalence of mycobacteriosis in striped bass, coupled with the potential health problems for people, has resulted in a widespread and ongoing cooperative research investigation among numerous agencies—not only the Virginia Institute of Marine Science (VIMS) but also the U.S. Geological

Survey, the Centers for Disease Control and Prevention, the U.S. Fish and Wildlife Service, the National Oceanic and Atmospheric Administration (NOAA), the University of Maryland, and the Maryland Department of Natural Resources (DNR).

Bacteriological studies conducted in Vogelbein's lab at VIMS have isolated and tentatively identified more than a dozen different species of mycobacteria in striped bass, some of which are new to fisheries science (probably because, Vogelbein suggests, before this disease outbreak nobody had ever looked). "That's a pretty broad suite of organisms associated with one disease," he adds. "But which ones are pathogenic and which aren't, we don't know. You need to understand that most of the previous work done on fish mycobacteria had been very superficial. . . . In some [stripers], we can isolate more than one type of mycobacterium. This is puzzling to us and has never been reported before. It may be that the skin infection we see is a separate disease from the internal one. The dogma has been, once you see skin lesions you're in the terminal phase of the disease—but that's not always the case. In certain instances, we've seen only the skin disease and no visceral one."

Trying to determine whether the causative agents are different mycobacteria is sometimes very difficult and can take months, according to Vogelbein. Equally confounding is the data surrounding one species, *Mycobacterium marinum*, a classic fish pathogen. It's one of the three most common species known to infect fish in aquariums. It's been showing up in less than 6 percent of the striped bass in Vogelbein's lab. It's the same species, however, that's known to cause "fish handler's disease." And in lab experiments, *marinum* is the primary species able to induce significant disease in striped bass.

That's not been the case with *Mycobacterium shotsii*, the most common isolate, which has been detected in about 75 percent of the stripers sampled. Here is the way Vogelbein explains his exposure studies, conducted in collaboration with the National Fish Health Laboratory in Leetown, Virginia: "We went into the literature to see what had been done. As far as experimental exposures, researchers typically expose fish to essentially lethal doses of mycobacteria, and cause pathology and death very quickly. We decided in our studies to attempt to mimic more the chronic disease we think we see out in the wild. So we went with a lower dosage through intraperitoneal [abdominal cavity] injections that wouldn't kill the fish right off the bat. Our goal was to study the progression of the disease over a period of about forty-five weeks.

"One thing that was puzzling to us was the skin lesions. We've not been able to induce those that we see in the wild fish. The visceral spleenic disease, yes. We did this with three different mycobacterial organisms—*marinum, shotsii,* and *gordonae,* which is thought to be a harmless organism present in the soil. *Marinum* is one we rarely isolate, but the one that whacks the fish experimentally. The real puzzle is *shotsii,* which is so prevalent. Under controlled conditions, there was infection but very little disease. So what's going on here? I suspect that, in our lab exposures, we just didn't create the right environmental conditions that foster its development."

Vogelbein is looking now to develop molecular probes specific to the various mycobacteria that have been isolated from striped bass; these would then be applied to fish tissues collected in the archive, in an effort to learn more about the pathological relationship.

So far, there has been no obvious die-off of striped bass. But could the infection be killing them slowly? Might they be dying in scattered places, to become food for passing birds or for crabs along the river bottoms? There are strong indications that this is indeed occurring. Those who assess the fishery have long assumed natural mortality to be relatively low, given that stripers are such a long-lived species. Fishing mortality has thus been the main indicator upon which to base calculations of allowable harvest rates for the various states in the ASMFC. However, two recent independent studies reveal, alarmingly so, that natural mortality in striped bass has been rising since 1998.

These analyses are based on what's known as tag-and-recapture data. On the spring spawning grounds in both Maryland and Virginia, fish are collected in either gillnets or pound nets. They are measured, scales are taken to determine their age, and then a small metal tag with date and location is inserted on their bodies before they are released alive. Later, when a fisherman catches one and notices the tag, he calls the U.S. Fish and Wildlife Service phone number on it. It's a time-honored method to help determine, from year to year, where the fish go, what proportion of tagged fish are being captured, and how well they're surviving.

In September 2003, Victor Crecco of Connecticut's Division of Marine Fisheries released his statistical analysis, derived from examining the rate of tag returns. He found that mortality from fishing had remained constant or even dropped slightly. But in Chesapeake stocks eighteen inches and above, natural mortality had *increased fivefold* over the previous five years.

Using a different model to examine the tag-recapture data from those same surveys, Desmond Kahn of Delaware's Division of Fish and Wildlife reached a similar conclusion in 2004. In both the Maryland and Virginia sectors of the Chesapeake, Kahn wrote, "survival has declined significantly, from 60 to 70 percent in the early to mid-1990s down to 40 to 50 percent in the late 1990s and early 2000s. . . . In each jurisdiction, survival has declined roughly 20 percent." Kahn's conclusion, like Crecco's: the change was "due to natural mortality," not an increase in fishing pressure. Notably, Kahn says, "This decrease in survival is not showing up in tagging programs in either the Delaware River spawning stock or the Hudson River surveys—only in the Chesapeake, where there have been reports of widespread fish health problems."

As Jim White has seen in Rhode Island, there is evidence that the disease is spreading beyond the Chesapeake. Recent reports based on the American Littoral Society's tagging program indicate lesioned striped bass showing up from Maine to New Jersey. It is not yet known if most of these fish exhibit internal granulomas as well, as White has also observed, and various studies in the Northeast are just getting under way.

In July 2004, I met with the research scientist Eric May at NOAA's Living Marine Resources Cooperative Science Center on the University of Maryland's Eastern Shore campus. His office was crammed with papers, books, and mounted fish of various species. The burly, redheaded professor told me that he'd grown up fishing in Oregon, "but fighting striped bass is as good as fighting steelhead any day." His Ph.D. was in comparative medicine, making him "a pathologist by trade" who teaches graduate courses in fisheries biology and fish diseases.

Some of May's current students are also trying to sort out what's going on with striped bass. "My statement to the Maryland DNR has been, ultimately these diseased fish will not survive. Once they get to the point of their liver, kidney, spleen, and heart being affected, I'll guarantee they won't. Even though they may go out into salt water where the skin lesions could heal, the internal lesions don't go away. It's a slow, insidious disease that takes a long time to kill, but it does kill."

Like his Virginia colleague Vogelbein, Maryland's May has puzzled long and hard over where the disease came from. "Prior to 1988—and there had been a helluva lot of fish pathologists crawling around the Chesapeake Bay before me—[there had been] no report of mycobacteriosis and no outbreaks. Over the next nine years, we isolated maybe one or two mycobacteria during that entire span. All of a sudden, 1997, the shit hits the fan. Why? A disease just doesn't get locked in. Something happened here."

May speculates that the change might have had something to do with well-meaning, but misguided, attempts at stock restoration when the wild striped bass population was at its nadir. "Number one, we allowed hybrids of striped bass and white perch stocked by DNR into the northern part of the bay. What impact did this narrowing of the gene pool have? I mean, striped bass were at such low levels, it was hard to conceive that a female could mate with more than one male at any given time. Just finding a 1-on-1 ratio was hard enough. Now they absolutely don't want this kind of 'stock enhancement' to happen anymore. But closing the barn door after the horses are out is a little bit too late.

"Maryland also had a very active hatchery program. When you do hatchery rearing of fish, you're not necessarily using those that are as robust and healthy as you find in the wild stock. So did we introduce something that was 'genetically inferior' into the entire system from hatchery propagation?"

Just as baffling to Eric May is the fact that Vogelbein's efforts to experimentally infect some striped bass with the ubiquitous *Mycobacteria shotsii* have not resulted in the eventual death of those fishes. "*Now* what the hell's going on?" May wondered aloud. "Either Wolfgang is looking at the wrong bug—and he's too good a pathologist for that to be the case—or maybe something else is required to tip the balance.

"This has led to my current thinking, which is quite different than what I was caring about a year ago," May continued. "That it is possible the infection is more widespread than we realize, and that it requires some other stressor for it to manifest. This comes from some things we know about lung tuberculosis. In certain areas of Western Europe right after World War II, the infection rate of people with the tuberculosis bacterium was anywhere from 50 to 70 percent. But not 50 to 70 percent came down with tuberculosis. It was only those who wound up being starved, homeless, essentially completely stressed, who came down with the disease. So the pathogen itself may be everywhere, but for the fish to actually contract the disease—I'm beginning to relook at the issue of starvation."

Previously, May had argued with people like Jim Price that starvation had to be considered independently of mycobacteriosis—simply because it is known that as the disease progresses and demands more and more energy from the fish while spreading through its body it will cause the fish to become very emaciated. Now May suspects that Price may have been right all along. He believes the infection itself is independent of starvation, but manifestation of the disease may be a function of starvation.

The previous fall, I'd paid a visit to John Jacobs, a marine biologist for the NMFS at its Oxford, Maryland, laboratory. Accompanying me were Jim Price and Joe Boone. In 1998, 1999, and 2001, Jacobs had been the lead researcher on a general health survey of striped bass in the Chesapeake. The findings, not yet public at the time, contained some startling results.

Jacobs explained that, during the fall of those three years, they had conducted random sampling of 320 wild fish. Body weight on average among the fish was very low in relationship to length. Indeed, between 70 and 80 percent of the bass had no visible abdominal body fat. When the fish were opened up to examine the muscles, researchers found little fat (lipid) storage but elevated moisture content. "That's important, because the two are inversely related," Jacobs told us. "As a fish begins to starve, it's going to take on water. Basically it's using up its fat storage, which is being replaced with water. That was fairly consistent throughout those three years."

But that wasn't the most telling part of the study. At the same time, food deprivation studies were conducted with forty additional striped bass at Maryland's Horn Point Laboratory. For a two-month period, the fish were contained in hatchery tanks at a constant temperature between 18 and 20 degrees Celsius (64 to 68 degrees Fahrenheit). In order to establish reference points, some were fed while others were given nothing whatsoever to eat. "What we found," Jacobs said, "when we took similar-sized fish from our survey in the wild and compared these to the fish in the deprivation studies, there was *no statistical difference* between them. In other words, the wild collected fish were not significantly different in chemical composition, weight at length, or mesenteric body fat than the fish deprived of food for two months."

That took a moment to settle in. Boone finally broke the silence. "Isn't that highly significant?" he asked.

"Well, there are some caveats," Jacobs replied. "Because a fish in general, when it's food-deprived for an extended period of time in a tank, is different than in the wild. It can't move as much, so its metabolism slows down dramatically. With that said, we did not expect to see this at all."

"That's amazing," Boone said. "Sounds like all wild stripers need is more food. It's that simple."

Jacobs added that the bass in the wild seemed to be in poor nutritional health independent of disease. They were seeing plenty of malnourished fish that exhibited no signs of mycobacteriosis. And they were seeing some healthier-looking fish that did have the infection. Whether these were really two separate problems, however, Jacobs couldn't say. "They could be the same problem, it's just that at our ability to detect this at low levels [of fish surveyed], we're not observing that."

Eric May is now looking at energy requirements in striped bass and finding that previous modeling of the fish's consumption needs may be alarmingly off-base. May told me: "Researchers out in the Midwest are finding that probably we are underestimating the energy required for growth, because we've been underestimating the energy required for excretion, either feces or urine." This means that analyses of how much striped bass need to eat may have been significantly underestimated. According to May, they may need 20 to 40 percent more food than had been thought.

As he diagrammed the revised bioenergetics model on a chalkboard, May continued: "Which goes back to the multispecies concept, or ecosystem level, of management. This means that when we are dealing with protecting menhaden, or looking at bay anchovy and other species that serve as prey items, we're going to have to do a lot bigger job than we anticipated. In my opinion, menhaden is a human problem, because we're intervening and taking them away from striped bass. So we're competing with the striped bass. . . . If this new model is correct, then our estimates of the numbers of menhaden needed to sustain a healthy striped bass population are also a gross underestimate."

Studies in the early 1990s found that menhaden then accounted for between 37 and 66 percent of the striped bass's diet. However, by 1998–99, menhaden consisted of only 12 to 27 percent of their diet, depending largely on the age of the bass. A report by biologist Anthony Overton on this development stated: "It is well-known that nutritional shifts have deleterious effects on fish, and in many situations affect the immune system. This is not to say that this is the only possible cause for the condition, but it is one of many possible origins of ulcerative dermatitis [skin lesions] in striped bass."

So while the striped bass population climbed toward a historic high due to strict conservation measures, the population of their favored food declined to near a historic low. The predator-prey relationship fell completely out of balance during the 1990s. Maryland biologist Jim Uphoff found that, in the upper Chesapeake Bay, the ratio of edible-size menhaden to striped bass declined by an astounding 97 percent between 1983 and 1998. Not surprisingly, a deterioration in the striper population's nutritional health ensued. Uphoff wrote in 2003: "Transmission of disease would have been aided by high density of striped bass in poor nutritional condition residing in degraded habitat (Chesapeake Bay was the most hypoxic [oxygen-starved] estuary in the mid-Atlantic region in the late 1990s)."

Analyses by Jim Price, although not peer-reviewed science, utilized numerous recent studies in coming up with some dramatic figures that he presented to the ASMFC's Atlantic Menhaden Management Board in December 2003. Between 1955 and 1959, the estimated population of menhaden up to the age of two averaged 795 billion fish and comprised fully 77 percent of the diet of stripers between the ages of three and six in the Chesapeake Bay. Yet during the 1998–2001 period, menhaden consumption by stripers the same age and in the same locality had fallen to 21 percent. In 2000, the menhaden population ages two and younger was estimated at 158 billion fish—an 80 percent drop over a forty-year period. Striped bass, by the time they reached the age of six, in 2001 consumed 38 percent less forage and weighed about 40 percent less than fish of the same age in the late 1950s.

Given the decline of menhaden in the bay, stripers have turned to alternatives, a phenomenon that is creating problems of its own. In the Chesapeake, the bay anchovy—not nearly as nutritious a food source as menhaden—used to be preyed upon only by small stripers. A recent study, however, found that the age-three stripers were now consuming five times more bay anchovies than in a study conducted eight years earlier. In 1994, the year after a failure of menhaden recruitment, the bay anchovy population in the Chesapeake entered a decline of its own—and has fluctuated ever since.

What else, then, might the bass be finding to eat? They are widely known to be opportunistic feeders and will pursue pretty much anything smaller when they're hungry. Increasingly along the coast, the bass have turned to eating small lobsters and, in the Chesapeake, sometimes blue crabs. This has brought cries of alarm from fishermen, including commercial cod fishermen and even recreational trout fishermen in upstate New York, that the striped bass are damaging their livelihood and enjoyment—"like roaches, all over the place, eating everything," as one has said. The answer? Cull the herd, some say.

But while striped bass are being blamed for overeating other prized species, the fact is that more than ever before, they are showing signs of malnutrition and disease.

In Maine, according to Barry Gibson, until recently an editor of *Salt Water Sportsman*, "the bellies of the bass we catch along the shore are stuffed full of little green crabs, most about the size of a half-dollar. That's not even a native species, and there are millions of those. [They arrived from Asia in the ballast water of ships.] However, I can't believe they could eat and digest enough green crabs to get the fat and protein that they need."

Alan Peterson, retired from the National Marine Fisheries Service, reports that in Cape Cod Bay, "You open 'em up and they were feeding on little sand dollars! Now what a bass can get out of a sand dollar, I don't know." The problem appears to be worsening. In a study undertaken by Jim Price's organization and East Carolina University early in 2005, analysis of two hundred migrating female striped bass from

Oregon Inlet, North Carolina—at the peak of their feeding season—revealed *half* to have empty stomachs and less than one-quarter to have eaten any menhaden, at a time when the baitfish are supposed to be plentiful.

In the face of all this, what's happening to menhaden becomes paramount for the future of the striped bass. In 2002, the commercial fishery took 384 million pounds of menhaden, to be ground up into fish meal and fish oil. The majority of those landings occurred in Virginia waters of the Chesapeake Bay—more than twenty times the catch of all other finfish species combined.

As Eric May puts it, "Striped bass are the most political animals you can invent. And menhaden are not too far behind them." I would come to understand what the scientist meant only after I moved—literally and figuratively—into the belly of the beast.

Chapter Twelve

The Town
That Menhaden Built

$\mathbb{N}$o other town in the Tidewater region of Virginia has such grand old Victorian homes as Reedville. Main Street is still known as Millionaire's Row, where The Gables (now a bed-and-breakfast) yet retains a ship's mast extending through its third and fourth floors. During the early years of the last century, Reedville boasted the highest per capita income of any community in America. Today, it remains the nation's third-largest fishing port in terms of pounds landed—375.3 million pounds in 2003, almost exclusively menhaden. Surprising as it may seem, fish meal and oil rendered from menhaden together constitute nearly 40 percent of the total U.S. fish export volume annually. And Reedville's Omega Protein Corporation, headquartered in Houston, is the most productive non-governmental fish processing facility in the world.

I'm enjoying iced tea on a screened back porch with Wendell Haynie, whose grandfather's house can be seen directly across Cockrell Creek. On the wall next to a painted red crab is a little plaque his wife had him tack up: "Me and My Old Crab Live Here." Now in his

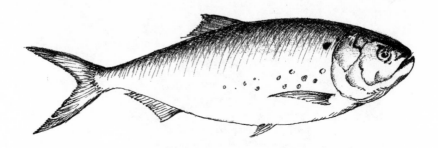

seventies, Haynie had said over the phone that he was "too old to get into anything controversial," but would agree to fill me in on Reedville's history. His family, after all, had lived in these parts for just about forever. One ancestor, John Haynie, had been the second person of European descent to settle among the Indians on Virginia's Northern Neck in the 1840s—well before that Yankee skipper Elijah Reed docked here in 1867. Reed arrived from Maine in a three-masted schooner, bringing along fishing gear the local watermen had never seen—seine nets, kettles, presses. He'd been following schools of menhaden down the western edge of the Chesapeake Bay. When he finally came upon this estuary of the Great Wicomico River, with its natural deep water and protected coves, it seemed a natural location to set up shop.

Atlantic menhaden (*Brevoortia tyrannus*) are bony, oily, inedible members of the herring family, their large heads distinguishing them from close relatives such as alewives or shad. Legend has it that Squanto first instructed the Pilgrims at Plymouth in 1621 on the art of fertilizing their cornfields with menhaden, or *munnauwhatteaug*, "that which manures." As a fellow named Edward Johnson wrote in his journal in 1628: "The lord is pleased to provide great stores of fish in the springtime, especially . . . menhaden. Many thousands of these they put under their corne, which they plant in hills five foot asunder."

While "scrap," or the dried fish, continued to be used for fertilizer, oil recovery from menhaden for industrial purposes began in the early 1800s. With the introduction of purse seine nets in the 1850s came large-scale fisheries. "At one point in time [1880–85], there were fifteen processing plants right on this creek," Haynie says. "Used to be between 4,000 and 5,000 people involved right in this area and the middle peninsula."

Distributed from Florida to Nova Scotia, menhaden were simply ubiquitous, because their food source—phytoplankton, the microscopic plant that supports the ocean food chain—is so widespread. They move through the water in vast schools, often following a single lead fish, their toothless, herbivorous mouths agape to consume huge quantities of microscopic plants. They are a critical species in the flow of energy and nutrients, billions of silvery sea-strainers that improve water quality and hold down algae growth. The capacity of menhaden to "filter" phytoplankton is unmatched by any other fish species (oysters play the largest such role).

They are also a remarkable commodity. As Rachel Carson once put it, "Almost every person in the United States has at some time eaten, used, or worn something made from menhaden." It's in Rustoleum and Friskies Fancy Feast and Pepperidge Farm shortcake cookies and Soothing Seas Aromatherapy body cream. The oil has been used in the manufacturing of soap, linoleum, waterproof fabrics, and certain kinds of paint. With its high percentage of polyunsaturated fats, menhaden oil has also been popular for many years in Europe as a cooking oil, as well as to make margarine and shortenings.

By World War II, due to menhaden's 60 percent protein content, the primary product had become fish meal for poultry and swine feed. That's when Haynie began working on the boats, which he did for eighteen years. "When I started, if you got 10 million fish for a boat, that was a good season. Now if you don't get 50 million a boat, it's not. But there are only ten boats. Used to be a hundred. All up and down the coast. They were all crewed from this area. Great big old menhaden fifteen to eighteen inches long, fulla roe.

"The fishing started going really poor about forty years ago; that's when I got out of it," Haynie continues. "Most of the people from here that stayed with fishing went to the Gulf [of Mexico]." (Government records show that the numbers of menhaden went from 8.1 billion fish during the late 1950s to a "severely depressed condition" of less than 3.9 billion fish in the 1960s.)

Today, Haynie says, the population of Northumberland County is about the same as it was a hundred years ago, 12,000-some, while Reedville itself numbers about 250. Besides the corn and bean growers in the area, the many retirees moving in, and a small tourist trade, it's pretty much a one-company town. That's since Omega Protein a few years ago bought up the only remaining local competition, American Protein (Ampro), founded by Haynie's great-grandfather.

In Reedville, the lowly menhaden is still king. And the fetid smell emanating from the factory stacks after a night's "cooking" of the catch is, the locals shrug, simply "the smell of money."

My closest view of Omega Protein's operation came on a Sunday, in July 2004, from the deck of Ferrell McLain's forty-two-foot charter boat. This summer, McLain has only been able to book customers maybe two days a week, if that. The numbers of striped bass are way down, especially finding any over eighteen inches long. A big part of the reason, he believes, is that "there's not enough for them to eat around here anymore." Thanks, he also believes, to Omega Protein.

Cruising out of Cockrell Creek toward the open bay, McLain points out a "snapper rig" that pursues menhaden for bait. These boats ship frozen "bunker," as striper fishermen call menhaden, through a seafood outlet to various parts of the East Coast. (Catching menhaden for bait comprises about 17 percent of the Atlantic coastal fishery.) "See those brick ruins there? That was an old menhaden factory," McLain says. "You'll see a bunch of abandoned ships rusting away around that bend; they were part of the Ampro fleet."

We close in on Omega Protein's operation. I count ten battleship-gray vessels, each about 170 feet long, names emblazoned along their hulls—*Smith Island, Atlantic Breeze, Coastal Harvester,* and more— former fleet supply ships that the company purchased from the U.S.

military. Early Monday morning, these boats will all be out, on courses charted for them by the company's spotter planes. The fish have a
unique characteristic known as "flipping" as they feed in schools so
large that they're detectable on the surface. Forming tight groups is
an evolved defense mechanism to reduce their vulnerability to predators; unfortunately, this makes the menhaden far more susceptible to
purse seining. Once the fish are sighted, a vessel will lower two boats
carrying a 1,500-foot-long seine net. As they approach the fish, the
boats separate and crew members pay out the net as they encircle the
menhaden, then close or "purse" it. Hydraulic power blocks are
cranked to bring the net aboard, at which point the "steamer" comes
alongside, lowers a giant vacuum hose, and pumps the fish into the
ship's refrigerated hold. A single setting of a seine may corral up to
300,000 menhaden.

Once shore-side, the fish are pumped out again into the processing
plant, where the day's catch is cooked and separated into fish oil and
solids. The solids are then dried into various grades of fish meal. The
factory isn't rendering today, but there's definitely a residual stench in
the air. "That's nothing," McLain says, holding his nose. "You're taking
something wet and basically cooking it till it gets dry, so that's what
makes the smell." The vile odor, I'm told, can sometimes carry as far as
Tangier Island and Smith Islands, way out in the middle of the bay.

I'd hoped to see the operation firsthand, but the next day Omega
Protein executives turned down my request. Over the phone, one of
the vessel captains told me: "I don't know what's going on. They told
us last year we could take anyone out on the boats. I'm an ol' country
boy who'd like to show somebody what we do. But it's been a total
turnaround. I'm not privileged to do a whole lot of talking. I've been
in this business forty years and never seen anything like this."

On the outskirts of Reedville, I turned right onto Menhaden Road.
There were a few spotter planes parked in a field, where the company
was nearing completion of a new $17 million fish oil refinery. It will
be able to process an additional 100 metric tons of fish per day, tripling
the existing production capacity. A half-mile farther down is the

existing factory that I'd seen from McLain's boat. As I approached the open gate, a security guard lowered it and then picked up a phone. I made a hasty U-turn.

"Hey, look," Toby Gascon, Omega's designated spokesperson told me by phone in a moment of candor, "we're fighting for our lives right now."

As they did with the proposed Storm King Mountain and Westway projects, once again striped bass are swimming in the way of "progress." At least that's how the corporate powers view the escalating controversy over whether menhaden fishing needs to be curtailed—if not temporarily shut down altogether—in the Chesapeake Bay. While the largest market remains meal for the animal feed industry, Omega Protein is also the world's biggest manufacturer of liquid fish protein for aquaculture. In 1989, the U.S. Food and Drug Administration ruled that fully and partially hydrogenated menhaden oil is also a safe ingredient for human consumption. In 1997, the same status was extended to refined menhaden oil. This means that not only is the oil a new ingredient in jars of spaghetti sauce and sticks of margarine, but a whole new health food and nutritional supplement market has opened up. In November 2002, the American Heart Association recommended that Americans consume fish oil daily—because omega-3 fatty acids appear to help protect against cardiovascular disease. Omega Protein has cornered the market as America's largest producer of what it labels OmegaPure. The company's new Health and Science Center was completed in October 2004, billed by Omega Protein president Joe von Rosenberg as "the only fully integrated fish oil processing facility in the United States." Currently, the company produces about 40,000 tons of meal and 20,000 tons of oil a year. Its fish oil sales in 2002 helped Omega boost its business by 18.5 percent, to an annual $117 million.

The company has a long and curious history. It is 60 percent owned by the Zapata Corporation, originally an oil-and-gas operation started in 1953 by former president George H. W. Bush. Some believe that Zapata's offshore oil rigs were used as a staging ground for the ill-fated Bay of Pigs invasion, which within CIA circles was known as "Operation Zapata" and whose landing ships were named *Houston* (Bush's domicile) and *Barbara* (his wife). Coincidence? Perhaps, but long after Bush sold Zapata in the mid-1960s, six years of filings on his later years with the firm were "accidentally" destroyed by the Securities and Exchange Commission (SEC) soon after he became vice president in 1980.

Soon after the FDA's decision to approve menhaden oil for human consumption in the first year of the elder Bush's presidency, in the early 1990s a multimillionaire named Malcolm Glazer started accumulating shares in Zapata. Glazer, son of a pawnbroker, had made his first fortune selling mobile home lots near Rochester, New York. Today he is the reclusive owner of the Tampa Bay Buccaneers football team, and his First Allied Corporation is a holding company for a family real estate empire that includes shopping centers in fifteen states.

Zapata is a publicly traded holding company, about half-owned by Glazer and his offspring. Early in 1998, not long after another favorable FDA decision on menhaden oil as a good source of omega-3, Zapata's share value nearly doubled following word that it planned to spin off Zapata Protein, rename it Omega Protein, and offer part of the fishing firm to the public. By late 2003, Zapata's only source of income was its 60 percent stake in Omega, and its value on the New York Stock Exchange had soared 184 percent in a little over a year. That November, Omega's share value jumped 9 percent after a financial services firm gave the company a "buy" recommendation based on the emerging market for omega-3 fatty acids as a human food supplement. Omega's 95 percent total return on investment was far higher than that of other food processors, including Archer Daniels Midland (25 percent) and Tyson Foods (20 percent).

Meanwhile, the SEC was said to be taking a look at the inner workings of both companies. An investigation by *BusinessWeek*, reported in its March 15, 2004, issue, "raises serious questions about whether sudden increases in the value of Zapata Corp. and Omega Protein Corp. were orchestrated. . . . Spikes in stock prices of both came after curious buyout offers." Bidding for Omega, out of the blue, was a company nobody had ever heard of, Ferrari Investments, from a sleepy coastal village in Argentina. Bidding for Zapata was a mysterious would-be corporate raider named Theodore Roxford, who later admitted he "might have been acting" on Malcolm Glazer's behalf. It was all tied in, the magazine indicated, with Glazer's effort to outmaneuver two horse-racing tycoons and acquire the British soccer club Manchester United, perhaps the most lucrative sports franchise in the world.

All of which is seemingly a long way from the quaint little town of Reedville where, with Omega Protein the only fish oil source to receive direct FDA approval, those pesky striped bass are once again a fly in the ointment. The town's attitude is aptly summarized in an exhibit at its Fishermen's Museum. After noting a long history of its menhaden industry being accused of "offensive odor," having "caught and/or scared away sportfish," and limiting beachfront access, it continues:

> Today's critique is not so much about public health concerns or aesthetic real estate desires, but focused on the issue of nature as commodity. Unlike earlier conservationists who viewed man as steward of the environment and responsible for its rational management, the protectionist complaint holds that we are responsible for protecting nature against *any* form of exploitation. For them, the commercial fishing industry's offense is a result of its view of nature as means for livelihood. The joining of protectionist with recreational sensibilities has gained ground in recent years. The two camps are now working to severely restrict, if not close, menhaden fishing grounds.

It is difficult to visit Reedville and not come away with an appreciation for what the menhaden industry has meant to it. The museum walls are filled with sections on the ages of sail and steam, pictures of the African-American chantey singers who rowed the seine boats and hauled in the catch, the spyglass Captain Reed used when he first saw the great potential for catching menhaden in the bay. Reedville is still a place with no town government, where the fire department is manned by volunteers. As one citizen told me, "You don't say anything bad about somebody around here, because they're all related! A lot of people don't mind that smell from the fish plant, because frankly it keeps outsiders away. We don't want to be like an Annapolis and have hundreds of sailboaters down here living in condominiums."

It is certainly true that Reedville's last remaining menhaden processor has been cleaning up its act. Air scrubbers have been installed on the drying machines, reducing the odor as well as particles in the discharges. Wastewater treatment systems have supplanted the practice Haynie remembers of "dumping everything in the water, so there was no oxygen at all, and fish and crabs wouldn't live in it."

Omega Protein also raises the point that a Virginia sector of the Chesapeake Bay is pretty much the only menhaden-rich territory it's still allowed to fish. Maryland outlawed menhaden seining in its waters nearly fifty years ago. In more recent times, under pressure from sports fishermen who don't appreciate the seiners getting in their way, various state legislatures along the Atlantic coast have evicted Omega's "steamers" from their inshore waters. "They stale the whole area with death," as one fisherman says.

What Omega doesn't say is that, as far back as 1965, the menhaden reduction fishery began concentrating its efforts in the Chesapeake. Consider the reported landings of menhaden in Reedville: 488 million pounds in 2001, 382 million pounds in 2002, 375 million pounds in 2003. "That's equivalent to more than a thousand pickup-truck loads of fish a day," says Jim Price. "It's equal to five times the amount of seafood that the entire Maryland commercial fishery is able to land— counting oysters, clams, fish, everything. One company, one fishing operation."

While spawning occurs mainly at sea, the menhaden larvae are transported by ocean currents into the estuaries. They use the bay as a nursery during the first year of life. In recent years, recruitment—the number of new menhaden hatched into the fishery—has plummeted. Between 1975 and 1991, average recruitment was estimated at about 4.4 billion fish a year. By 2001, recruitment was calculated at some 500 million, the lowest figure ever recorded.

Clearly, something is awry with the menhaden spawning stock biomass, the total weight of fish in a stock that are old enough to spawn. Menhaden reach sexual maturity at around age three. Given the decreased numbers of younger fish, the percentage of those larger menhaden landed by the reduction fishery has increased threefold since 1993, compared to the previous thirty-year average, according to calculations made by Price. Since those are the same size menhaden that increased numbers of the larger striped bass also forage upon, Price sees the combination of harvest and predation as reducing the menhaden's spawning biomass "to a level which may cause the stock to collapse." This is more acute, considering that there are also about one-third as many age-two menhaden as in the 1950s, with about 45 percent of the remaining ones being caught annually by the purse seine fishery before they ever reach spawning maturity.

"The menhaden modelers are telling us that the spawning biomass is still considered to be robust, or at least enough to sustain the species," Maryland biologist Eric May says. "Yet they're also telling us that the juvenile numbers are way down. You can't have it both ways."

Other marine scientists, however, counter that menhaden are very resilient. One or more large year classes each generation could stabilize the population. Also, because menhaden migrations are extensive and their fecundity great, only a portion is susceptible at any given time to the intensive seining off Virginia and North Carolina.

Yet what's happening to striped bass would seem to bear out a steep decline in certainly the Chesapeake Bay's menhaden population. Stripers love menhaden for their fat content, which is about 20 percent, approximately four times the fat concentration of most other fish. "It's almost like they know they're eating some kind of rich chocolate, as opposed to, say, a carrot," says New York biologist John Waldman. It may also be that just as fish oil can be an important supplement to the human diet, the same is true for the bass. Perhaps even more so, especially as they get older.

Price has also calculated that, prior to the mid-1960s when the reduction fishery began focusing on Virginia, the bass consumed three times more menhaden in the bay than they do today. Nor are the bass all that's apparently being affected by the menhaden shortage. Maryland bird biologist Paul Spitzer has long kept track of flocks of loons that make a fall migration stopover across a sixty-square-mile habitat on the Choptank River. Until recent years, the loons could be observed feeding on large schools of menhaden. No more. During the 1989–99 decade, Spitzer's loon count fell steadily—from between 750 and 1,000 during a three-hour period of observation to between 75 and 200. A typical flock went from between 100 and 500 birds down to between 15 and 40.

Spitzer observed a similar problem in ospreys, which had been making a resurgence since DDT was banned in the 1960s. "Migratory menhaden schools formerly arrived in May, in time to feed nestlings," Spitzer told *Discover* magazine. Now that menhaden are largely missing, there has been a decrease in osprey chick survival, along with a decline in active nests on Gardiners Island from seventy-one to thirty-six, during the period 1995–2001. "The collapse of the menhaden means the endgame for Gardiners Island ospreys," Spitzer has said, noting a similar pattern in other areas of the Eastern seaboard such as Sandy Hook, New Jersey, and Plum Island, Massachusetts.

With striped bass and menhaden, the predator-prey imbalance is not limited to the Chesapeake either. Barry Gibson, a former editor of *Salt Water Sportsman* and a summer fisherman in Maine, reports that the last year the menhaden came there in any numbers—and there had been big schools for twenty years before that—was 1993. That year,

Gibson caught sixty-one striped bass more than thirty-six inches long. In 1994, when the menhaden disappeared, he caught a mere four striped bass with equal effort. Previously, as he worked his way upriver, the bass could always be found underneath the schools of menhaden.

For its part, the menhaden industry claims it's not their operation —but those overabundant striped bass—that is responsible for depleting the menhaden. It's undeniable that the fishery and the fish are competing for the same resource. And the bass may have more effect on the menhaden population than anyone thought. But the fact is, Omega Protein (along with the Menhaden Resource Council and the National Fish Meal and Oil Association) has, on at least one occasion, done its best to block scientific examination of the food web in the Chesapeake. Back in 1994, the Virginia Marine Resources Council (VMRC) was considering a proposal, backed by the nonprofit Chesapeake Bay Foundation and to be funded through recreational license fees, to explore the repercussions on other fish species of the commercial taking of menhaden in the bay. "We got our ass handed to us, on what should have been a slam dunk," the foundation's fisheries program manager, Bill Goldsborough, recalls. "Because the industry opposed it, the commissioners caved in."

The Atlantic States Marine Fisheries Commission does have a fishery management plan for menhaden, first approved in 1981 and twice revised since. For years, however, its Menhaden Management Board was dominated by representatives from the industry. It was the *only* such ASMFC board that allowed industry members a direct vote in the decisions. Insiders referred to the head of the board's technical committee as the leader of the "Menhaden Cheering Society." For years, the board and the committee passed the buck between them, stalemating any real regulations. In 2001, both the board and the technical committee were finally restructured. And some recent signs suggest that the group may take a harder look at what the industry is doing to menhaden stocks, and at those repercussions.

As late as the ASMFC's annual meeting in December 2003, however, the modelers were maintaining that, based upon the 174,000 metric tons landed in 2002, the coastwide Atlantic menhaden population was "not overfished, and overfishing is not occurring." They maintained, too, that it was simply impossible to isolate what might otherwise be happening in the Chesapeake Bay, where nearly two-thirds of that Atlantic catch originated. Meanwhile, other marine scientists were beginning to question the numbers of the menhaden technical committee. According to Desmond Kahn, "Some scientific research suggests that forage species like menhaden need to be fished at a lower level than managers might think, to allow for their high natural mortality [from disease and predation]."

In the summer of 2004, Jim Uphoff was finishing an analysis of the National Marine Fisheries Service's menhaden stock assessment, which emanates from a lab in Beaufort, North Carolina, also the home base of Omega Protein's only remaining competition, the Beaufort Fisheries Company. Uphoff's conclusion? The assessment is "grossly wrong." He explains: "The big thing is, for as important a species as menhaden are, there is virtually no monitoring of the population. Everything is run off of catch data. But what you've got is what's known as 'inverse catchability.' In other words, as the population goes down, the fishery becomes more and more efficient. They take a larger fraction [of menhaden] per shot. Because these guys aren't out fishing randomly. They have airplanes and, when they set a net, they set it on a school of the right magnitude. That would indicate that the fishing mortality rates on menhaden are going way up, as opposed to going way down."

Also, as the population shrinks, juvenile fish aggregate with the older ones for protection, making it more difficult to avoid catching the younger ones while giving a misleading impression of how healthy the stock is. Catches can be sustained, even as the numbers of fish decline. Indeed, you might not *see* the decline until the stock collapses.

Boston University biologist Les Kaufman adds: "The thing is, menhaden are like passenger pigeons. It could be over before you'd ever know that they were overfished, because their populations are extremely volatile and very responsive to environmental flux. Their response to a profound change in the availability of food, or anything like that, is to crash."

Other marine scientists argue that Uphoff's concern about "inverse catchability" would only be applicable if all menhaden were in the Chesapeake; however, as we know, they are scattered up and down the coast. To get an accurate picture we would need to have good population and migration information and then be able to measure accurately rates of local depletion.

Still, at long last, due to the unprecedented demand from environmental and fishing groups that something be done, the ASMFC began to take notice. This came on the heels of a March 2004 meeting at which an interstate panel of scientists concluded that current menhaden management measures were inadequate. Shortly thereafter, Uphoff presented an important paper at an Ocean City, Maryland, symposium in which he concluded that total menhaden demand had increased sevenfold over the past generation. A low ratio of supply that began in the mid-1990s for striped bass "coincided with the outbreak of lesions and decreased presence of Atlantic menhaden in diets." The findings of another scientist, Kyle Hartman, were similar to Uphoff's, although Hartman's analyses suggest that a shift in diet from menhaden to anchovy had only a minor effect on striped bass growth.

Then, in October 2004, the ASMFC convened a three-day scientific workshop on menhaden in Arlington, Virginia. A standing-room-only crowd was on hand. At the close of the proceedings, recommendations for action were supposed to take place. Since Omega's boats took five times the biomass of menhaden in May 2004 than they had the previous May—coinciding with the time when striped bass were spawning in the Chesapeake—it was suggested by Bill Goldsborough, the Chesapeake Bay Foundation's representative, that the menhaden fishing season be delayed a month, until June 1. Goldsborough also said that it would be prudent to place a cap on the current level of men-haden harvest, perhaps at an average of the past five years.

Although these suggestions hardly seemed extreme, the industry's response was swift. Lawyers for Omega Protein threatened a lawsuit against the ASMFC should the managing body seek to restrict the men-haden catch in the bay. It would be "arbitrary and capricious," they claimed, to take precautionary action or to institute ecosystem-based management—because the existing fisheries laws don't offer grounds for doing so. In response, conservationists pointed out that the ASMFC had in fact already taken a precautionary move to set limits on taking horseshoe crabs, even when there hadn't yet been a stock assessment.

The science presented during the workshop convinced the NCMC's Ken Hinman that, more than ever, something needed to be done quickly to curtail the menhaden harvest. Yet no scientist was will-ing to make specific recommendations. Angler and environmental groups responded to the lack of movement by forming a new alliance called Menhaden Matter. Spokespeople from the Chesapeake Bay Foundation, the Coastal Conservation Association, Environmental Defense, and the National Coalition for Marine Conservation gathered at a press conference on October 26, 2004, to release a case study demanding immediate proactive measures to protect "the most impor-tant fish in the sea"—and insisting that the ASMFC "lay the founda-tion for an ecosystem-based approach to fisheries management in the future." The industry retaliated by setting up a website of its own—Menhaden Facts—seeking to disparage the new coalition with state-ments like, "These fanatical big game angler organizations appear will-ing to go to any lengths of deception and defamation in their attempts to expand the sportfishing industry at the expense of the centuries-old, sustainable harvest of menhaden." The fight went on, all too reminis-cent of the earlier striper wars.

Yet what the industry wants most—a booming market for its products—may not necessarily come to pass. According to *National Fisherman*, "On the industrial side of the fishery, where menhaden is processed into feed for poultry and pigs, the demand for fish is depressed by a surplus of soy, which serves the same purpose." All that ground-up menhaden, it seems, could be readily replaced by ground-up soybeans. As for omega-3 vitamins for people, other companies are already seeking ways to produce these from alternative sources—such as algae, although algal populations in the bay are already so large that they indicate an ecosystem with far more nutrients than are consistent with productive habitat for striped bass and other species.

While the ASMFC's menhaden board decided that no additional management measures were necessary at its November 2004 meeting, the board agreed to have scientists look into claims that "localized depletion" might be occurring in the Chesapeake Bay. Some members of the technical committee were, in fact, slowly coming around to Jim Price's conclusion that the resurgent striper population's impact on the menhaden had not been properly factored into its menhaden population modeling.

At last, in February 2005, the menhaden board overwhelmingly approved development of an addendum to their management plan that would place a 110,400 metric-ton cap (an average of the last five years' landings) on the menhaden fishery's Chesapeake Bay catch in 2006 and 2007. The board also proposed beginning an immediate research program to examine the menhaden's status in the bay and to consider other fishing constraints during the six-month addendum process. Larry Simns, president of the Maryland Watermen's Association, told the menhaden board: "I hate to go against my [Virginia] watermen brethren, but I think it's the right thing to do." Maryland Governor Robert Ehrlich affirmed his state's commitment "to being a leader in menhaden management in the Chesapeake Bay as these filter-feeding fish are vital to its sustainability." For his part, as he once did for striped bass, Price has decided to petition the National Marine Fisheries Service for designation of Atlantic menhaden as a "threatened" species.

It's about a five-hour drive from Reedville back up Virginia's Northern Neck, across the Bay Bridge, and south again down Maryland's Eastern

Shore to the little town of Oxford. There, from his picture window, Jim Price casts his gaze beyond the backyard at the Choptank River flowing past his wooden dock. "I've lived on this river all my life," he says. "I should be able to look out this time of year and see schools of menhaden with terns working them. I haven't seen a one. I haven't caught a single eighteen-inch striped bass this year in the Choptank. This river's just about dead."

Joe Boone has just arrived, driving down from his farm a couple of hours away on the western shore. Jim Uphoff lives in nearby Easton, and our plan is to meet him at the dock where Price keeps his boat. On our way out the door, Price pauses to show Joe and me his aquarium. He points out a couple of mud crabs. "That's a lot of what you find in rockfish [striped bass] stomachs these days," Price remarks. "They even eat that pipefish you see there."

Loading the gear into Price's twenty-eight-foot Bertram, the three of us head out into the Choptank River for a morning of fishing. Ultimately we're bound for Tilghman Island, about eight miles down the river toward its mouth, site of one of the last remaining large-scale charter-boat and commercial fishing centers in the upper bay.

The oldest private ferry service in the United States, running from Oxford to Bellevue since the seventeenth century, passes in the near distance and sounds its horn. For a moment, all feels timeless. A few gulls soar overhead. "There used to be thousands," Price says. He notes that three-quarters of the pleasure craft we're seeing on this Saturday morning are sailboats. Fishing boats number maybe 10 percent. A decade ago, it was precisely the opposite. "You'd see easily twenty to thirty crab workboats dotting the shoreline," Price adds. "Now you might see two or three."

It's too windy to do much fishing. We pass under a drawbridge at the northern end of Tilghman and out the other side toward Poplar Island. Price describes his role on one of the state's seven "tributary teams" involved in the long-range plans to restore the beleaguered bay. Farm runoff, it's been determined, is the most important issue for the Choptank River basin. "The most important areas are the nurseries where the fish spawn," Price says. "This area is dying not from the bay *in*, it's dying from the rivers *out*."

I've got a plane to catch, so there isn't much time for any casting. There are no signs of fish anyway. No birds, no bait, no fish. Joe Boone puts his fishing rod back in a holder and muses: "You know, menhaden filtered out millions of tons of algae, turned it into flesh, and went out of the bay with it. That's how the bay evolved over 10,000 years or so. Now that the oysters are gone, too, all of a sudden you don't have anything removing these massive volumes of algae, which of course is made even worse by increased fertilization from agriculture. The algae dying contributes to another problem, oxygen depletion in the water, which means fish start to suffocate. This was finely tuned by nature, but it's like removing one little pin and then the whole system starts to collapse."

As we head back toward the dock, I am left pondering how Bill Goldsborough of the Chesapeake Bay Foundation summarized the situation over a recent lunch. "You've got the menhaden industry harvesting hundreds of millions of pounds out of the bay, of a filter feeder that should be eating algae," he says. "Then the menhaden are being ground up and processed into a feed that's going to chickens. The chickens are producing all this manure and nitrogen that ends up back in the water, stimulating more algae growth. And that can stimulate disease outbreaks—the prime victim of which is menhaden!"

I could think of no better definition for the phrase "vicious circle." The intricate web that nature has woven into and around the Chesapeake Bay ecosystem—where what happens to algae, menhaden, striped bass, and chickens is all interrelated—human practices can rapidly rend asunder.

Chapter Thirteen

Upriver

Jim Uphoff had come a long way from his beleaguered early years as a Maryland fisheries biologist, forced to file charges of harassment against his own superiors in 1984. The new chief in 2004 of the state's Tidewater Fisheries division, Howard King, was himself a former marine scientist who spent his first years with the department "walking every tidal tributary around the bay" to identify spawning runs and ecological problems. King was a strong admirer of Uphoff. "Without Jim," he told me, "we wouldn't be anywhere near where we are in considering the importance of the menhaden issue."

The "menhaden issue" involves more than overfishing. Although spawning occurs along the coast, menhaden spend their first year of life in estuarine waters. In the Chesapeake Bay, they are extremely vulnerable to environmental conditions. Blooms of toxic algae, after being ingested by menhaden, have been known to cause widespread fish kills. The appearance of these blooms is often correlated with increased pollution running off the land into the water, from both

urban and agricultural sources. Fish kills also occur when schools of menhaden enter enclosed inshore bodies of water in such large numbers that they consume all the available oxygen and suffocate. High water temperatures, which increase the fishes' metabolic rate, accelerate oxygen depletion. So does pollution.

Striped bass, of course, are not only stressed by having less available prey—but similarly by what accumulates in their habitat from runoff upstream, often in the form of waste from sewage treatment plants and farm animals that creates excessive nitrate pollution. Where there has been considerable human settlement around their nursery areas, stripers are pushed to the margins rather than able to roam in a wider area where food is more abundant. This habitat squeeze will only be exacerbated should urban sprawl—with its resulting pollution—accelerate in the coming years.

These are reasons why King's fisheries agency has embarked on a major assessment of human land use on fish habitat in Maryland's part of the Chesapeake, with Uphoff being placed in charge of one study called the Impervious Surface Project. This refers to paved roads, rooftops, compacted soils, basically anything that can't absorb water. As one might expect, the resulting runoff affects the water quality of bay tributaries much more in areas of greater urban—or suburban—development. King's hope is that, by comparing impacts on fish populations in places where the degrees of development differ, counties might learn to divert growth away from sensitive locations in the future. "Unless we start to do this," King says, "fishing as we know it will go by the wayside."

As I rode with Uphoff en route to meet several colleagues for a seining-and-trawling survey, the bearded biologist spoke of how the human population in this sector of the mid-shore Chesapeake is expected to double within the next decade or so. "With so much development, the impervious, or hard, surfaces will increase and so will the runoff," Uphoff explained. "Streams then don't function in the gentle way they normally would. They become much more prone to flooding and drying. The water can be warmer and laden with much more in the way of nutrients and contaminants. It's a steady progression. You may get plenty of nice houses, but meanwhile the [fisheries] resource went to hell."

For Maryland fishery managers, this raises questions of how to offset loss of habitat by either decreasing the rate of allowable harvests or increasing the size limits at which striped bass, for example, are allowed to be taken. Uphoff says they're trying to come up with "biological reference points," or benchmarks of what a fish population needs in order to maintain itself. It's all part of the movement toward ecosystem-based planning.

The Miles River, which we'll be exploring today, empties into Eastern Bay on the eastern side of the Chesapeake Bay. Right now, it is one of the least-developed and healthiest of the seven rivers that Uphoff and his crew sample. Around the Severn River, Annapolis's river on the western shore, where a surge in new homes, shopping centers, and businesses has brought the impervious surface to about 20 percent compared to 4 percent surrounding the Miles, a study of yellow perch is indicating severe problems with young-of-the-year survival, something Uphoff hopes doesn't "spill over" into striped bass.

In 2002, according to *National Fisherman*, the Chesapeake Bay still produced nearly 500 million pounds of seafood worth $172 million. The shallow waters that make the bay region so productive, however, also mean less capacity to absorb pollutants, which in a deeper system would largely sink out of harm's way. Relative to its volume of water, the Chesapeake has more land draining into it than any other bay on the planet. With 16 million people already living in the watershed, vast amounts of pollutants—primarily nitrogen and phosphorus—are carried into the 150-some major rivers and streams in the Chesapeake drainage basin and thence into the bay itself. About 40 percent of this excess of nutrients comes from agricultural runoff and sewage treatment plants. Other contributors include lawn fertilizers, septic tanks, and nitrogen compounds released by automobiles or even power plants hundreds of miles away, which fall into the bay system when it rains.

When that rain is excessive—as it was in the spring and summer of 2003, bringing near-record river flows after several years of drought—the resulting runoff is devastating. That year, an estimated 353 million pounds of nitrogen (up from 130 million pounds in

2002) and 30 million pounds of phosphorus (up from 6 million pounds in 2002) washed into the tributaries. These literally fertilized the water, triggering massive blooms of algae that blocked life-supporting sunlight from getting down to the underwater grass beds. Those beds play a crucial role in the life cycle of juvenile fish, providing food and habitat. Submerged aquatic vegetation, or SAV as it's called, also helps to protect shorelines from erosion and traps sediments that otherwise cloud the water. In 2003, nearly one-third of the Chesapeake's SAV died.

That wasn't all. When the algal blooms themselves die, they decompose in a manner that consumes large amounts of dissolved oxygen (DO). Relatively small amounts of oxygen dissolve in water, compared to the atmosphere, which is why the gills of fish must be so efficient in order to extract it. Fish, after all, need to breathe, too. But the proliferation of dead algae can cause conditions of hypoxia (water with low DO) or, worse, anoxic waters with no oxygen at all. When these conditions exist, marine life either is displaced in the first instance or dies from suffocation in the second. The EPA's Chesapeake Bay Program reported that about 40 percent of all the water in the Chesapeake lacked adequate oxygen in 2003, resulting in what are called "dead zones," devoid of marine life. This is a problem in highly developed regions worldwide; the Gulf of Mexico, for example, suffers from a 7,700-square-mile "dead zone," a result of pollution flowing down the Mississippi River. In the Chesapeake, typically starting twenty to thirty feet below the surface, in 2003 the dead zones extended for 150 miles, from above the Bay Bridge in Annapolis down to the mouth of Virginia's York River. During the two summer weeks when the "dead zone" was at its greatest extent, watermen spoke of blue crabs dying in their pots, of red tides wiping out oyster beds, of striped bass disappearing from their customary habitats.

In July 2004, the problem was not as widespread, but still extensive: 35 percent of the bay waters had oxygen levels low enough to be stressful to marine life, and in some sectors the anoxic water was more widespread than the previous summer. Virginia biologists working in the lower Rappahannock River detected as little as 0.8 milligram per liter (or parts per million) of oxygen in their water samples. The general habitat requirement for striped bass and other fish is five milligrams per liter; otherwise they are forced to abandon the area. Crabs require three milligrams. Bottom-dwelling worms and other invertebrates need one milligram; otherwise they will die.

At the same time, in June and July water clarity in portions of the Chesapeake showed a sudden improvement. Turbidity lessened; SAV bloomed. There were few clues as to why. Long strands of macro-algae were so thick in places that they kept watermen from crabbing, but perhaps the macro-algae were also trapping sediment and thereby keeping the waters less murky. So it was conceivable that what looked "better" might really be another sign of ecological distress. As biologist Eric May put it: "We've gone from essentially a system of bugs and critters on the bottom, burrowing and renewing, to a system that's now dominated by bacteria and algae. When you go into that shift, you change the entire nature of that bottom and what it can provide. Because large patches of the bottom are now anoxic, striped bass are being pushed into the shallows in order to find food. These are also areas where the water temperature is higher, meaning you no longer have the kind of cooling refuges that a bass likes to go to. So you have these fish forced into places they've never been before and eating things like grass shrimp and polychaetes off the bottom, instead of their traditional foods."

At the loading dock on the Miles River, Uphoff introduced me to his veteran team. Jim Mowrer has been with the fisheries agency for thirty-four years, Rudy Lukacovic for thirty-one years, Margaret McGinty for fifteen years. Unhitching their twenty-one-foot workboat from its trailer, they stowed the gear as a light rain began to fall. "Guess we'll go upriver and work our way back," Uphoff said.

Standing at the wheel, wearing green shorts and a green bush hat, Uphoff started the engine. "They were talking yesterday about getting a seat for me, but I said no, I'll wait till I get my wheelchair," he said. The other crew members smiled. We left the harbor and passed under the Route 33 bridge, dark clouds on the horizon. We went by numerous osprey nests atop the channel markers and, along the oak-lined banks, big white colonial homes set back from the boat docks. "No poor people on the Miles River," Uphoff said.

The first survey of the day would employ a bottom trawl. Uphoff cut the motor, and Margaret and Rudy fed out the wooden-weighted net by hand. Mowrer simultaneously took the water's temperature, also metering its salinity, conductivity, and DO levels. "The trawl is a sensitive technique," Uphoff said, "because if you don't end up with any oxygen, you don't bring in any fish. If there are less than three species in your trawl, you've got very low DO."

He steered us slowly upriver for a six-minute tow, past the shoreline cordgrass. When the trawl was hauled in, the first "catch" was a six-foot-long stick. "See what happens when we don't let Jim do it?" Margaret said. Eight different species got counted and released. Ctenaphores, or comb jellyfish, were the foremost collected. These can easily tolerate water with low dissolved oxygen, Uphoff explained. Their larger brethren, the sea nettles that customarily eat them, have for unknown reasons largely disappeared from the system, leaving the ctenaphores to "go crazy." This is worrisome, because those proliferating comb jellies are highly efficient consumers of fish larvae and zooplankton.

A foul smell turned out to be a dead deer floating in the tidal river. An eagle soared overhead as Uphoff made toward the shoreline. "With all the riprap and bulkheads that people have built," he said, "we've had a harder time locating a beach to seine on. Only found three good sites along the Miles." A blue heron waited on the beach. An empty tennis court dominated the background. After we dropped anchor, Mowrer and Rudy waded out to make a big circle with the seine net. It was the same procedure, indeed the same type of net, I recalled Boone and Uphoff using on the young-of-the-year surveys twenty years earlier.

"Did you notice the water color? I wonder if this is an algae bloom dying off," Margaret said. I peered down at a muddy, plantless bottom. A log arrived in this haul, too. "We're building up a lumber pile, that's all," Uphoff said. Margaret bucketed water into the net to drain the mud, while the other scientists started counting fish. "Fair number of fingerling stripers this time," Uphoff said. Striped bass weren't known to spawn in this river, and he guessed these young ones had swum up from either the head of the bay or the Choptank River. "When they start getting mobile, doesn't take much to get 'em moving around." Then he added: "I'm more interested in some of the things that aren't here." Like crabs. Like menhaden.

Uphoff explained that the results of inshore seining could some-times be misleading. When water quality has deteriorated, "our seine catches may actually go up—because the fish have been driven into these shore-zone habitats where they can breathe. That's why we also do the trawling and the water quality scans."

The morning proceeded, with a total of four trawls and three seine hauls along a ten-mile stretch, before we headed back around noon. Mowrer and Rudy had kept the count in their heads, then dictated the numbers to Uphoff: seventy-six white perch at one site, thirteen young-of-the-year stripers at another. Rudy came up with an arrow in one haul. "Miles River Indians?" Uphoff wondered. On other trips they'd pulled in a cap, a pair of shoes, and more apparel. "Jim's outfitting his family," Rudy deadpanned.

"One thing we didn't see today was menhaden," Uphoff said. "Not a one. Used to be . . . menhaden would make up 60 to 70 percent in the hauls. It's down to about 10 percent or less now. We don't see that much of anchovies or spot either, which should be very common in the trawls.

"Everybody was so happy to have plenty of striped bass; the intel-lectual curiosity disappeared real fast. I remember one of the managers saying, 'We're slapping ourselves so hard on the back that we're liable to break our arms.'" He paused. "Makes it very difficult for people to let go of the success story. That's part of what's going on today—they can't afford to believe it may not go on forever."

As governor of Maryland from 1979 to 1987, Harry Hughes not only declared the striped bass moratorium but was the prime mover behind the landmark Chesapeake Bay Program initiated in 1983 among Maryland, Virginia, Pennsylvania, the District of Columbia and the federal government, with its stated goal to "implement coordinated plans to improve and protect water quality and living resources" in the estuarine system. Now in his mid-seventies and retired with his wife to a ranch-style home along a branch of the Choptank River, Hughes remains an active champion of protecting the bay. When I stopped in to see him, I raised the disquieting point that so many things didn't seem to be getting any better.

"No, they're not," Hughes responded. "The only thing one could say that's comforting is, it would probably have been a lot worse without the Chesapeake Bay Program. When we first got the legislation passed, we kept saying, 'Don't expect the situation to change overnight.' It took 200 years, starting with all the deforestation to grow cotton and tobacco, to put the bay in this condition. So it's going to take at least ten years to start turning it around. Well, it's twenty years now and that's still not happening.

"The one thing nobody can do anything about, I guess," Hughes continued, "is the number of people that keep moving into the watershed. The only answer is spending a *lot* of money—on the order of $18 billion—to, among other things, upgrade the sewage treatment plants and control the farm runoff. But the political leaders have got to take hard positions, like we did with the rockfish [striped bass] moratorium, and I don't see it happening."

For certain, some improvements have taken place. A ban on using phosphate detergents that took effect in the bay states between 1985 and 1990 has helped stabilize phosphorus loads. Prodded by the Chesapeake Bay Foundation (CBF), in the spring of 2004 the Maryland legislature passed the so-called flush tax—a $2.50 monthly surcharge on all homeowners, expected to generate around $750 million to pay for upgrades at sixty-six wastewater treatment plants. This followed a CBF report that only 10 of 300 such plants baywide are using the right equipment to remove nitrogen.

The Chesapeake cleanup efforts are plagued by "the politics of postponement," CBF director Will Baker has noted. This may be partly attributable to the huge overstatement of cleanup effectiveness over the years, which the Chesapeake Bay Program acknowledged in 2004. EPA computer models had indicated that the flow of nutrient pollution from rivers into the estuary had nearly met the goal of a 40 percent reduction since 1985. An investigation published on the front page of the July 18, 2004, *Washington Post* revealed otherwise. Its analysis of U.S. Geological Survey water monitoring data from the mid-1980s through 2003 found that nitrogen and phosphorus levels showed *no decline whatsoever*. An EPA spokesperson said there had been "no intention to mislead. Our models continue to be refined over time." But as described by Professor Howard Ernst (author of *Chesapeake Bay Blues*), the annual claims that conditions were improving "sucked the public outrage out of the system" and eliminated any incentive to make necessary changes. Following the *Post*'s exposé, three U.S. senators (Barbara Mikulski and Paul Sarbanes of Maryland, and Virginia's John Warner) sent a joint letter to the U.S. Government Accountability Office, asking it to assess the Chesapeake Bay Program's real progress.

In November 2004 the CBF adopted a more aggressive strategy, filing suit against the EPA for its failure to enforce the Clean Water Act and reduce nitrogen pollution in the bay and its rivers. Specifically, the environmental group accused the EPA of foot-dragging on requiring sewage treatment plants in the watershed to impose permit requirements limiting nitrogen output. After four Maryland water-watchdog groups followed with another lawsuit, a potentially landmark agreement was announced on January 3, 2005. The EPA had brokered a six-state, $4-billion pact aimed at keeping about 17.5 million pounds of nitrogen and 1 million pounds of phosphorus out of the Chesapeake Bay each year. Citizen actions seemed to be producing results.

Of all the pollution choking the Chesapeake, a July 2004 report by the CBF relates, "the largest source is agriculture and, increasingly, from the manure produced by livestock, which now outnumber the watershed's human population by 11 to 1. Most of that manure is spread on the surface of nearby cropland, and studies show that within two years as much as half of its nutrient pollution washes out of the soil and into rivers and streams or seeps into groundwater."

The Delmarva Peninsula (short for Delaware, Maryland, and Virginia) is a swath of land, 180 miles long and up to 70 miles wide, that encompasses the sixth-largest poultry-producing area in America. In 2001, more than 532 million broiler chickens were raised here in nearly 5,700 chicken houses. Poultry in this region accounts for nearly two-thirds of farm output value; in addition, virtually all of the corn and soybeans grown in the region is used for local chicken feed. A big percentage of Delmarva poultry is being raised in tightly confined quarters under contract to Perdue Farms, Inc., a family-owned business with annual sales of $2.7 billion, headquartered near the Eastern Shore town of Salisbury, Maryland. Today, Perdue Farms has about 20,000 employees, ranking third in the poultry industry behind Tyson and Con/Agra.

The way the industry is set up, the farmers don't own the birds. Perdue is the only company with its own genetic stock, and it supplies the flocks to the chicken houses. It also sells farmers the antibiotics and feed for their chickens. While the big integrators like Perdue also handle all the processing and marketing, the farmers are left to deal with the waste. The Chesapeake Bay Program calculated that, in 2003, poultry manure excreted in the bay watershed generated more than 185 million pounds of nitrogen and over 51 million pounds of phosphorus. While some is spread on farm fields to fertilize crops, vast amounts also end up polluting rivers and streams. The CBF spearheaded an effort to change this equation through then Maryland Democratic governor

Parris Glendening. A new rule established "manure liability" for Perdue and other corporations, as part of a requirement to obtain water quality permits needed by their processing plants. The companies fought back, and in June 2003 new Republican governor Robert Ehrlich Jr. rescinded the rule, saying his predecessor had overstepped the state's authority.

I decided it would be worth a shot talking, to be blunt, chicken shit—its uses and abuses—directly with Jim Perdue, the fifty-four-year-old heir to the business his grandfather began in the family's backyard back in 1920. I'd heard that Perdue, in fact, had started out as a fisheries biologist, working for two years at a fish hatchery in Maryland before obtaining a Ph.D. in aquaculture from the University of Washington/Seattle. To my surprise, he personally returned my phone call and we made arrangements to spend a morning together. I was particularly interested in seeing his new Perdue AgriRecycle operation, which removes manure from the chicken farms and turns it into pellets of organic fertilizer. "I think they're seriously aware of the problem," the CBF's Alan Girard had told me, "and are trying to take proactive measures to deal with it."

As we sat down for breakfast in a local restaurant, Perdue's first words were: "Is striped bass still classified as *Morone saxatilis*?" If he was trying to impress me, he did. Yes, I replied. "Well, I know they've changed it a few times," he said. He was right. Striped bass had over time bounced back and forth between the genus *Roccus* and *Morone*, and had various designated species names as well. Back when the bass were "beginning their downhill slide in population," Perdue said, he was working at the Horn Point Laboratory. "We got some stripers from the upper Choptank, trying to figure out what was going on and to spawn them."

Did he do much fishing himself? "Well, my sons tell me that, for someone who majored in fisheries, I'm terrible at it." Still, he relished telling the tale of the day he went trolling from his windsurfer, landing a big bluefish that also hooked *him*, then heading into shore "bleeding like a stuck pig—people were astounded I still had the fish."

Two miles away from this restaurant, Perdue recalled, "I grew up on our farm. We literally have the Perdue office across the street from where my dad was born, and right next to it is where I was born." His grandfather, Arthur, had left the railroad at thirty-five and started a table-egg business with a flock of fifty-three backyard layer hens. That was the same year his son Frank entered the world, later to become TV-famous with his nasal, high-pitched voice advertising his family's chickens.

Jim had shied away from the family business initially, but he found that in the aquaculture industry, "making a go of it economically was tough" and he "didn't like the politics [of teaching] at universities." So the prodigal returned to poultry, though he's kept his hand in water matters. Perdue is a member of the Chesapeake Bay Commission, which represents the legislatures of the three bay watershed states in evaluating ways to improve the bay's condition, as well as the Oyster Recovery Project, aimed at bringing back a population whose harvests have declined an astounding 98 percent since the mid-1950s.

"When John Smith came here, the bay was already opaque," he said, putting on his biologist's hat. "Why? Because it's a wonderful place to grow food. One of the definitions of an estuary is a eutrophic environment, meaning highly nutrient-rich, mainly because there is a lot of marsh drainage. Now there is no doubt we have increased this dramatically with our sewage plants, our agriculture, our lawn fertilization, you name it. We don't have anything sucking up the algae now. The problem is, it's all interrelated, like a body with the joints."

Perdue is part of the effort to figure out how to get the $18 billion that former Governor Hughes and others believe it will take to clean up the bay. (In late October 2004, a panel studying the bay's financial support concluded the price tag could be closer to $28 billion.) The big challenge, Perdue believes, is getting Pennsylvania more involved, since about half of the fresh water entering the bay flows down from its Susquehanna River basin. There, Pennsylvania's Lancaster County has the second-highest agricultural production of any county east of the Mississippi, and it ranks fifth in livestock production nationally. A significant portion of Pennsylvania's animal waste—almost 77 million gallons of liquid manure and more than 58,000 tons of solid manure annually—is spread on fields not covered by any nutrient management plan. "Out of sight, out of mind," according to Perdue.

Not only does Pennsylvania have two-thirds of the region's dairy cows, but its factory hog farms are booming. And when it comes to hogs, Perdue reminded me, the majority of the waste is liquid. Hence, with the containment lagoons that hog farmers build, the walls can collapse and send the urea into the watershed. This is in contrast to chickens, whose uric acid waste is predominantly dry. "People forget about this, but there's a huge difference," Perdue says, "because chickens are not mammals like you and me—and pigs."

Still, chicken shit is not to be sneezed at, in terms of nitrate pollution. The day before my meeting with Perdue, the CBF had released a new report, "Manure's Impact on Rivers, Streams, and the Chesapeake Bay." According to the study, "Poultry manure is higher in nutrients than cow manure, and the poultry industry has been expanding in the region, while milk and beef production have declined." The report named the three "hot spots" in the Chesapeake region: Pennsylvania's Lancaster County; Virginia's Shenandoah Valley, the largest turkey producer in the United States; and the Delmarva Peninsula, in particular Sussex County, Delaware—birthplace of the modern broiler industry, with some 200 million chickens raised annually, making it the most concentrated poultry-growing county in the world.

Such a concentration creates a situation in which far more manure is produced than local farmers can use to fertilize their crops. Since the cost of transporting the heavy, wet manure to other agricultural regions is prohibitive, most farmers see no alternative but to overapply it to their own land. This ends up releasing additional pollutants into the air and water. Just as clear-cutting of forests results in the loss of salmon spawning grounds in the Pacific Northwest, so does poultry production contaminate a watershed that striped bass and other fish must use as nurseries.

Asked by the *Washington Post* for a response to the CBF study, Jim Perdue was quoted: "We haven't been sitting around not doing anything." Which was precisely what he wanted to show me today. As we got into his SUV, Perdue said, "It's not like we're absentee landlords, you know." People expect businesses to be "corporate citizens" today, he believes. In a bottom-line world, however, businesses often need a push. In Perdue's case, it came from Maryland's implementation of new nutrient management regulations for farmers in 1998. Those regulations came on the heels of the 1997 scare over the *Pfiesteria* microbe and a series of fish kills in the Pocomoke River that a commission appointed by Governor Glendening determined had been caused by runoff of poultry manure. "Cut me a break!" exclaimed Perdue, still rankled by what he considered "a sham."

The day before, Judith Stribling, a professor of biology at the University of Maryland at Salisbury and a Nanticoke River activist, had told me about the situation at the time. "Farmers embraced poultry farming because it was a way for them to stay in business. You can't support yourself anymore doing traditional corn and soybeans unless you have 5,000 acres. They also embraced chicken manure because it was better [to use on their fields] than inorganic fertilizers. But in the early '90s, farmers' fields were basically being used as waste disposal sites. They were putting this stuff out there all year round, big piles sitting in the rain for months."

Both nitrogen and phosphorus are fed to chickens to make them healthier—nitrogen being a primary component of their protein, and phosphorus being needed for bone growth. Until 1998, farmers had been told they needn't worry as much about phosphorus. However, huge buildups of phosphorus can travel laterally off a field, rainfall carrying it into creeks and rivers. After the *Pfiesteria* outbreak, the new nutrient management rules decreed that if a farmer's soil tested too high in initial phosphorus content, he couldn't add any poultry litter to it.

This first motivated Perdue to seek to provide alternatives. He had an enzyme called phytase added to the company's feed mills, which cuts the chicken's phosphorus output in half. He considered burning the manure to produce energy, a standard practice in Great Britain for years. But Perdue felt that "nitrogen going into the air was just going to create more pollution, and what are you going to do with the left-over ash, which is high in phosphorus? We decided the safest route was to recycle the litter into organic fertilizer." (Some critics contend Perdue fought a British company's proposal that would have burned 50 percent of the watershed's chicken manure, mainly because he didn't want competition with his recycling effort. Perdue did admit to the *Washington Post* that starting the recycling operation was also a way to circumvent his company's having to accept any liability for the manure generated on its poultry farms.)

We crossed the Nanticoke River, heading west toward the little town of Hurlock—"this is classic Eastern Shore farming country, best grain crop in years," Perdue said proudly—and then to a poultry farm known as Beulah Winds. This is one of Perdue's big contract growers, four chicken houses that raise about 600,000 birds a year. Parking alongside a 500-foot-long, 50-foot-wide aluminum shed, we walked to a rear entryway. Until one o'clock the previous morning, this shed had housed some 25,000 birds. Now they were all en route to a Perdue processing plant. Just inside, a pair of bulldozers were shoveling manure onto a conveyor belt, which can load twenty-four tons in about ten minutes onto a waiting sealed truck. Perdue's company takes the litter off growers' hands for free. Delaware and Maryland together pay half the transportation costs.

We set out again behind the truck, heading another twenty-or-so miles east into Delaware. Just outside the little community of Blades stands Perdue's three-year-old AgriRecycle Micronutrient Facility, the planet's largest manufacturer of pelletized organic fertilizer. Its own railroad line crisscrosses the perimeter, since about half of the raw fertilizer is transported to the Midwest. The same nutrients vexing the Eastern Shore are in short supply out there, where farmers need to replace the nitrogen and phosphorus that their crops absorb from the soil.

Inside the cavernous plant—two and a half football fields long—Perdue and I stood on a catwalk thirty-five feet above a concrete floor and its interlocking series of conveyor belts, pipes, and holding tanks. The manure is processed here into a commercial-grade soil additive for parks, highways, and landscapers. A negative-air system keeps dust and odor from escaping into the environment.

After our tour, AgriRecycle president Tom Ferguson waited to greet us in his nearby office. He's a tall, mustached, garrulous fellow who's been in the business of buying and selling manure for more than twenty years. "You know, manure was the first fertilizer, and it's still the best," he said soon after our introduction.

"That's a great bumper sticker," Perdue enthused.

"Then somebody built a fence and somebody else learned how to concentrate and feed chickens and cattle," Ferguson went on, "and first thing you know you've got a little too much of a good thing in one place."

He leaned back in his chair and offered a brief history: how pelletizing and marketing chicken litter began in 1956 in South Perth, Australia, after the country passed a law that banned chemical fertilizers on public walkways. After the Midwestern farmers, Perdue's next biggest market is golf courses, many of which are built on streams or ponds and are now under nutrient management programs that require less use of chemical fertilizer. "Golf courses had used sewage sludge for years, with lots of problems," Ferguson continued, but Perdue's organic variety offers a new alternative for your favorite putting green.

The Perdue pellets are also nourishing the grass surrounding the Statue of Liberty and being sold in lawn-and-garden centers under brand names like Cockadoodle Doo and MicroStart60. Delaware governor Ruth Ann Minner uses the product in the orchids on her night tables. A New York horticulturist percolates it into a manure tea for his grand-champion African violets. Some are brokered to mushroom growers in Pennsylvania.

So far, Perdue AgriRecycle is still operating in the red. It's turned a profit in only two months of its three-year existence, and at 60,000 tons a year being processed, is currently at only about two-thirds of the plant's capacity. Along the Delmarva Peninsula alone, Ferguson estimates that about 600,000 tons of manure are generated. Assuming at least half of that gets used in a traditional manner to grow properly managed crops, it still leaves an excess of between 200,000 and 300,000 tons of litter in need of a home, lest it potentially run off and pollute local waterways.

Concerns exist, however, about arsenic and other heavy metals found in poultry manure. Arsenic goes into poultry feed to stimulate the chickens' weight gain, to help them process food with greater efficiency, even to make their meat look more attractive. Researchers in Alabama detected a buildup of toxic metals in the soil when manure is repeatedly applied to the land. Since arsenic can pose a human cancer risk, new studies by Johns Hopkins University and the U.S. Geological Survey are under way to determine what ultimately happens to arsenic in manure—including Perdue's pelletized variety.

The Chesapeake Bay Foundation's July 2004 report called for Maryland to reinstate an Animal Waste Technology Fund, committing $5 million to initiate competition among private enterprises "for the most cost-effective, environmentally friendly alternative uses of manure." The CBF has proposed that consumers pay a penny-a-pound user fee on meat, similar to Maryland's "flush tax," that could be used to fund solutions. And Perdue is working with the University of Delaware on developing buffer zones for farmers' fields near bodies of water, planting Leland cypress and other types of trees that are good at taking up nitrogen. Cover crops, planted in winter months, can absorb nutrients and prevent them from leaching into the water table during the nongrowing season. All these farsighted ideas, if they come to fruition, could go a long way toward improving the habitat for striped bass and their piscine brethren.

Ultimately, there is the overarching issue of land-use development. The American Farmland Trust ranks the Eastern Shore area as America's ninth most threatened by development pressures. "Development is eating up farmland," Perdue said simply. We had left AgriRecycle, and soon we were easing down a rural road beside acres of corn and soybeans. "As long as farming is done the proper way, it is environmentally I think the best use of the land," Perdue said. "And don't you think this is aesthetically more pleasing than development after development?"

Spoken like a true country boy. Which, in a sense, he still is. I came away from our morning together believing that Perdue does indeed want to make a difference—provided, of course, that doesn't mean he gets swallowed up by Tyson's. In a telling remark over our breakfast, the chicken man had said: "We've continued to reduce costs and be more efficient, but you've gotta get bigger to do that. That's one of the unfortunate things about capitalism, I think. Because capitalism breeds efficiency, and efficiency oftentimes has to do with size."

Size, in one form or another, is what's overwhelming the Chesapeake Bay region. Things are at such a pass that, in July 2004, the Maryland Watermen's Association sent letters to all its members, asking their opinions about whether to file a class-action suit against the bay's worst polluters. Association president Larry Simns said they wouldn't go after farmers, but that doesn't necessarily rule out marketers like Perdue, although it would prove difficult to show how much of the pollution going into bay waters is his company's responsibility— as long as the onus of liability for the manure continues to be with the poultry farms themselves.

Lawsuits may be what it will take to make a real difference. As charter-boat captain Jim Price told Robert H. Boyle, in concluding an *Audubon* article about the Chesapeake: "We can have a good, healthy bay. But the problem is that the bay is being mismanaged by politics and not managed by science. And sooner or later that's going to cause disaster."

For the striped bass, whether they're swimming through polluted runoff from houses, industries, or farms, it's a soup they can hardly be impervious to. Nor is their race to stave off the impacts of humankind confined to the Atlantic coast. In another bay region some 2,800 miles across the country, once again stripers are a sentinel of ecosystem degradation—and their future is very much at stake in a contest between powerful economic interests and conservationists.

Chapter Fourteen

California Stripers

It was New York writer–activist Robert H. Boyle who first alerted me that the striped bass's mystique—and the fish's uncanny ability to show up in human politics—was not confined to the Atlantic coast. Just as the Hudson River stripers had interceded to block the Storm King Mountain plant and the Westway Project, their transplanted brethren were making quite a splash out west, Boyle told me over a drink at his 133-acre farm near Cooperstown, New York. The issue in California wasn't overfishing, but rather the diversion of water out of the striper's estuary in the San Francisco Bay–Delta—pumping it south, along with millions of fish eggs and larvae, to agribusiness enterprises and municipalities.

Striped bass first swam into the consciousness of California in 1879. Centuries earlier, Spanish explorers had described the enormous Northern California estuary as a "great inland lake that stretched farther than the eye could see, abounding with game, fish, and fowl of all kinds." Nourished by runoff from the Sierra Nevada mountain range and northeastern California, California's two largest river systems—the

Sacramento and the San Joaquin—merged with the salt water at the northern arm of San Francisco Bay to create the biggest estuary on the western coast of the Americas.

By the late 1870s, local salmon populations had been pretty well decimated by a combination of hydraulic gold-mining and overfishing. Hoping to compensate for their loss, S. R. Throckmorton, chairman of the California Fish Commission, put in a request to import new species to Livingston Stone, secretary of the American Fish Culturists' Association in Washington, D.C. Originally from New Jersey, Throckmorton knew a good location to scoop up striped bass, in the Navesink River not far from the mouth of the Hudson. About 135 of the species were netted for him, mostly juveniles not more than three inches long. The fish were placed in tanks that, according to a report, were "filled with water, half from the river, high tide, and half from a spring with sea salt added." Then the stripers journeyed west in a baggage car along the recently completed transcontinental railroad and were finally released into California's Carquinez Strait.

Eleven months after that first planting, Throckmorton received "a very handsome striped bass taken in this harbor [San Francisco], measuring 12^1/$_2$ inches in length and weighing one pound." The commissioner had preserved in alcohol the first shad successfully transported to California a few years earlier. Not so with the striper, it seems. His report described its flavor as being "fully up to the best specimens of the fish of the east." With the Fish Commission concerned that such a small number of immigrants might not be enough to establish the species, a second introduction of about 300 more striped bass from the same New Jersey estuary took place in the lower Suisun Bay in 1882.

Most of the native anadromous species that spawned in the Sacramento River, including the salmon, deposited their eggs on the bottom. These were easily smothered by the vast loads of sediment washing down from the gold-mining operations in the Sierra Nevada. Striped bass, however, had semibuoyant embryos and pelagic larvae that floated right on through unscathed. The bass also seemed to have

abundant prey available at all life stages, with few rivals as quick or as voracious.

They proliferated. Only two decades after their arrival, the commercial striper catch alone was averaging well over a million pounds a year. This became, as Boyle put it, "the greatest fishing anyone ever had on the doorstep of a major municipality." From April through mid-June, like their East Coast counterparts, the stripers spawned in fresh water—about two-thirds in the upper Sacramento River, the rest in the San Joaquin River delta—then moved into brackish and salt water for the summer and fall. The Pacific Ocean was so cold, most stripers didn't venture out of San Francisco Bay. During the late autumn and winter, many moved upstream again to fresh water in the Delta and the lower Sacramento. A few made their way elsewhere. West Coast striped bass have been found in the ocean all the way from southern British Columbia to twenty-five miles south of the Mexican border in Baja California. A small resident population came to establish itself off Coos Bay, Oregon.

Recreational anglers soon demanded stronger management to protect the popular bass from overfishing, and in 1935 the California legislature outlawed commercial fishing for stripers altogether. Boyle, in a 1995 article for *Outdoor Life*, reported quite a scoop of an interview about that. "My father used to fish for striped bass with gillnets," Joe DiMaggio told him. "Then the sport fishermen came along and took it [the netting] up with Fish and Game." Giuseppe DiMaggio, a fixture along San Francisco's Fisherman's Wharf, was out of the bass business. "But what the hell did I know then? I was a snot-nosed kid playing baseball," son Joe recalled.

A sportsfishing boom lasted for decades. Then things suddenly began to change for the stripers and the sportsmen who fished for them. The catalyst was a severe drought in 1976–77, which resulted in vast amounts of fresh water being diverted from the Sacramento and San Joaquin rivers to huge pumping facilities of the State Water Project and federal Central Valley Project in the south Delta. From there, the water continued flowing south through an elaborate network of dams and pumps to the big cotton growers and other mega-farms in California's dry Central Valley. (Water from the Bay-Delta system irrigates about one-quarter of the state's 29 million acres of farmland. Most of that water is delivered at cheap prices, if not for free, under contracts between the governments and the growers.) Water from these rivers was also supplied via the California Aqueduct to municipalities to the south, causing many Northern Californians to accuse Southern California of stealing their water.

The trouble was, the volume of these diversions changed the hydraulics of the striper's estuary dramatically. In 1972, scientists monitoring the population of young stripers for the California Department of Fish and Game had published a landmark study relating the index of summer striper abundance to the late spring and early summer freshwater flow through their larval habitat in the Bay-Delta estuary. Yet that flow was being steadily altered. In average years, about half of the water destined for the Bay-Delta was diverted from its natural course; in dry years, more than 60 percent. Annual water exports doubled from two million acre-feet in the 1960s to four million acre-feet in the 1970s. As Boyle wrote: "The combined suction of the two pumping stations proved strong enough to reverse flows in the delta and turn the ebb tide into a flood tide. The massive disruption confused fish attempting to move upstream to spawn and, more directly, annually killed or entrained (sucked in) an estimated 200 million young fish."

California's striped bass population began to plummet. During the early 1960s, it had been estimated at between 2.3 and 3 million adult fish. By 1993, that number had fallen to some 574,000. Between 1959 and 1976, the average young-of-the-year index of juvenile stripers was 66.6. Since 1996 it has averaged less than four juveniles per summertime haul in all but one year. This precipitous drop in both adult abundance and juvenile recruitment coincided with water exports that increased to six million acre-feet in the 1980s.

I'd rarely seen anything as unexpected as the California Delta country. Turning south at the outskirts of Sacramento on a mid-May afternoon in 2004, I soon found my car about the only one cruising along an amazingly remote thousand miles of winding waterways. Fed by five major rivers including the Sacramento and San Joaquin, this unique freshwater system consists mainly of hundreds of fertile agricultural islands—sitting below sea level, protected by levees, often connected only by ferryboat.

It's easy to get lost navigating all the sloughs, cuts, and canals. That's just what happened as I tried to make my way to Rio Vista, where the main drag was said to feature a neon sign aiming visitors toward the

Striper Café. Alternatively, I thought I might find Steamboat Slough, a palm-treed resort reputed to be set up like a Hawaiian club. Instead, when I stopped at dusk to seek my bearings, I was told I wasn't far from Isleton, population 830. Across a drawbridge I could find a room there for the night, along with "probably the best prime rib in California." And, I was told, perhaps run into some of the state's estimated 300,000 recreational striper fishermen, many of whom fish the Delta waters.

Entering Isleton, I felt like I'd been plunked down somewhere in the deep South, not less than a two-hour drive from Silicon Valley. Working my way inside the Hotel del Rio past what looked like a Harley bikers' convention, I found myself bellied up to the bar along-side Gary and Omer ("without the H"). "First thing you gotta realize is, we're on Delta time here," Gary told me. What does that mean? I asked. "Well, it takes us two hours to watch *60 Minutes*," Gary said. "And about twenty minutes to cook Minute Rice," Omer added.

It turned out they'd both come in from a day of striper fishing. From across the bar, a young fellow in a Timberland T-shirt asked: "Fish today, Omer? Catch anything?" After a pause, Omer said only, "I fished," then looked at me and added, "We call 'em CPR fish—catch, photograph, and release." In California, an eighteen-inch minimum size limit was in place, but some sports fishermen apparently let all sizes of fish go. The bass population simply wasn't what it used to be. Gary and Omer still had their tackle shop across the street (the Master Baiter), and they still participated in bass tournaments, but it had gen-erally been a lousy year. The next morning, over breakfast at Elmer's, I heard a similar tale from other locals: there were no stripers around now. They didn't know why. Things had been going downhill for years.

"It was interesting that the striped bass decline happened at pretty much the same time on both coasts," recalls Lee Miller, a biologist who recently retired after more than thirty years with the California Department of Fish and Game. "Back East they fished 'em before they could spawn; out here we 'fished' 'em through the pumps before they could even grow up," he commented when I visited him at his farm south of Sacramento. He remembered that saltwater intrusion during the

1977 drought had made the Delta water quality so poor that the pumps had to be shut down. "When it rained around mid-December and they turned them back on, they sucked up probably a million young-of-the-year bass that had accumulated there. Never seen anything like it."

Unseen are hundreds of millions more eggs and larvae that typically vanish into the big pumps when they're running in April and May. Either the newborn bass are chopped up going through the system or, if they do make it as far as the irrigation canals or reservoirs on the other side, they are locked away from the Delta, never to return. Even fish that are deflected off a sometimes successful screening system, and manage to survive in the vicinity of the pumps, can't make it back into the bay system because the force of the flows is too great.

Tom Cannon, an estuarine fisheries ecologist, had been technical director for a 1970s Hudson River study that determined that a substantial impact on the striper population was caused by eggs and larvae that were being drawn into power plant cooling systems like Indian Point's and then pumped back out (a process called entrainment). Since 1977, he has worked closely with California Fish and Game officials on similar problems in the striper's West Coast habitat. It was clear that Pacific Gas & Electric's power plant was killing at least 10 percent of each new year-class of stripers. The fish population's loss to water diversions, however, was "far greater than anything I had seen on the East Coast," Cannon wrote. Another 20 to 40 percent was taken by the south Delta pumping stations and, through hundreds more smaller diversions, Delta agriculture was taking 10 to 20 percent more. "That was a lot [of hits] for any population to sustain, even for the prolific striped bass that each spawned five to ten times over as many years, with two million to four million eggs each time."

Nor is falling victim to these snares the only effect of the water exports. In low-flow years, when water gets either drawn toward the pumps or diverted to irrigate the abundant fields of rice and corn in the Delta itself, there is less dilution of toxic runoff. Industrial chemicals flow into the waterways from big oil refineries around San Francisco Bay. Pesticides pour down numerous agricultural drains, onto farmland

and into holding ponds, and eventually re-enter the striper's river habitats. During the most recent extended drought, which ended in 1991, it was found that the preferred first food for larval striped bass (the copepod *Eurytemora affinis*) suffered a drastic decline in abundance and was largely replaced by less energetically desirable species.

In his book *Inland Fishes of California*, Peter Moyle, a renowned professor of fish biology at the University of California at Davis, described another factor likely influencing the size and strength of the striper population. "The San Francisco Estuary has been labeled the 'most invaded estuary in the world' because of the hundreds of species of alien invertebrates, plants, and fish that have become established in the past 150 years," Moyle wrote. "Indeed, the rate of invasion, mainly by species carried in the ballast water of ships, has increased during the period of sharp striped bass decline. Some of the invaders have significantly affected the food supply of larval and juvenile bass." Extraordinary numbers of an Asian clam that arrived in the 1980s, for example, severely reduced the densities of plankton customarily eaten by these young stripers. The tiny opossum shrimp, once the food of choice for somewhat older bass, are a plankton feeder whose population has collapsed since the Asian clam carpeted the bottom of the striper's spawning grounds in Suisun Bay.

With such ecosystem assaults plaguing the region's most popular gamefish, it's not surprising that the annual young-of-the-year index of striped bass was long a thorn in the side of the water interests. "The information we developed about striped bass was instrumental in decisions by the State Water Resources Control Board to limit the manner in which the state and federal water projects could operate," recalls Dave Kohlhurst, another retired Fish and Game biologist. "In the early 1980s, water quality and flow standards were basically set to try to protect striped bass."

While water export limitations were agreed upon between Fish and Game and the state's Department of Water Resources in 1982, this was not to hold true for long. First, that November came the culmination of a fight over the proposed Peripheral Canal. This project, designed to carry Sacramento River water around the Delta to the pumping stations, was expected to benefit the striper population because fewer would be caught up by huge water flows. As a ballot initiative, it would also have written extra protections for the Delta area into California's constitution. Agribusiness giants, especially the cotton empire of a Central Valley man named J. G. Boswell, didn't like it. According to a 2004 biography of Boswell (*The King of California*), agribusiness executives were afraid the bill would reduce rather than increase the amount of water they received: "Taking care of wildlife, not farmers, seemed to them the legislation's primary aim." Oddly enough, the Boswell interests forged an alliance with part of the environmental community, whose concerns were that the Peripheral Canal would deliver too much water south of the Delta. What another author, Marc Reisner, called "the oddest alliance since the Hitler-Stalin Pact" managed to defeat the ballot initiative and stop the canal.

In 1986, an agreement was signed by the directors of California's Fish and Game and Water Resources departments, committing them to find ways of offsetting adverse fishery impacts caused by water diversions. Soon after that, however, stripers went from being the poster child in California's water wars—"the canary in the coal mine," as one fisheries scientist puts it—to playing a definite second fiddle. In the early 1990s, as biologist Kohlhorst observes, "There was an active attempt on the part of the water users and, to some extent, the water agencies to make striped bass seem less important or potentially even detrimental to the ecosystem." In fact, they began to suggest that stripers were responsible for the decline of other fish species.

Over recent decades, fish with similar life histories adversely affected by water operations have also seen their abundance fall drastically. During the 1990s, the Chinook salmon and the delta smelt were added to the federal endangered species list. As an "introduced" species, striped bass aren't provided with the same protections that native fish receive under California law or the federal Endangered Species Act.

Suddenly, then, the conservation emphasis shifted. Indeed, when it came to salmon, stripers were now considered by some fisheries managers as the villain, even if it was the region's water diversions that put the salmon much more at risk.

In certain circumstances, particularly in closed systems, stripers can exert predatory pressure. The water export system has caused some baby salmon moving downstream to pass into a small reservoir called Clifton Forebay, into which water flows en route to the pumping stations. Here, striped bass wait in what is basically a giant swimming pool. It's been estimated that between 63 and 99 percent of these particular juvenile salmon are consumed by striped bass.

Outside of situations such as those of Clifton Forebay, generally speaking, stripers do not prey that frequently on little salmon. Nor do they eat many delta smelt, which don't form into schools and are not easy for the bass to find. As Peter Moyle comments, "Besides, delta smelt [a vegetarian] probably taste like cucumbers, and what self-respecting striped bass would eat a vegetable?" They are simply opportunistic feeders whose favorite food, as they start to grow, are threadfin shad that coexist on the spawning runs and are carnivores full of tasty fish oils.

The view of striper as villain also suddenly called into question a program to raise stripers in hatcheries, in hopes of expanding their dwindling population. The program moneys came from a special Striped Bass Stamp purchased from Fish and Game by sports fishermen for $3 a year. Over a period of ten years, private aquaculturists contracted by the state had produced 11 million fingerlings and yearlings, which tagging data suggested came to comprise as much as 30 percent of California's bass population by 1990. Then, in 1992, a new Fish and Game director, Boyd Gibbons, canceled the program. In his view, if even a single winter-run Chinook salmon were eaten by a hatchery-reared bass, this would put the hatchery in violation of the endangered species laws. Planting of striped bass resumed in 1993, following Gibbons's abrupt departure from the agency, but, Dave Kohlhurst recalls, "we had to mitigate for their impacts on listed [endangered] species." The permits could be issued only so long as the bass population stayed below a certain level. In 2001, after this limit was exceeded, the program was halted.

Discontinuance of the program was a sore point for Fish and Game biologists like Kohlhurst. While there were divergent opinions within the department, those who'd worked actively for years on striped bass science and management thought stopping the hatchery program was "nonsense," says Kohlhurst. "Striped bass, salmon, and delta smelt had coexisted together since 1879, and the declines of those three species had occurred simultaneously. They all did fine under good conditions in the estuary, and poorly under bad conditions. Well, those who disagreed with us finally admitted that the bass didn't cause the other fish populations to decline—but they still claimed the bass were 'impeding their recovery.'"

Fishermen with the California Striped Bass Association, started in 1974 and now with several thousand members in nine chapters, were similarly disturbed when the Striped Bass Stamp was replaced with a Bay-Delta Enhancement Stamp. Recreational anglers now had to pay an extra $2 a year into state coffers, with the funds no longer being dedicated to striped bass. Meanwhile, state officials were discussing raising the minimum size limit, imposing a maximum size limit, or placing seasonal restrictions on fishermen. Jack Chapman, a member of the association's Sacramento chapter, says: "All those approaches are a band-aid to try to slow down the decline of the fish, as opposed to fixing the real problem—which is water diversion."

Further muddying the waters was a new hypothesis about the rise and fall of striped bass populations, put forward by research ecologist William (Bill) Bennett. He seemed eager to explain his notion the morning we met at U.C. Davis. First off, Bennett said, it needed to be understood that he was originally from Massachusetts and had grown up observing the Atlantic striper migration. "So something always bothered me about the striped bass issue here in California," he continued. "The fish were never thought to leave San Francisco Bay, or just peek out around the corner and come back. Of course, back East most all our fish migrate from the Chesapeake or the Hudson. Why should it be so different out here?"

Then Bennett began reading about climate change. Because the Pacific waters around Northern California had entered a warming phase in 1976, coinciding with the striper's decline, Bennett hypothesized that many stripers had left the bay for the open ocean and couldn't find their way back again. The food supply along the coast was more abundant, but the species also became more vulnerable to sports fishermen. When weather patterns shifted back to a colder regime in 1999, Bennett's modeling work indicated, more adult striped bass began appearing in the bay again. What had appeared to be a large population decline due to the water exports, in his view wasn't that at all.

Bennett's mentor, Peter Moyle, supports his younger colleague's theory. "This idea that what happens to striped bass outside of the estuary is important was just in nobody's thinking before," he says. "For Bill to even suggest that was really disturbing to the basic mind-set," one that blamed the pumps for the striper's troubles. Bennett's theory has indeed sat poorly with many experts. Some have questioned the fact that initial funding for his study came from the U.S. Bureau of Reclamation, the federal agency overseeing the water projects, and that urban water agencies in California have provided him with subsequent funding.

Biologist Dave Kohlhurst is particularly critical of the notion that older striped bass can't find their way back from the ocean. "That would be pretty damned maladapted, if you think about it, for a species not to be able to home in on its spawning grounds again. Bennett will tell you there are consistent catches of striped bass all up and down the California coast. I was never able to document those. We don't get tag returns from those locations."

Recent research conducted by U.C. Davis marine pathobiologist David Ostrach also casts doubt on the Bennett theory. Ostrach's analysis of the otoliths, or ear bones, of thirty-one female striped bass has been able to determine the salinity the fish live in during different life stages. His analysis indicates that *none* of the fish he studied were going out into the ocean and not returning for several years.

The way Kohlhurst's colleague Lee Miller sees it, the whole situation adds up to this: "The water export people have gotten off the hook, because striped bass have become irrelevant. I don't understand this, but you've got a lot of people running around espousing this purist idea that striped bass were introduced out here, so who cares if they go away? Well, the ecosystem is so changed in so many ways, it's beyond ever getting back to anything pristine! So what in the world are they talking about?"

Who are the people only too pleased to see the striped bass become politically incorrect in the Bay-Delta? There are, of course, many legitimate champions of bringing back the endangered Pacific salmon, but powerful interests on other fronts are at work as well. To learn about the land they came from, I headed south into the arid Central Valley west of Fresno, the beneficiary of much of the water exports. Robert H. Boyle had put me in touch with Lloyd Carter, president of the California Save Our Streams Council, a crusading journalist turned lawyer, and we'd arranged to meet on a Sunday afternoon in the heart of J. G. Boswell's cotton country. In his mid-fifties, Carter is now a prosecutor and appellate attorney for the California Attorney General's Office in Fresno. He's also the water law instructor at San Joaquin College of Law. The use of taxpayer money in effect to subsidize the vast water needs of agribusiness and the control of water by what he calls "the hydraulic brotherhood" continues to outrage him.

Carter had started out as a newsman with UPI, then moved to the *Fresno Bee*. "Nobody wanted to cover the farm beat," he recalled. "Who wants to do a story about cotton futures? For the first ten or fifteen years of my career, I bought hook, line, and sinker whatever the hydraulic brotherhood put out, just unquestioning regurgitation of the press releases." Eventually, it was the tens of thousands of migratory birds dying from botulism and avian cholera outbreaks that, in Carter's phrase, "lifted the scales" from his eyes. A drainage plan that dumped toxic agricultural wastewater into the Kesterson National Wildlife Refuge had been the culprit. "They referred to this as 'agricultural return flow,' or an even more benign euphemism, 'summer water,'" Carter said, shaking his head. "This valley 150 years ago was an absolute paradise—sturgeons weighing hundreds of pounds, ducks and geese literally by the millions that sounded like thunder when they took off," he added wistfully.

Carter had begun delving into "the dark side of the 'green revolution,'" becoming a prize-winning journalist for his exposés about the selenium poisoning of Kesterson Refuge. Eventually he became too controversial for the editors in Fresno and headed off to law school. His last major published diatribe was a piece for *Penthouse* in 1999 zeroing in on "the uncrowned king of Agropolis," J. G. Boswell II, "a reclusive, unassuming man who calls himself a simple cowboy. In fact he grows more cotton than any other individual in the world. No one knows how rich he is."

Boswell's empire was, as former Supreme Court Justice William O. Douglas once put it, a "corporate kingdom undreamed of by those who wrote our Constitution." A man who helped launch the careers of three California governors (including Ronald Reagan), Boswell and "a handful of other big growers," Carter wrote, "got the U.S. Army Corps of Engineers (funded, of course, by the American taxpayer) to build four 'flood control' dams on rivers flowing out of the Sierra Nevada so Boswell *et al.* could safely farm the bottom of what was once the biggest body of water west of the Mississippi River, the legendary Tulare Lake. Boswell now controls rights to public water that could well be worth nearly $1 billion. His property in California alone is estimated at 225,000 acres."

For years, ever since bottling up those rivers and routing their flow into a thousand miles of ditch bank, lake farmers like Boswell had been watching stripers and other fish die, victims of algal blooms and too little oxygen, as Mark Arax and Rick Wartzman write in *The King of*

California. In the late summer of 1997, along a twenty-five-mile stretch in Boswell's cotton kingdom called the Homeland Canal, a defoliant sprayed nearby resulted in a fish kill estimated at ten million threadfin shad, along with the death of thousands of birds. The investigation, an anonymous Fish and Game staffer had told the two authors, got "buried by scientific incompetence and politics."

Agribusiness has been estimated to consume more than 80 percent of the surface water used in California, and Carter has come to call cotton "the Great Satan of crops." The cotton industry is, Carter continued, "by far the state's biggest water consumer—three million acre-feet a year, enough water to meet the domestic needs of thirty million people! It's an extraordinary waste of a precious resource, in a desert, to grow a crop that consistently loses money. It costs seventy cents a pound to grow this stuff in a world market that has fluctuated, in most years, between thirty-five and fifty cents a pound. But because it's all subsidized by the taxpayers, American cotton farmers can flood the global market and control it. Think about the impact in West Africa, where about ten million people depend on cotton for a subsistence existence but can't compete, which is literally causing starvation, misery, and poverty."

California's Central Valley region is paying a different sort of price. Cancer is epidemic among the Hispanic laborers who spray the fields with pesticides ("we use more poisons to drench the landscape than anywhere else in the world," Carter says). Nearly one-fourth of the populace in Kings County lives below the poverty line. The valley has also become the number one source of methamphetamine in the United States, with Mexican drug gangs running the illicit rural laboratories. About 20,000 trucks a day emit diesel fumes along Highway 99, as do 15,000 more that travel Interstate 5. Surrounded by mountains, the valley now has the worst air quality in America; the number one cause of absenteeism in Fresno's public schools is asthma.

J. G. Boswell II, I was told, had retired at eighty-two to an estate in Idaho. But Carter suggested that I at least make the fifty-mile drive south from Fresno to where the Boswell empire began, in the town of Corcoran. "You need to get a feel," Carter said, "for the desolation out there."

Deep into a desert where millions, if not billions, of egg, larvae, and juvenile striped bass have been pumped down the state and federal canals without any accounting, a big sign appears on the outskirts of Corcoran along two-lane Highway 43: "Farming Capital of California." The town is also home to the maximum-security Corcoran State Prison, whose inmates include Charles Manson and Sirhan Sirhan. Main Street is not easy to find and doesn't last for long. Gone are the many stores of Corcoran's boomtown era, replaced by a few fast-food restaurants and merchandising outlets. About 60 percent of the town's 15,000-some residents are Hispanic, and the jobless rate here has hovered around 16 percent for over a decade.

Below Corcoran, 150,000 of the acres controlled by the biggest farmer in America stretch for fifty miles. You can travel for half a day here, as Boswell's biographers have noted, and never leave his farm. Lines from an old song, "Summertime," come to mind: "Fish are jumpin', and the cotton is high." But I see no evidence of either on this Sunday afternoon, and shortly, thinking more of the Dylan title "Desolation Row," I've lost my desire to see what remains of the Tulare lakebed. I've reached the ecological dead end for stripers, and the only thing to do is turn around and seek out those fighting for the fish in California's striper wars.

It's a relief to be back on the water. Antioch, forty miles as the crow flies below the state capital of Sacramento, is called the gateway to the Delta—the juncture where the San Joaquin and Sacramento rivers meet. At 7 a.m., I meet four biologists from California's Department of Fish and Game at the Antioch marina, where their forty-three-foot boat *New Alosa* is docked. They'll be conducting a spring survey: setting and retrieving drifting gillnets, then inserting small metal ID tags into adult striped bass before releasing them again in hopes that anglers, when they catch one of these fish, will provide the relevant data to the department—similar to the Atlantic coast "mark-recapture"

studies, as they're called, that help estimate the current striper population size.

The wind is kicking up—to my surprise, there are even whitecaps on the water—as we pass under the Antioch Bridge. Nina Kogut, chief biologist aboard, tells me that we'll cut through Broad Slough into where the lower Sacramento juts into Suisun Bay before putting the net out. "We've not seen as many striped bass this year as in 2000," Nina informs me, "but then again, they're bigger." According to Mike Silva, a bearded fellow who's been laying out the 100-fathom-long gillnet on these surveys since 1984, it's likely that the stripers either spawned or started moving earlier due to record-breakingly hot spring weather.

We motor past a PG&E power plant, a sheet metal facility, a valve manufacturer where a Weyerhaeuser lumber company used to be. Mike points out the Mount Diablo range, veiled in clouds today, on whose other side is the northern edge of San Francisco. "From that vantage point, before there was pollution," he says, "supposedly you could see both the rivers, up into the Sierras, all the way to the Farallon Islands."

The boat has only a five-foot draft but, when it's blowing like this, Ken Flowers at the helm needs to be extra cautious to avoid sandbars and other shallow spots. Closing in on a marshy shoreline thick with bulrushes and cattails, we anchor and the crew starts spooling out the gillnet from a hydraulic winch to a depth of twenty-one feet. When the net is hauled back in, it contains a lone shad. "Well, sometimes you make a few skunk drifts before you start catchin' something," Mike says.

As the morning progresses, things do pick up—a little. Nina has a set of disk-dangler tags, found to be the best for long-term use, in a tray on the deck rail. As a striper gets plucked out of the nets, maybe a twelve-incher, Ken or Mike—both wearing gloves and yellow slickers —quickly hand the fish forward. If they're coming fast, Nina and her assistant, Tricia Lippe, keep the waiting ones in a small tank. Nina carefully aims the tiny needle—"you just poke 'em a little bit, feed the wires in through the back side below the upper gill, and pull 'em out," she instructs—using a metal spacer as a substrate to wind the tag on and scrape off a few scales. The scales will go to a lab technician, who will use them to determine the fish's age. Nina gives a slight push with her thumb to see if any male milt gets released, then returns the striper overboard. All this is accomplished in just over thirty seconds.

But there aren't many stripers today: all males predominantly less than twenty inches long, between three and five years old. "No sea lions around here," Mike says. "What does that tell you? No fish." Indeed, a burgeoning population of sea lions often wreaks havoc on the surveys. The fabric of the nets is designed to be lightweight, because the scientists want any fish snared in them to remain alive. But when sea lions are around, even their playing with a net can damage it. "Seems like they're always on us now, no matter where we go," says Nina. "It starts with maybe one in the morning, but by the end of the day we'll have five hanging onto the net and ripping holes in it, taking the fish out."

The day ends around 1 p.m. with 100-plus fish, of different varieties, taken in six sets. A lone white pelican passes above us. We're on our way back when Nina says: "People talk about how fresh water used to go all the way down to the C&H sugar plant in San Pablo Bay. Now you see all this water diverted, and slowly the saltwater intrusion is coming in farther; you just don't have the flow you used to. Even though it was only seasonal, at least it flushed things out. You didn't get the festering that you do now."

That "festering" is likely taking a toll on newborn striped bass in their spawning areas, according to new studies by U.C. Davis's David Ostrach. The eggs from twenty-one female striped bass were removed prior to fertilization, then analyzed in order to obtain an accurate picture of what chemicals might already have bioaccumulated in the adult fish and be passing on to their offspring. In every sample were found "biologically significant levels" of pesticides, PCBs, and the first reported finding in the estuary of polybrominated diphenyl ethers (PDBE) contamination and its transfer to a new generation. PDBEs are chemicals found in flame retardants used in house paints and other products.

A known hormone and endocrine disrupter, they are now as ubiquitous in the striper's environment as PCBs.

Apparently due to some or all of these pollutants, Ostrach also detected abnormal larval development in every striped bass examined from the Delta river systems over the three-year study. At five days, the progeny had very little or no remaining yolk material—similar to what Massachusetts lure maker Bob Pond had observed in Chesapeake stripers in the 1970s. This meant they were lacking any stored energy reserves to help them as they went out to seek their first real food. In addition, liver growth accelerated during the first three days of life, but then the organ shrank in size and became much less functional by day five. A healthy liver should be full of glycogen, another form of stored energy, but these were not. The combination of these two problems raised critical questions as to whether similarly impaired striped bass could survive beyond the early larval stage.

Already, "because of elevated levels of mercury, PCBs, and other chemicals," California health officials advise adults to eat no more than two meals per month of striped bass taken in the Delta; pregnant women, nursing mothers, and children under age six are cautioned to eat no more than one such meal a month and no striped bass over twenty-seven inches, as greater amounts of chemicals bioaccumulate in larger fish. Ostrach's study may raise the level of alarm, since PDBEs, for example, were not previously known to be so ubiquitous.

"One advantage the striped bass has over us," Ostrach points out, "is that these types of compounds are fat-soluble. A female striper with, say, 100 grams of fat would burn 40 to 60 percent of it for energy, while the rest of the contaminant load for that year gets passed on to the eggs. So, technically, striped bass can wipe their bodies fairly clean of PCBs and PDBEs yearly. We humans don't have the luxury of clearing those fats every year. They continue to bioaccumulate and get passed on to a developing fetus through the mother's blood supply."

Enabling more fresh water to flush out the system would make a huge difference in the contamination levels of the estuary and its fish. The Natural Resources Defense Council (NRDC), similar to its legal efforts on behalf of fishermen in the Hudson River, has for fifteen years been representing the California Striped Bass Association (CSBA) and a number of other fishermen's organizations in a lawsuit concerning Friant Dam on the San Joaquin River. Over time Friant Dam has captured so much river water for use in agribusiness that sixty miles of the San Joaquin are now bone dry, decimating its once-prolific salmon run. In 1998, and again in 2004, federal judges ruled that the Bureau of Reclamation needs to obey U.S. laws for protecting fish and the environment—and allow more natural flow into the river. The case is now under appeal from agribusiness interests claiming that any San Joaquin restoration program won't leave them enough available water.

Whitey Rasmussen, board president of the CSBA, has been striper fishing in the Delta for fifty years. Asked to compare how things have changed, he says, "It's like walking from broad daylight into a fog where you can't see nothing. Last year, in downtown Stockton, which is at the head of the channel on the San Joaquin, it was so algae-laden no fish could live in there. But when you get good fresh water in here, you're gonna pull stripers in. Water quality is the key to it all."

In October 2004, Tom Cannon—the former Hudson River analyst, now a volunteer advisor to the CSBA—completed an extensive review of the effects of water diversions in the Bay-Delta. Cannon is highly critical of the Cal-Fed Bay-Delta Program, established in the mid-1990s as an umbrella bureaucracy, charged with both solving environmental problems and ensuring a more reliable water supply for agricultural and urban users. The consortium, in the view of Lloyd Carter and others, is "a self-perpetuating bureaucracy to do more studies, and basically stall-and-delay."

While some $500 million has been spent on ecosystem restoration in the watershed since Cal-Fed's inception, in 2001 and 2003 annual water exports also reached record annual levels of 6.3 million acre feet. According to the Cannon report, this "results in the loss of over 50 percent of larval and juvenile striped bass from the Delta. These losses directly affect recruitment of three-year-olds into the adult population and thus reduce subsequent adult spawning populations and egg production. Reductions in the spawning population directly translate into fewer young produced."

Cannon says that Cal-Fed (recently renamed the California Bay-Delta Authority) "has lost sight of why the estuary and its fishes have declined and is being driven not by scientists, but by people with the most money and political power." The latest ploy to influence the authority's decisions on managing the Delta was the so-called Napa Agreement, which arose after a convocation of water interests in Napa, California, in the summer of 2003, a meeting that excluded environmental and public-interest groups. A deal was cut that proposes to increase maximum pumping limits from the state's massive Harvey O. Banks facility near Tracy by fully 25 percent—from the current 6,680 cubic feet per second to 8,500 cubic feet per second.

Barry Nelson, formerly executive director of the Save San Francisco Bay Association and now co-director of the NRDC's Western Water Project, provides some background. "One of the breakthroughs in the Cal-Fed planning process was when Southern California said, 'We don't need more water from the Delta, we just need that water to be reliable.' We said, 'We can work with you on that.' Since then, the south-of-Delta water interests have completely abandoned that common ground. It's because of what's happened to the Colorado River, where the drought has brought severe limitations on what can be taken out of it for Southern California. So their Metropolitan Water District has abandoned the core deal of Cal-Fed. That's where the Napa Agreement came from. The Boswells of the world always wanted more water and were pretty careful never to strongly endorse Cal-Fed—until now, because they'll get additional water, too."

As Lloyd Carter wrote in an op-ed piece for his old paper, the *Fresno Bee*: "Critics claim the new plan could increase exports of Northern California water by up to one million acre-feet of water, enough water to cover a million football fields a foot deep, and devastate Delta recovery chances." Pumping all this additional water south of the Delta will mean draining reservoirs more frequently, resulting in higher river temperatures and literally no water at all in dry years.

Carter wondered how the Napa Agreement's boosters can "justify the use of precious Northern California rivers to grow surplus subsidized cotton on more than 700,000 acres of land in the western San Joaquin Valley . . . [and] so developers in Southern California can jam yet another tacky subdivision onto a landslide-prone barren hillside."

None of this bodes well for the striped bass. The Cannon report describes "a reluctance to provide further striped bass protections despite laws that require it, and a subtle if not overt attempt to sacrifice striped bass . . . [for] increased water exports." The NRDC's Barry Nelson adds: "Ten years ago, we were seeing many fish species headed toward extinction, and striped bass was one of them plunging fast. If the Napa Agreement is implemented and Interior goes forward with its new policies, we'll go back to those days."

Indeed, NOAA fisheries biologists in Sacramento concluded in August 2004 that the new pumping plan "is likely to jeopardize the continued existence of Sacramento winter-run Chinook salmon and Central Valley steelhead" as well as spring-run salmon. Internal agency documents subsequently obtained by the *Sacramento Bee* indicate that NOAA administrators in Long Beach not only overruled those biologists but revised the plan to conclude that the Delta diversions wouldn't harm the fish that much. It turned out that the Bureau of Reclamation had offered "suggestions" on the changes. According to the *Bee*, this "removed the last major obstacle" to sending more water south.

This could not come at a worse time for the striper population as well. Although the latest California Fish and Game estimate is that 1.4 million adult fish remain in the Bay-Delta system, the big old bass—the twenty- and thirty-pounders—are now extremely rare. Egg production, calculated as high as 434 billion in 1975, had fallen to 70 billion by 2000. And the young-of-the-year index of juvenile abundance in 2004 was a mere 0.8, the lowest figure in the summer tow-net survey's forty-five-year history.

Nelson summarized the situation this way: "The American public has an image of the small family farmer and a guy standing by a barn with forty acres of corn. You could not find anything more different from that than the folks who are hurting the striped bass. The area south of the Delta is probably the most industrialized piece of agriculture in the world. These are massive corporations, and we should think about them the way we do oil companies. The myth of the American farmer has been exploited by the despoilers of the Delta. And it's a big lie."

As if to bolster Nelson's argument, recent research reported by the Washington-based nonprofit Environmental Working Group indicates that Central Valley growers reap up to $416 million a year in taxpayer-funded water subsidies. In 2002, the largest 10 percent of those farms got 67 percent of the water—not the small farmers that the Central Valley Project was originally designed to benefit.

The California Striped Bass Association, however, has no intention of giving up the fight. In a letter sent in October 2004 to the state Department of Fish and Game as well as the Department of Water Resources, the CSBA's board wrote: "The decline or lack of recovery in the striped bass population does not bode well for the Bay-Delta ecosystem." It called for the two agencies to adhere to existing legal protections and prior commitments—such as the export limitations they originally agreed upon back in 1983, but implemented only selectively. It also called for new fish protection facilities and improved salvage systems at the south Delta pumping plants.

Cannon's seventy-three-page report accompanied the CSBA's letter. "I wrote it," he says, "because the powers-that-be have written off striped bass and are allowing them to be wiped out of California. Well, once we proved that power plant entrainment of larvae was having a substantial effect on the striper population in the Hudson River, some major alterations were made to these facilities to reduce those impacts. There's no reason why that model can't still be applied here."

Back on the Hudson, however, the early model that Bob Boyle, Tom Cannon, and others fought to implement is coming under siege once more. But striped bass have gained some new champions in their corner.

Chapter Fifteen

Keepers of the River

"It's a sacred fish here," he said, speaking of the Hudson River. "Stripers in many ways define the culture, the same way the salmon do on the West Coast. In the Hudson Valley, the Indians lived off of them. They were a major cultural and subsistence asset of the Algonquin and Delaware communities. Then the first Europeans to visit the valley were dumbstruck by the stripers. There's a great passage in the log of Robert Jewett, who was Henry Hudson's pilot, describing the thick schools that the prow of his ship cuts through like a comb. He calls them salmon, but we've never had salmon in the Hudson River. The Housatonic was the most southerly range for Atlantic salmon, so Jewett was looking at striped bass."

It was Memorial Day weekend in 2004. Robert F. Kennedy Jr. and I were speeding toward a boatyard in Ossining, New York, to pick up his seventeen-foot Boston Whaler before dark. "Stripers are the flagship species for the Hudson," he continued. "Their life cycle is intertwined with the biological integrity of virtually every part of the river." (The bass swim up into the headwaters of freshwater tributaries to spawn, move down into nursery grounds at the widest part of the river in Haverstraw Bay, then winter and grow in the New York Harbor area.)

Kennedy's father, the former U.S. attorney general and senator from New York who was assassinated during his 1968 campaign for the presidency, had taught him how to fish. Striped bass had been "kind of a mystical fish for me since I was a little kid. This was the icon fish, the one everybody wanted to catch. I'd be standing on the dock with a bucket full of scup, watching these fishermen getting off their boats with big striped bass. I'd watch while they cleaned them on the pier. Those guys were my heroes. They had the king."

One of his earliest memories was watching his older brother Joe arise before dawn at the family Hyannisport compound on Cape Cod to head for a jetty on nearby Squaw Island. "There's always stripers there. I remember how proud my brother was when he brought 'em home. Dinner was a big deal in my house. During the summer we had seafood almost every night—scallops and clams that we'd catch or rake ourselves. We always had lobster pots. Still do. My uncle Teddy checks the pots every day, and my brother Max keeps his own. I had a fishing rod in my hand as soon as I could walk. And there was nothing better than striped bass, a fish big enough to feed our whole family."

Darkness was setting in when we arrived at the boatyard on the Hudson. While he maneuvered the car, I hooked his thirty-year-old Whaler, the *Lickety Split*, onto the back bumper. In the morning, we'd be taking it upstate for a camping trip on Magdalen Island on the Hudson. He figured there ought to be some stripers around.

In recent years, Bobby Kennedy has emerged as one of America's most charismatic environmental activists. He continues to teach at the Pace University Environmental Litigation Clinic, which he founded in 1986. He's president of the Water Keeper Alliance—an international coalition that now numbers more than 125 grassroots groups—and a senior attorney for the Natural Resources Defense Council (NRDC). He coauthored the best-selling book *The Riverkeepers* in 1997 and, in 2004, published another best seller, *Crimes Against Nature,* a powerful study of the Bush administration's environmental policies. He writes frequent op-ed columns for many newspapers and speaks to audiences all across the country.

Through it all, the Hudson River has remained a fulcrum of his many crusades. "My father took a trip up the Hudson in 1965 and said that the pollution he saw was a disgrace to our country and we had to clean it up," he said. The next year, Robert H. Boyle mobilized the coalition that became the Hudson River Fishermen's Association, to fight for "the ecological betterment of the watershed." But it would be another eighteen years before Bobby Jr. began fishing and working with Boyle.

Of all the Kennedy kids, he'd always been the nature boy. "My mother told me that when I was in the crib, I'd spend hours watching ants," he said. "As soon as I reached mental competence, wildlife was my obsession. All I thought about was how to catch an animal." Author and longtime family friend Jack Newfield recalled: "I first met him in 1967, when I was writing a book about his father. Bobby was thirteen years old and had a menagerie in the house. It was actually the morning of his father's first big speech against the Vietnam War on the Senate floor. Young Bobby had a coatimundi that had bitten his mother, Ethel, on the leg. So Senator Kennedy had to tend to the wounds, making him late to give his speech.

"Bobby has a deep intellectual, philosophical, and emotional sensibility. He has not just a lawyer's intelligence, but an artist's—the novelist's perceptions, the poet's understanding. I think he sees things like his father did, in an unusually profound way that cuts to the core of any issue or problem."

It did not come easily. After his father's death when he was only fourteen, for a time learning falconry (still one of his passions) helped keep him focused. Kennedy had gone on to obtain a law degree. But, as he would write in *The Riverkeepers*: "Everyone, a friend has observed, is given something to overcome. My biggest battle was with addiction to drugs."

Following his arrest for heroin possession in 1983, he set out "to retrace my steps to that point where I had started off on the wrong path." This led him to volunteer work with the NRDC, which was then representing Boyle's organization as a legal counsel in the Storm King Mountain fight. "Two things made me fall in love with the [Hudson] river," Kennedy would write. One was reading Boyle's book, "which advocated an in-the-muck, waders-on environmentalism. Second, I took Boyle's advice and acquainted myself with the river first-hand." After Boyle founded the Riverkeeper group in 1984, Kennedy began working closely with John Cronin—the first appointed Hudson "keeper"—tracking down polluters and eventually filing lawsuits to hold them to account. Over the years, Riverkeeper, through their work and that of others, has won more than 300 legal actions and forced polluters to spend about $3 billion in remediation.

Although Boyle, Cronin, and he came to a parting of the ways with one another a few years ago, Kennedy continues to regard Boyle as "a real hero of the environmental movement. He basically spent his life working to round up fishermen and hunters, sportsmen, to mobilize and deploy them as a political force. Today our Water Keeper Alliance is based upon the idea that the people who use local waterways are the most important constituency for protecting those waterways. That was all Bob Boyle."

Striped bass were often at the epicenter of Bobby Kennedy's struggle to preserve the river, just as they were for Boyle. Indeed, the first time Kennedy ever cross-examined a scientist on the witness stand, stripers were the main topic, he said as we climbed into his car for our two-hour trip with his family up into the Hudson Valley.

The year was 1988, and Riverkeeper was suing to shut down the Chelsea pumping station. Constructed back in the 1940s on the Hudson about sixty miles north of Manhattan, it was designed for use in drought emergencies, capable of sucking out a hundred million gallons a day of Hudson River water, injecting it into the Delaware aqueduct, and sending it on to New York City. The city now wanted to activate the station after a severe drought had caused a substantial drop in its reservoir capacity. The trouble was, the intake pipe off Wappingers Falls was located at a point in the river that recent studies had revealed to be a locus for spawning of striped bass and some other migratory fish.

At the permit hearing, a crucial question was what would happen to the fish once they had been whisked into the aqueduct, which runs for nine miles underground before dumping into the West Branch Reservoir. The municipality's Environmental Impact Statement (EIS) recycled an argument used by power plants in the past: that the adult fish, once inside the aqueduct, would never be forced all the way to the reservoir. They'd be able to sense the impending approach of a screen, look around for an escape hatch, somehow locate a little scoop placed on the moving screen by the pumping people, be lifted up the belt, and then be carried over by jets that would send them safely back to the river.

It might sound good on paper, but Kennedy and the Fishermen's Association weren't convinced. Ian Fletcher, the expert biostatistician who'd been so instrumental in stopping the Storm King and Westway projects, came to Kennedy's house to go through the EIS with him line by line. It took three days. Fletcher was of Scottish descent and, Kennedy recollected, he liked to wear a kilt and play the bagpipes. He was also brilliant.

"The point Ian made," Kennedy said, "is that fish in the Hudson River, unless they have contact with the bottom, don't even *know* that they're moving with the tides. Not if they're in the middle of the water column. It's the concept of relative motion, like us when we're in an airplane and feel like we're not moving. Down there it's dark; the fish only have six inches of visibility. So once they'd hit that screen and turn around, they'd stay there just swimming around against the current until they became completely exhausted. Ian had observed fish behavior in a darkened flume at his lab and by dropping marked carp into one of these intakes. Those were experiments never tried by the utility scientists, and he proved that all their assumptions about fish behavior were wrong."

The Chelsea pumping station's advocates had hired their own "expert" to defend the theory, who'd signed off on sworn testimony that was little more than a cut-and-paste job taken from old power plant applications. "We had him on the stand for a couple of days," Kennedy remembered, unable to conceal a gleeful smile. "I asked him,

'Well, if the fish could sense the approaching screen, how do they do that? Do you mean like a bat can sense things?' He started to become very uncomfortable. He said, 'Yeah, like a bat.' I said, 'Well, a bat has sonar; do fish have sonar?' At that point, I think he realized he'd gotten himself in a jam. He said 'No, but they can sense the electrical stimulus in the water.' He just started making up stuff. And I said, 'Well then, why don't they avoid fish nets?' Right? Because a screen's no different—but they'll swim right into a gillnet on the river. Fish do have the ability to sense electrical activity. But if there's a structure that stands still in the water, they can't see ahead like a bat can. So we took the case to federal court—and stopped the pumping station."

Above Poughkeepsie and maybe ten miles from Woodstock, with the Catskills looming in the near distance, a sizable contingent of folks was already waiting for us beside the Hudson River. This wasn't to be what you'd call an "intimate" camping trip. For the past twenty years, each Memorial Day weekend Kennedy had been bringing together many longtime friends and their families, gathering for the past nine years on the otherwise deserted Magdalen Island. It was only a five-minute boat ride across the Hudson. Nonetheless, ferrying across this season's sixty-eight guests, along with their tents and other gear, was to take up much of the first day.

Upon arrival at the thickly forested island, all of us waded ashore on slippery shale, climbed a little hill to where the main campsite would be, then fanned out along a ridge to locate a decent spot to pitch our tents. I found a place under a big red oak. A small campfire was going by the time I reconnected with other humanity. Kennedy and Bill Robinson, an avid striper fisherman and a raptor biologist who used to train animals for Hollywood, were sitting together on a log. A salt-of-the-earth kind of guy, Robinson had a couple days' beard going.

He launched into a story of "an amazing phenomenon" that he'd observed here ten years ago. This was the Middle Hudson, and striped bass had always been known to spawn farther downriver. That particular spring, suddenly stripers gathered in the channels right off Magdalen, where twenty-foot-wide shoals known as Hogback and Saddlebacks had remained intact since the glacial lakes emptied centuries ago. "These shoals were nothing but boiling water," Robinson marveled. "For every pod of females you saw, there were four or five males under them. Never seen anything like it, before or since."

Robinson, who lives nearby at Overlook Mountain, had been the first to use live herring to catch a striper in this region, "across the way there by Shithouse Bend" (the locals' name for a sewage treatment plant). Now everybody did it. In these same waters around Magdalen, Kennedy said, for three consecutive years a record-breaking Hudson River striper had been taken by recreational anglers, the most recent being a fifty-four-pounder. Due to PCB contamination, commercial fishing for Hudson stripers was still banned—and the fish had proliferated. "Commercial gillnetters hate the bass," he added, "because they rip huge holes in their nets and eat all the shad."

A month ago, Robinson said, you'd see fifteen boats fishing for stripers here with live herring. But he had some bad news. The bass had disappeared a week ago, apparently migrating south. Bobby's face fell. Mine did, too. "Sorry about the stripers, Dick," he'd say several times during the weekend.

I accompanied him to put some gas in the *Lickety Split*, leaving the river by the Old Saugerties Lighthouse and cruising slowly past the wild iris into Esopus Creek. This creek, upper portions of which are a world-renowned Catskill trout stream, had been the focus of a lawsuit brought by the Pace Environmental Litigation Clinic on behalf of Riverkeeper and four fishermen's organizations. The creek is part of New York City's reservoir system, which transfers drinking water from under mountain ranges between one watershed and another. For more than a decade, the city had been discharging highly turbid water into Esopus Creek without obtaining a federal permit as the Clean Water Act required. With four Pace University law students conducting most of the arguments, in January 2003 a U.S. District Court judge handed down the largest penalty ever awarded in a citizens' suit against a municipality ($5.7 million) and ordered the city to obtain a discharge permit before proceeding any further.

After gassing up, heading back toward the island we scooted past a huge barge—one of many, from all parts of the globe—moving down-river from the Port of Albany forty miles north. Located near there was the General Electric plant Robert Boyle had helped expose, the plant that for thirty years had discharged those estimated 1.3 million pounds of PCBs that had contaminated the river's striped bass, before the chemicals were outlawed nationwide in 1977. Sports fishermen were currently allowed to keep one striper a day from the Hudson, but because of persistent levels of PCB toxicity in the fish, in recommen-dations similar to California's, New York health officials suggested that people eat no more than one a month, and none at all for children and women of childbearing age.

"The resident population of stripers is still highly contaminated. They're outnumbered by the migratory fish, which are pretty clean, but there's no way to tell the two apart by looking at them." Kennedy's voice intensified as he went on: "General Electric knew during the '50s and '60s that PCBs were dangerous—we have letters from Monsanto Corporation, which was the manufacturer, warning GE not to let any of it go into the environment. It would have been an easy thing for them at the time to collect the PCBs and recycle them. But they could get away with dumping them into the river, so that's what they did."

In December 2000, after a ten-year battle between GE and the federal government, the EPA ordered the company to spend $500 million over five years to dredge PCBs embedded in the river bottom north of Albany. The plan calls for removing 2.6 million cubic yards of sediment that hold about 100,000 pounds of PCBs, dredging this from thirty-three "hot spots" along a forty-mile stretch of the Hudson. For its part, GE asserts that it's better to leave the PCBs undisturbed where they lie—something, of course, that would save the company millions.

"The science indicates that if we don't dredge, then most of the fish in the Hudson will be inedible for over a hundred years," Kennedy was saying as we approached our camp. "If dredging happens, most fish will be edible within two years."

What about GE's argument that dredging would stir up the sediments in a detrimental manner? "That's GE's propaganda," Kennedy responded, his eyes flashing. "The sediments get stirred up anyway during storms, so that stuff is constantly recirculating in the river. GE's been *ordered* to do this. One of the aims behind the federal Superfund program has been to force recalcitrant polluters to clean up, because it gives EPA the power to charge them treble damages. That's what was happening during the Clinton administration, and more sites got cleaned up in his first three years than they had in the previous twelve. But [President] Bush has let Superfund go bankrupt. So GE's supposed to start dredging in 2006, but there's a good chance they'll never do it."

He aimed the *Lickety Split* toward a cove about a quarter-mile from the camp. We'd tie up here to some trees on the bank. But Bobby wasn't quite done with the topic of PCBs: "Besides being a probable carcinogen, the real problem with PCBs is that they're very potent endocrine disrupters. That means they disrupt the formation of nervous tissue, so when fetuses are exposed, the babies are more likely to have cognitive impairment. Even small amounts of PCBs have the potential to produce a number of problems in our children. And every woman between Oswego and Albany has elevated levels of PCBs in her breast milk."

His voice trailed off momentarily. Then he continued: "I don't let my kids eat any fish from the Hudson. Frankly, I myself tend to be risk-averse, and I'm not gonna let the polluters get between me and the fish. It's probably not a smart thing for me to do, but I eat fish anyway. I had myself tested. In my body I've got two to three times the amount of PCBs as in a normal human being my age. For a man of fifty, not intending to have any more children, the risks are more internalized. But we don't feed our kids tuna anymore either, or other freshwater fishes, because those are loaded with mercury."

Everyone had been gathering wood for the fire, and we sat down for a dinner of beef stew. Tonight would be the annual "talent show." She wasn't in attendance this year, but in the past, Brenda Boozer had been on hand. A board member of Riverkeeper, she is also a mezzo-soprano who has sung for the Metropolitan Opera. And, Kennedy said as night descended, her performance of *Carmen* had resulted in his favorite Hudson River fishing story.

"It was a very still night," he remembered, "and a full moon lit up the sky. Down the river maybe half a mile, we were out fishing—Bill Robinson, my son Bobby, and a couple of his buddies. All of a sudden, here came this beautiful music off the island, her voice carrying all the way downriver. Brenda could crack glass with that voice. Later I heard she'd been dancing on top of the coolers with castanets. And in the hills, the coyotes started to sing, because they heard her, too. All this was echoing back and forth across the river. That's when we hooked into a twenty-eight-pound striper! Bill was fighting it the whole time, while the air was filled with this incredible aria from *Carmen* and the howling of the coyotes. We brought the fish back and cooked it on the grill. It was a magical night."

The 2004 talent show started with the kids, all of whom were eager to perform their songs, recitations, and walking handstands. Bobby Kennedy closed the night with a telling of "Rip Van Winkle," which he noted was set right across the way in the Catskills. I recalled his saying about the story's author: "If you look back in history, probably the first great environmental battle of American history occurred here in the 1840s, when Washington Irving fought efforts to build railways along the Hudson's shore. He ultimately lost, but delayed the project for over twenty years."

The next day, on the way to explore nearby Cruger Island, Bill Robinson spoke of how ubiquitous zebra mussels have become in the upper Hudson. "You can't throw a baited hook over without bringing up a big mess of them," he said. They were an alien species, having arrived first in the Great Lakes on ship's ballast, then somehow working their way up into the Hudson in the early '90s. John Waldman, then a senior scientist with the Hudson River Foundation, had told me that an estimated *550 billion* zebra mussels had colonized habitats with hard sediments in northern sectors of the river, becoming the most abundant animal in the freshwater reaches there. They now consumed some 85 percent of the phytoplankton that fish depend upon during their early life cycle. Shad seemed to be suffering the most but, according to researchers at the Institute of Ecosystem Studies, striped bass have also decreased in growth and abundance since the zebra mussel invasion. Luckily, in the fresh water, there was also a detritus-driven plankton food chain that's able to support the young fish. But the situation was highly unstable.

"Invasives" in a human form, who have encroached upon the river, have created an equally tenuous situation. "The biggest issue is urban sprawl, the destruction of landscapes," Kennedy said as we headed back onto the highway the next morning. In 1995, he had led negotiations that resulted in a landmark interagency agreement to protect the Catskill watersheds that provide most of New York City's drinking water supply. Now, pointing out the car window in the direction of the Catskills, he continued: "The Hudson is not just a ribbon of water. Any activity in the watershed ultimately affects the productivity of the ecosystem." He cited a Chesapeake Bay study, which had provided impetus for Jim Uphoff's latest research there, indicating that if as little as 10 percent of a watershed was paved, there could be permanent harm to its biological integrity. "Stopping the growth of these impervious surfaces was the real objective of the [New York] watershed agreement—to get a grip on the big threat, the pressure being put on hundreds of planning boards all over the Hudson Valley from thousands of developers and property owners, and to encourage the towns to preserve the pristine landscapes around them."

Today he felt there was cause for optimism, that the land ethic in the Hudson Valley had begun to change. "We're holding our own. Five years ago we were absolutely despised by government officials, who basically were doing the developers' bidding. There are probably fifty or sixty major developments that would have occurred in the watershed had we not stopped them." He paused momentarily, then continued: "But when you get an administration like the one in Washington now, you have to go back and fight rearguard actions all the time."

Although I hadn't been able to catch any stripers on the Hudson trip, the opportunity would come that summer on Martha's Vineyard. On the seventeen-foot fishing boat, having four rods going simultaneously wasn't easy, but somehow nobody hooked into anybody else. As we cruised up the Vineyard's North Shore, Kennedy watched as his sister's brother-in-law, fashion designer Kenneth Cole, helped six-year-old Finn Kennedy reel in an eight-pound bass. The little boy was shown how to release the fish—"you know it's going to swim away when the dorsal fin comes up," Cole said. Soon Kennedy caught and let go an eight-pounder. "See how light-colored they are from living on that sandy bottom?" our captain, Mo Flaherty, marveled as the fish swam free again.

For a while we hung around a big rock at Cape Higgon that he knew the stripers liked. Then an odd thing happened. Numerous terns were fishing in the same vicinity. Kennedy had just finished recounting a story about casting out one day and accidentally knocking a wing off a tern—"which got me very worried about my future karma"—when, on Cole's next cast, one of the birds mistook his lure for a baitfish and snared its wing. "Every good tern deserves another," he quipped. Then, as Cole reeled the bird in, Kennedy said quietly, "Let me get it loose." Which he proceeded to do, deftly, placing his finger in the beak as the tern desperately flapped its wings—but didn't bite him. Cutting away the section of line that connected the lure to the tern, Bobby carefully unwrapped the struggling bird. Then he handed it to Finn to set free. The little boy lifted the tern high with both hands, and it soared heavenward.

It was all catch-and-release striper fishing, an ethic that had taken hold among sportsmen all along the coast in recent years. Passing an area known as the Brickyard at the end of a long morning, Kennedy said impulsively: "Wild in there, let's take a cast." Immediately, a fish struck. In a flash, it was a triple hookup. Environmental activist Laurie David, screaming with excitement, brought in the first big fish she'd ever caught—a nice-sized blue. She had "caught the bug," it seemed, and was working her way up to a striped bass.

On the way back to land, I thought of a passage Kennedy had written in *The Riverkeepers*, a theme he'd returned to many times since. "We aren't fighting for the Hudson just for the sake of the sturgeon, striped bass, and shad, but because we believe that we and our children will have richer lives in a world where these fish thrive, where fishermen on the river wait for the tides and repair the nets, where their families and others can eat their catch. In such a world, our children can connect themselves to 350 years of New York State history and feel that they are part of something larger than themselves—a community."

Two months later, the Hudson Valley was ablaze in autumn leaves. Every fall, Kennedy brought a few students and friends along to Senasqua Beach, on the Croton River. "You get a good mix of fish because it's in the salt line," he told me. It was also a big spawning area for striped bass. His aim was to net a few fingerlings, along with perhaps some other unusual species, to stock in the big aquarium that graced the entryway to the Pace Environmental Litigation Clinic.

When I arrived at his house early on a Saturday afternoon, he was going on three hours' sleep. He'd gotten in at 4 a.m. after giving talks in Las Vegas and Phoenix, doing everything he could to expose the record of "the worst president in American history for the environment." We carted the hundred-foot-long seine net out of the basement, affixed it to his roof rack, and drove north toward Croton Point.

He was furious about a recent ploy by the Bush White House to undermine a legal ruling that would otherwise force power plants into using the best available technologies to minimize their fish kills. He'd originally brought the case fourteen years ago against the federal EPA. Back in the 1970s, thanks initially to Bob Boyle's efforts, the EPA had put regulations in place to require all new power plants to install closed-cycle cooling towers. "With once-through cooling, you're constantly sucking new water from the river—and killing obscene numbers of fish in the process. With closed-cycle technology, you're saving 90 percent of the water, so killing far fewer fish." The EPA, however, made a technical mistake when promulgating the new rules and, under pressure from the utilities, never corrected it. So nothing had moved forward.

A fellow named Brendan Kennedy (no relation) had discovered this. "He's an interesting character," Bobby continued. "A Vietnam vet, Special Forces paratrooper. He'd come back and gone to work for Con Edison at the Indian Point nuclear plant. He'd also continued his hobby, which was parachuting. He'd parachute into this parking lot in Yonkers at night, but one time the police nailed him. He wouldn't reveal who'd dropped him. The headline in the paper said something like 'Con Ed employee refused to cooperate with police.' So Brendan got fired and had to look for a new career. He went to law school, first in his class, then walked into our clinic one day and said, 'I want to sue Con Ed.' So we gave him a job, and he dug up this old statute— 316(b)—that we used to win our case."

Power companies had tried to get around the expense of installing closed-cycle cooling with so-called restoration measures. Rather than protecting the fish at the intake pumps, they proposed "replacing" the dead fish by stocking hatchery-raised ones, building artificial wetlands, or simply controlling runoff. This, Kennedy's team argued, was an absurd trade-off; you were talking apples and oranges. In February 2004, a U.S. Court of Appeals agreed with Riverkeeper that this was not to be allowed.

Two weeks later, when EPA staff circulated a proposed rule to require the closed-cycle technology at the largest power plants on America's most sensitive water bodies, an obscure agency inside the Office of Management and Budget scuttled it. The Office of Information and Regulatory Affairs, Kennedy says, "may be the most antidemocratic institution in government. It operates in secrecy, reviewing proposed regulations under an executive order." The revised rule, as it appeared in the *Federal Register*, merely requires all facilities to demonstrate reductions in the numbers of fish being killed in their intake pipes.

This means, for example, that New Jersey's Salem Nuclear Generating Station, along the Delaware River, will continue to do business pretty much as usual. In recent years, although PCBs in stripers remain a problem, the river's bass population had been making a strong comeback after surrounding sewage treatment plants were upgraded, thus largely eliminating the oxygen depletion problem in the river. However, a study by marine scientist Desmond Kahn in 2000 found that the Salem Nuclear plant—with no plan to install cooling towers— killed, on average in a worst-case scenario, 32 percent of the river's larval and juvenile stripers. His follow-up assessment in 2001 concluded that the striper's spawning stock productivity "may be severely restricted" by the power plant.

"The Salem plant's own consultant, Martin Marietta, admits they kill more than 300 billion [larval and juvenile] Delaware River fish each year," Bobby said, shaking his head. "Now all these plants will keep on using their antiquated technology—and reducing local fish populations."

It was discouraging, to say the least. But Kennedy was battling a corporate mentality—and a government—that, as the subtitle of his new book says, is "plundering the country and hijacking our democracy." When it all felt hopeless, there was a sense of renewal on days like this one, as we turned in by the Croton Yacht Club and parked by a chain-link fence along the Hudson.

He grew animated again as he described the natural wonders of this particular spot. "You get such an amazing variety of fish around here. Not only the anadromous species—stripers, shad, sturgeon, herring— but saltwater fish like mackerel, flounder, pipefish, Atlantic needlefish. Also a lot of shrimp and blue crab—stripers love the shrimp. You get freshwater fish like big catfish and pumpkinseeds, and brackish fish like killifish and mummichogs. There's also the hogchoker, which has what looks like fur on the back of its body. A little bit offshore the bottom of the Hudson is paved with them.

"Where Long Island juts out into the ocean, it breaks off cells of the Gulf Stream and the fish circulate up the Hudson River. So you even get a lot of tropical fish who think they're in Belize or Costa Rica—sea horses, clownfish, occasionally angelfish, and these exotic stargazers which can give you an electric shock—though they die off in the winter. You also get permit, a tropical fish that people are shocked comes in here! Bob Boyle actually once netted a mangrove snapper, right here on Senasqua Beach." (The first such ever seen in the Hudson; in the western hemisphere these fish typically range from Florida south to Brazil.)

Kennedy had earned his beach-seining "fins" here, in the company of Boyle. "In fact, Bob Boyle gave me his net. One of the first times I met him. He'd had an old net for fifteen or twenty years with holes in it. But I used it a lot. I still have it."

A few friends had shown up with their children. Kennedy had waterproof sneakers for everybody in the back of his car. I'd borrowed a friend's wetsuit, but it turned out to be a one-piece variety that proved extremely challenging to fit myself into. Perhaps it was a matter of our differing sizes. At any rate, Kennedy and other company were making for the water by the time I got my legs in. "Dick's probably still gonna be here when we're finished," I heard Kennedy say. Well, the heck with the arms. I let them flop away at my sides, and went to join the party.

Kennedy walked into the river holding onto a pole at the outer end of the seine, which had a bag at its midpoint for gathering the fish, while his photographer friend Bill Abranowicz stayed near shore hanging onto the other end. As Kennedy reached shoulder-high water and turned to make a circular tow back toward the beach, he called for me to come help with the hauling. I could feel my wetsuit filling up with the Hudson as I waded toward him. At least the water was still relatively warm. While we advanced with the seine, a practice that hadn't changed much since Biblical times, an Amtrak train came whizzing past along a nearby commuter rail.

"Keep that lead-line down, or the fish'll get out," Bobby shouted over at Bill as we reached the shore. The kids all came racing over. Mixed in amid the seaweed were hundreds of wriggling little fish. Nearly all of them were alewives, a close cousin of menhaden. But there was a catfish, and a couple of hickory shad, that the children placed in a bucket of water to be taken back to the Pace aquarium. There was, as well, a beautiful fingerling striped bass. Held up to the sun, it glistened with all the colors of a rainbow.

We made two more seine hauls. A huge carp splashed in, then out, of the bag on the next one. The last time, Kennedy decided to try going out deeper. The water was up to my neck when we made the turn. "We've just gotta make it to that barge there," he said as we pulled, somehow not stumbling over tree limbs and logs that lined the bottom. Grabbing onto the structure with one hand, he used it as leverage to guide us in. The story was pretty much the same: alewives and more alewives. "Sometimes we catch the amount of stripers that we just saw these guys," Kennedy said.

Bill and I folded up the seine and carried it back to the car. It seemed like half the Hudson came pouring out of my wetsuit when I changed. As we hit the road again, Kennedy pointed offshore. "Look at those seagulls—they're hitting the alewives," he said. The Croton River appeared over the first rise. "It's an amazing river," Kennedy said. "Used to be all marsh, with Indian middens [refuse heaps] forty feet high of 6,000-year-old shucked oysters." He pointed again, to the left of the road. "Over here are some forty-foot holes that I've been scuba diving in. Right below the falls, all the fish get bundled up into the aerated water. I'd go to the bottom and grab a big rock to stabilize myself. I'd see big goldfish, catfish, yellow and white perch, brookies—just watch the pool fill up with fish. There's an alewife run in the spring that makes the whole river look like quicksilver. And giant striped bass come up to feed on them, just gorging on the alewives. It's such an experience to see all those different kinds of fish in the same place."

That experience is what he was fighting for. "In the '60s, the Hudson River was reeling against the ropes; it barely had a pulse," he continued. "Now, even though we've still got a long way to go, it's again one of the richest fisheries in the North Atlantic. The threat to the Hudson today is not a single project, as it was in the old days—a highway built along the banks, or a power plant on Storm King Mountain. Today it's the death of a thousand cuts, irresponsible development in the watershed that's got to be kept in check."

That night, above a beautiful walnut desk in his study, I noticed that a small hammerhead shark was mounted. Paintings and sculptures of falcons looked down from the walls. Another sculpture, clearly of Native American origin, was marked as having been a gift to Senator Robert Kennedy. At the entry to the living room were two coffee tables, one with many pictures of his father, the other with framed photographs of his late brother Michael and cousin John Jr., along with one of Bobby himself, shown with a falcon on one arm, his daughter Kyra standing beside him holding a little nest.

"My father brought us to some of the most beautiful and wildest places in America," Kennedy said. "He taught me that wilderness was the source of our values and our virtues and our character as a people, and that we had an obligation to protect it for the next generations."

As he paused, bathed in the glow of lamplight, the momentary resemblance to his father was almost eerie. "I believe we still have time to preserve a planet for our children that provides at least some of the opportunities for dignity and enrichment as those our parents gave us. My job is to be able to look myself in the mirror and say that I spent my short time on this planet trying to make it a better place for my children. I have to look my children in the eye."

Those children, and grandchildren, Kennedy hopes will also be able to know the thrill of pursuing a striped bass. In northern New England, one man is determined to make that legacy a reality.

Chapter Sixteen

Stripers Forever?

At the base of Brad Burns's spacious backyard in South Portland, Maine, the Presumpscot River is but a few miles from its mouth at Casco Bay. Burns keeps his twenty-foot Aquasport, the *Sea Beagle*, anchored a little offshore from a wooden dock. On this mid-September afternoon, he was carrying a spinning rod for me and a fly rod for himself. "John Cole gave me this," he said as he hopped into a canoe and paddled out to retrieve his boat from its mooring.

The author, whose book *Striper* had helped launch the campaign to rescue the bass in the early 1980s, had been Burns's mentor in conservation. After reading the book, Burns had sought out Cole in nearby Brunswick. The two men became close friends—fishing together, going to dozens of meetings, even mounting a restocking effort of striped bass into the Kennebec River. In the curious way the fish has of uniting kindred spirits, Burns and Cole had gathered fingerling stripers from Bob Boyle's Hudson River group and then set up a rearing pond at the federal hatchery in Attleboro, Massachusetts, right next to Bob Pond's operation.

During the nineteenth century, striped bass larger than fifty pounds had been common in the Kennebec. Burns had heard stories of year-round resident fish being brought up through holes in the winter ice. But like most of Maine's rivers that once teemed with life, the Kennebec had become a void because of dams that blocked fish migration and oxygen-choking pollution from paper mills. During the 1970s, after Maine citizens pushed through legislation to force a much cleaner environment, a remnant migrant population of striped bass had once more been observed. Still, many believed that Burns and Cole's native restoration plan was quixotic. However, only a few years after their restocking efforts began, in 1987 came the first verified spawning success for stripers in the Kennebec in more than fifty years. It continues today, with young-of-the-year being found there every year. Not in huge numbers—this isn't the Chesapeake, but a cold northern river—but there nonetheless. As Cole once wrote (he died in 2003, at the age of seventy-nine), the striper's return was proof positive that "given half a chance, the fish will make the most of every available environment."

Partly at Cole's impetus, the Presumpscot—site of Maine's first dam and first paper mill—had, like the Kennebec, seen a dramatic restoration since the days when the river ran different colors and chemical fumes took the paint right off nearby houses. As we headed upriver, Burns said: "This is a small tidal river, but at times we've had terrific striper fishing. The bass that come up here are Chesapeake Bay fish. Back when I first moved to this house in the early '90s, I caught three in a row one evening that all had tags from the Chesapeake." His was all catch-and-release fishing. "I haven't killed a striped bass in years," he said, "because I love 'em too much."

We started casting below a natural waterfall, where the Smelt Hill hydroelectric dam had been abandoned in 2001 following a devastating flood. It was early evening. An auburn light glimmered down the hemlock-carpeted slopes. A statuesque blue heron stood on a log maybe fifty yards away. Burns reminisced about the long-ago days when as many as one hundred thousand Atlantic salmon would thrash their way along here, en route to Sebago Lake and their spawning grounds in the Crooked River twenty-odd miles north. "Can you imagine what a spectacle that must have been? All these thirty- and forty-pound silver bullets leaping out of the water everywhere! The first accounts of Europeans described how the entire surface of this river, for a foot deep, was all fish."

He's hoping there might be a few stripers looking for alewives along the bottom here today. "But bass chasing baitfish all over the place is just gone," Burns cautioned. "It's been really downhill for the last five or six years. A lot of the biologists scoff, saying the coastal fishery is at an all-time peak. Well, it isn't."

So strongly does Burns believe that the striped bass are in jeopardy that early in 2003 he formed a new organization called Stripers Forever. This came in the wake of the Atlantic States Marine Fisheries Commission's decision in 2002 to raise the annual commercially allowed harvest of Atlantic stripers by 43 percent. Stripers Forever now has some 4,500 members. What they're seeking is nothing less than banning commercial catches altogether, making striped bass a recreational gamefish like marlin or tarpon. This is something Burns managed to achieve in his own state, through the Maine legislature, in 1984. Connecticut, New Hampshire, New Jersey, and Pennsylvania had also enacted no-sale laws on stripers taken from their waters. Now, Burns felt, it was time to make it coastwide.

Could such a thing possibly be justified, at a time when the latest ASMFC stock assessment listed an estimated 4.7 million striped bass landed in 2003, only 1.1 million of those recorded as taken by commercial fishermen? (By comparison, when I joined the striper wars in 1982, Atlantic commercial fishermen landed 428,630 fish, recreational fishermen another 217,256.) Even though I personally had long felt that striped bass are too special a creature to allow their being caught up by the thousands in fishermen's nets, their amazing comeback made it difficult to sanction an outright commercial fishing ban. At least, so I thought upon first hearing about Brad Burns's organization. In the two years since its founding, however, warning signs about the health of the striper population had escalated almost as dramatically as their previous resurgence.

It wasn't only the startling rise in mycobacterial infections among the Chesapeake's population, or simply the overfishing on menhaden, that was curtailing the striped bass's primary food source, important as both those alarming developments have been. Since the ASMFC's rush to increase the commercial catch, the latest stock assessment in August 2004 showed significant overfishing the previous season on every age-class of stripers that fishermen are allowed to catch (a consistent twenty-eight-inch minimum size limit along the coast, but an eighteen-inch minimum size limit in the Chesapeake—where Maryland accounted for half of the total 2003 commercial catch). This was the first time such a decline had been observed since strict regulations began to be implemented in 1985. The first sign that too many fish are being taken is usually the disappearance of older ones—"growth overfishing," as marine science calls it. Fishing mortality rates on stripers ages eight through eleven was well above the threshold established by scientists for overfishing in 2003, and a 77 percent increase over 2002.

The fact is, the ASMFC management plan (Amendment 6, as it's called), permitting far more striped bass to be taken commercially than in the 1990s, makes no allowance for two critical factors: black market sales and striper bycatch taken in other fisheries. Some estimates are that the illegal commercial catch may be as high as 50 percent of the legal one. "Even though New Jersey has a no-sale law," says Tom Fote of the Jersey Coast Anglers Association, "we figure at least 200,000 pounds a year are being landed here and sold into the Fulton Fish Market in New York. We know people take coolers of fish to Philadelphia's markets, too."

In April 2004, a two-year-old Virginia program called Operation Back Door announced arrests at thirteen neighborhood fish markets and restaurants in the Richmond area for buying illegally harvested striped bass. The effort was modeled after a similar sting operation two years earlier in Hampton Roads, which generated about thirty arrests and temporarily dampened the black market in the state.

George Mendonsa's old outfit, Tallman & Mack, had been busted along with two other Rhode Island companies for admittedly having falsely labeled and illegally trafficked tens of thousands of pounds of striped bass while shipping them to other states and Canada. In 2003 the companies paid a $75,000 fine while also donating more than $50,000 to the nonprofit Fish and Wildlife Foundation.

Mendonsa had sold the trap-netting company five years earlier, but had continued to work for the new owners. When I interviewed him early in 2004, I asked whether he'd had any idea about what was transpiring. "Oh sure, I was on the boat. I knew it was goin' on," he told me. "I was encouraging them to do it." You were? I asked. "Sure I was." Why was that? I persisted. "Why the hell should I cooperate with them"—I presumed he meant the fishery regulators—"after what they done to me? Screw 'em!"

He added: "The Mendonsas are the ones that always caught bass, and the only ones *know* how to catch 'em! And after I'm gone, ain't *nobody* gonna know."

The more things change, the more they stay the same.

In the landmark article "Ecosystem-Based Fishery Management," which appeared in the journal *Science* in July 2004, reducing excessive levels of bycatch—defined as "killing of nontarget species or undersized individuals of a target species"—was listed as one of the key goals of the ecosystem approach. "Globally," the seventeen marine experts wrote, "discards in commercial fisheries have been estimated at 27.0 million metric tons, accounting for about one-fourth of the world's fish marine catch." Bycatch of striped bass in trawl and gillnet fisheries may be an even bigger problem than the black market. In the Hudson River, even though sale of stripers is not allowed, thousands of small, pre-spawning fish are still trapped and die in the nooselike mesh of gillnets being set for shad. Off the south shore of Long Island, according to

Charles Witek of the Coastal Conservation Association, "we see horrible shows of dead bass every fall, when the trawlers dump their nets. Mostly they're targeting weakfish." When the big female stripers mass off North Carolina's Outer Banks each winter, waiting to spawn in the Chesapeake, untold numbers—perhaps a million pounds or more—get swept up in gillnets staked to catch and sell dogfish. All these bass, too, are discarded dying or dead, and so not counted against the ASMFC's annual commercial striper quota.

Massachusetts, once in the forefront of striper conservation, is today the greatest offender in terms of bycatch. The problem surfaced in 1997, when the Cape Cod Commercial Hook Fishermen's Association distributed a letter about vast numbers of striped bass— "sometimes in excess of 10,000 pounds for a single boat in a single day"—being killed and wasted by another offshore gillnet fishery for dogfish. Any targeted netting of striped bass in Massachusetts waters has been outlawed since the 1940s. Yet at a meeting in 1999, two commercial net fishermen conceded that as much as 1.2 million pounds had been thrown back dead the year before, at a time when the state's entire annual commercial quota was 750,000 pounds! Striped bass simply can't survive being strangled in gillnets overnight.

The numbers taken by mobile gear dragging their nets along the ocean bottom, such as that employed in the multispecies trawl fishery off New England, can be even worse. In late August 2004, the environmental group Oceana filed a petition with the National Marine Fisheries Service (NMFS), requesting emergency measures to limit bycatch of stripers in groundfish trawls southeast of Cape Cod. As Oceana's Gib Brogan relates the story: "A couple of years ago, some fishermen told us about seeing 'slicks' of dead striped bass in the Great South Channel in the fall, when the bass were either migrating south or coming together to feed on herring schools. We chartered a boat and chased the trawlers for more than a day, trying to get pictures. We couldn't find anything, but heard stories of people who'd been seeing this for years and wondering what would show up in the observer records of at-sea scientists."

Under a Freedom of Information Act request to the NMFS, which is charged with keeping records on commercial fishing in federal waters, Oceana received data with rough estimates indicating to Brogan "that 2.5 million pounds of striped bass a year were being discarded over the stern by the draggers." If accurate, that would represent about half of the allowable coastal landings in state waters.

Eventually, managers may need as well to look at bycatch in the menhaden fishery: the numerous other species that are swept up in Omega Protein's purse seine nets. Only about 1 percent of its catch, the company claims, is anything other than menhaden. Still, when you're talking about 1 percent of a couple hundred million pounds of landings, that's not an insignificant number. "Under certain conditions, like when they set in shallow water where those nets scrape the bottom, it's got to be much higher than 1 percent," says Bill Goldsborough of the Chesapeake Bay Foundation. "My understanding is, they had a heads-up every time one of the scientists from the Virginia Institute of Marine Science was coming aboard, and control over the reporting of the information gathered, such that you can't look at those results and find any accounting for specific sets. You can only find average numbers, which come out to tell the story the industry wants."

At first glance, when you consider the vast increase in legal harvest over the past twenty years and then add in the thousands more stripers taken illegally and as bycatch, it might appear on the surface that there can't really be a shortage of bass. However, as fishing pressure escalates on the older fish, the younger ones are becoming increasingly susceptible to disease. What seems to be a stable—even booming—population may be far more fragile than we know. Things could change radically, in the span of a few seasons. It has happened before.

There has been a moratorium in place since 1990 on any fishing for striped bass in federal waters. These are regarded as any waters beyond three miles offshore, officially known as the exclusive economic zone

(EEZ) of the United States out to 200 miles. The EEZ is considered to be a refuge for the largest and most prolific breeding striped bass. That's where the trawlers are taking and wasting them. In the face of the already existing bycatch situation, it seems ludicrous that the National Marine Fisheries Service is simultaneously considering opening the EEZ again to striper fishing, allegedly for recreational purposes. Massachusetts, in fact, is the foremost state pushing for this, with state fisheries director Paul Diodati maintaining that "our traditional fisheries, especially in the island communities" want to be able to pursue those bigger stripers and claiming that the existing commercial quotas are adequate to protect the bass from suddenly being exploited in the EEZ. It is argued, too, that keeping a striper to eat or sell is better than simply tossing it back overboard as bycatch.

However, allowing striper fishing again in the EEZ would have many ramifications. Even though Massachusetts doesn't allow sale of stripers taken in nets, could it legally prohibit sale by the trawlers of fish taken beyond the state's territorial jurisdiction? Might a rod-and-reel-only fishery suddenly turn into a trawl fishery? And what about recreational fishermen being allowed to prey upon these premium offshore fish? Currently in Massachusetts, they're limited to bringing home two striped bass a day twenty-eight inches or longer. Conceivably an increase in a practice known as "high-grading" would result—meaning that fishermen end up selecting only the biggest fish caught, while throwing back dead the smaller ones that were caught earlier on the trip.

Opening up still more territory to fishing for striped bass can only exacerbate fishermen's greed. Maryland conservationist Jim Price underscored this early in 2004, when he called me about something he'd just witnessed out of Oregon Inlet, North Carolina. Price had been on his way back from Florida when he decided to see what the stripers looked like along the coast. He'd joined other fishermen on a charter. The ocean was teeming with female, soon-to-spawn bass. "You could catch them no matter what you threw in the water," Price related.

Sighting about fifty more boats in the same vicinity, Price realized to his horror that they were all fishing probably six miles offshore in the

EEZ, where striped bass are forbidden to be taken. The captain merely smiled when Price asked him about this. Back onshore, he demanded to know "why the Coast Guard doesn't enforce the law." They were understaffed, Price was told. The only federal officer available to police fifty miles of coast was away at a Homeland Security meeting in Texas. The next day, following Price's vehement complaint, a Coast Guard vessel did ticket twenty-eight boats with $900 in fines and confiscate eighteen illegally caught fish. This fishing practice was apparently routine off North Carolina, but enforcement was an anomaly.

Failing to raise any fish upriver, Brad Burns aimed the *Sea Beagle* in the direction of Casco Bay. A spot called "The Saddle" had "good moving water where the stripers sometimes hang out, looking for baby menhaden along the shoreline," he said. There was zip. We forged on, toward some pilings below the Interstate 95 bridge, "a prime spot where it's not unusual for gulls to be everywhere and the bass chasing pods of herring." Again, nothing. "I can assure you," Burns said as we started upriver again, "if this was September 1995, there would be bass all over the surface."

I couldn't help but wonder, though, how much of a part sports fishermen like ourselves have played in the bass's re-decline. Boat-builders had come to design and sell their latest fiberglass products specifically for striper fishing. The number of directed striper fishing trips grew from roughly one million in 1981 to seven million by 1996. The ASMFC calculated that the catch figures for recreational fishing along the Atlantic coast, including the Chesapeake, had been rising steadily since 1989—to roughly 1.8 million striped bass caught in 2002 and another 2.4 million in 2003.

That's a remarkable number of fish. More remarkable is the fact that *90 percent* of those fish were released alive. This catch-and-release ethic among sports fishermen is an amazing transition from a nearly 100 percent kill rate in the earlier days of striped bass

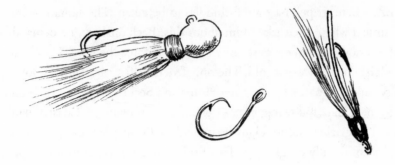

abundance. It has taken decades for people to heed what Zane Grey stated in 1919: "If we are to develop as anglers who believe in conservation and sportsmanship, we must consider the fish—his right to life and, especially, if he must be killed to do it without brutality." To a large degree, that consideration has been shown during the period of the striper's return. The joy of fishing has come to far outweigh the interest in keeping, certainly, more than one fish to bring home for the table. How many fish survive once they're put back in the water, though, depends entirely on how conscientiously the fish were reeled in.

"There certainly *are* problems with recreational catch-and-release angling," Burns said when I brought this up. "A lot of people don't treat the fish as they should, and much of it is just plain ignorance." Numerous studies, on a variety of species, have shown that the location of the hook wound is the single most important factor in whether a released fish will survive. What Burns successfully fought to establish in the Kennebec River was a ban on using live bait, which a striper will almost always swallow, with the hook penetrating its vital organs. Even using artificial lures, Kennebec bass fishermen are limited to a single hook. The fish is still more likely to be snared in the jaw and so more easily released.

The type of hook used is crucial, too. A traditional J-hook penetrates deep and, studies on fish kept in net pens have shown, will result in mortality for about 50 percent of released striped bass. By contrast, circle hooks—primarily applicable for bait fishing, as opposed to lure fishing—are now being manufactured with a curved point that often penetrates only the lip. (Prehistoric fishing artifacts made of bone, stone, and shell are similar in design.) Using this variety, studies by both Massachusetts and Maryland reveal extremely low mortality rates, an average of 1.2 percent of the fish released. Burns uses a barbless hook on his fly rod, the least painful to the fish and the easiest to release.

Survival also depends, of course, on whether the fish is brought in relatively quickly or "played" for too long and on the period of time it's kept out of the water after being caught. Also, interestingly enough, it depends on the air and water temperatures. One especially hot summer in Maryland, when the water temperature was about 80 degrees and the air temperature went as high as 105, even a fifteen- to twenty-second exposure to the air found mortality on shallow-hooked fish soaring from 2 percent to around 15 percent.

Fishermen also need to rein in their predatory instincts. As David Policansky of the National Research Council has written: "Catch-and-release anglers, probably believing that they are not affecting the fish population adversely, often catch and release dozens of fish or even more." This well-meaning, but overzealous, approach can result in mortalities of 18 percent and higher, according to studies in the Hudson and the Chesapeake regions.

As things stand, the ASMFC has estimated that probably 8 percent of striped bass that are caught and released along its range end up dying. That's based largely on a study conducted in Massachusetts where, Burns pointed out, the fish were encouraged to swallow the bait, then released into a confined environment at the peak of the summer. "Sure, if an inexperienced angler baits up a big old J-hook with a piece of clam, like the charter boats often do in Chesapeake Bay, mortality is going to go a lot higher than 8 percent," Burns said. "But other studies have shown it's as low as 1.5 percent. It's possible that recreational fishing mortality is being greatly overestimated, while commercial fishing mortality is badly underestimated."

In considering the future of the striped bass, we need to realize that its spawning habitats in the Chesapeake, the Hudson, and the California Delta are far less resilient today due to human pressures. For example, scientists already attribute many of the Chesapeake Bay's woes to a population spike since the 1950s that has coincided with burgeoning agribusiness and use of commercial fertilizers, along with the need for more sewage treatment plants. Population in the Chesapeake water-shed is expected to jump another 20 percent by 2030. Across the con-tinent in California, excessive nutrients from fertilizers, agricultural manure, and human sewage are likewise creating vast algal "dead zones" in waterways downstream from cities and farms. And there is little rea-son to expect less demand for water supplies in booming Southern California, a situation that will likely increase the exports from the striper's northern realm.

Electric power plants that burn fossil fuels are another big supplier of excess nitrogen to the bass's aquatic systems, in addition to sending carbon dioxide into the atmosphere. And this brings another element into the equation—the onset of climate change. The waters of the Chesapeake Bay have already risen by a foot over the last seventy years. Donald Boesch, president of the University of Maryland's Center for Environmental Science, has said that we may see an additional eight-inch rise by 2030. In a recent paper, Boesch and two coauthors write: "As relative sea level continues to rise, erosion of coastal margins will likely increase turbidity and may decrease the extent of submerged and emergent vegetative habitats. This would have far-reaching effects because these habitats are important nursery areas" for striped bass and other species.

At the same time, the Chesapeake's water temperature is likely soon to resemble that of North Carolina Sound. "Increased temperature would likely result in increased utilization of Chesapeake Bay by sub-tropical species such as brown shrimp, and decreased utilization by cooler water species like striped bass," the Boesch paper continues. Global warming, as marine scientist Eric May points out, increases metabolic rates in fish; even a 0.1-degree change creates a greater bioenergetic demand for striped bass. Another marine expert, Douglas Rader of Environmental Defense, has noted that as rising waters drown marshlands that are essential to the development of young menhaden, this will also pose a serious problem for striped bass.

Bob Wood of NOAA's Chesapeake Bay Program, a coauthor on the Boesch paper, puts it this way: "When you talk about climate change, there are multiple processes that could target the striped bass at multiple breakneck points—spawning, the egg stage, their development into juveniles. They are a cooler water species. . . . If salinity goes upstream and dissolved oxygen goes downstream, you've just lost every opportunity for that fish to have a range. Things can happen that we can't observe very easily or quickly in many components of the ecosystem. We can't treat fish like they're a stand of trees that someone has set aside on a few acres, that you can watch all the time and know exactly how many you can cut down at what time stage."

The impacts of global warming on the striped bass migration are already being seen beyond the Chesapeake, along the Atlantic coast. Until recent years, the first "keeper" bass, twenty-eight inches or bigger, caught in the Cape Cod Canal used to always occur around May 20. Now those fish are showing up at least three weeks earlier each spring. Standard wisdom also held that the first northerly winds and cold rainstorms of September would begin to trigger the fish's movement south. Now this often doesn't seem to happen until well into October. Many fish are hanging around New England into November, even sometimes spending the winter in the back reaches of rivers. As a result, more stripers are being caught there than in earlier years.

The prospect of intensified global warming is yet another clarion call for the implementation of ecosystem-based management. "This means erring on the side of caution in setting management targets and limits when information is sparse or uncertain," write the authors of the paper in *Science*. It "may simply involve using natural history and general knowledge to develop precautionary set-asides or safety margins, such as reduced catch limits or large closed areas."

Marine expert Les Kaufman of Boston University believes: "Striped bass are the Komodo dragon of the East Coast, emblematic of ecological conditions in our estuaries. Taking an ecosystem approach means that people beyond the East Coast are partially responsible for the welfare of striped bass, and they should be held accountable. The hydrology—in terms of peak flow—of the rivers up to almost the Canadian border, and as far west as Virginia and Ohio, [is] a determining factor in how well the Chesapeake bass are going to do. This should make enforcement of all land-use regulations an instant priority."

It's conceivable that just as striped bass have managed to cope with pollution and acid rain in the past, they may again prove all doomsayers wrong. Look at what's happened in the Delaware River. By the mid-twentieth century, a once-thriving striper population had largely succumbed to nutrient pollution and industrial toxics. But after large sewage treatment plants from Trenton to Wilmington upgraded their systems, coinciding with a moratorium on striper fishing in Delaware, a small remnant of the fish stock saw its young-of-the-year abundance increase a thousandfold between 1980 and 1993. In recent years, striped bass have become a normal part of the fish community in the Delaware River—and the region is now believed to contribute up to 15 percent of the coastal migratory stock. However, in November 2004, a tanker (the *Athos I*) spilled an estimated 265,000 gallons of crude oil into the river across from Philadelphia on the New Jersey side; how striped bass and other species will fare in the wake of the spill is not yet known.

Robert Boyle once wrote: "From both the Atlantic and the Pacific we have proof that the striped bass is a highly adaptable species, possibly (and fortunately) the most adaptable on earth. In the face of the fiercest assaults, the striper has managed to survive. For that we must be thankful, but we must also recognize that there are lessons to be learned from the close calls the species has suffered." Quoting H. G. Wells—"History becomes more and more a race between education and catastrophe"—Boyle went on to say: "That race is clearly being run in the oceans of this planet for the striped bass."

What we know for sure, Brad Burns says, is this: "I remember those big long discussions in the 1980s. There were biologists even blaming the fish's disappearance on sunspots, and others who simply felt it was the striped bass's time on earth to become extinct. Everything imaginable got theorized. But as soon as we stopped commercial fishing for striped bass, they came back. That's all there was to it."

Darkness was closing in on the Presumpscot River. The tide had dropped, and we were giving the area upstream of Burns's house one last shot. So far today, there had been no sign of striped bass. We motored past a few black ducks. I could see cormorants up ahead. "This was barren mud when I started fishing here fifteen years ago," Burns said. "You can see that the grasses have all come back." He angled his fly rod backward and then forward, once, twice, three times, and released. My own lure made a hissing sound as it fell just shy of the riverbank.

When he surveyed the membership of Stripers Forever recently, Burns said, the majority heavily favored the idea of the federal government creating some kind of Striped Bass Stamp, as had existed in California. As with the stamps purchased by duck and pheasant hunters, the moneys raised from sportsmen's contributions would go into a dedicated fund. The striper fund would initially be used to buy out the commercial striped bass fishery. This ought not to cost anywhere close to the $40 million allocated by the government a while back to retire a substantial portion of the groundfish fleet that had wiped out the cod on Georges Bank. In Massachusetts, for example, only about 350 fishermen (out of 4,587 commercial license holders) reported landings of more than twenty striped bass for the entire 2002 season. A relative handful of men continued to reap an income from striper fishing, and only a modest seasonal one at that.

In Congress, New Jersey Representative Frank Pallone had introduced a bill in March 2003 calling for no-sale, gamefish status for striped bass. Given the numbers of Chesapeake fish suffering from bacterial disease and apparently not finding enough to eat, Maryland marine physiologist Eric May believes: "There are some serious problems with the fish, and I think we're just seeing the tip of the iceberg. From a physiologist's standpoint, I don't know how long we can wait for this to blow up in our face. I think they have to look seriously at stopping the commercial fishing of striped bass."

As we have seen, the precedent for such a move was set seventy years ago in California where, despite periodic pressure from commercial fishermen to reopen the fishery, striped bass have retained their special gamefish status. The irony is that today, on both coasts, the fish that's long held such an archetypal status is being scapegoated by different special interests as being detrimental to the ecosystem. As I learned on my trip west, federal fisheries officials—with the tacit blessing of California's water barons—now maintain that, as an "introduced" species (more than a century ago) in the Bay-Delta region, stripers not only are less important than salmon and delta smelt, but are supposedly largely responsible for the decline in baby salmon. Surely this idea becomes ludicrous by comparison with the *real* invaders of the striper's ecosystems, such as California's alien clams or the Hudson's zebra mussels. In the Chesapeake, the menhaden industry points the finger at striped bass—and away from itself—as the prime suspect in the menhaden's decline. Along other parts of the Eastern seaboard, crabbers and lobstermen disdain the bass's comeback and claim that too many stripers are feasting upon their livelihood. It's a role reversal that, without fishermen still willing to fight on behalf of this most special of creatures, would end up with striped bass paying the price once again.

Commercial harvests, of course, do provide fresh-caught striped bass to seafood markets for consumers and, while curtailing these catches would help keep the striper population high, assuredly it would not solve all the problems. Sports fishermen might have to become accustomed to being able to pursue stripers only at certain times of the year. In addition, Maryland fisheries biologist Eric May would like to see the "trophy" sportfishing of stripers curtailed, "because we're now harvesting our bank reserves. We may have to say that anything bigger than twenty-eight to thirty inches should be off-limits." The National Research Council's David Policansky strongly agrees with May. Policansky believes that encouraging recreational fishermen to kill the big female stripers—as Maryland's so-called trophy season does—is a mistake. In 1993, Policansky published two papers that showed how fishing could affect evolution in fish populations. "That idea was new to many fishery biologists at the time," he says, "but it has become more widely accepted since. If there is any genetic component to the ability to survive and grow large . . . the best way to eliminate large fish from the population . . . is to selectively take the large ones and leave the small ones."

Some supporters of making the bass a no-sale gamefish, such as now-retired NMFS regional director Alan Peterson, want to see it happen for ethical reasons. "I just feel striped bass are a fish with a unique character, available to the common people," Peterson told me. "The whole nature of the fishery changed when people started making money out of it. It went from something you did for fun to something that turned into a job."

On a mid-September evening in Maine, I watched as a man dedicated to the enjoyment—and conservation—of striped bass finally, reluctantly, turned his boat toward home again, without our so much as glimpsing a fish. Brad Burns, carrying on the legacy bequeathed to him by John Cole, had, like myself, started out fighting for a higher minimum size limit in his home state and found himself swept into a battle bigger than anything he'd anticipated. He'd poured money from his photocopier business into his passion for preserving this particular fish.

"I'll tell you what I never dreamed of," he said now as we trudged through the brush back up the hill toward his house, "was that we'd be back in this fight again."

I thought of others who'd emerged over the past two decades as champions for the striped bass. They were men and women I would never have had the privilege to meet were it not for the bass. Respect and admiration—yes, call it love—for the striper cut across all class and ethnic boundaries. This was a truly American fish, from the days when John Smith and Henry Hudson came upon them in their coastal explorations—today as much a part of the workingman's lore and tradition as of those wealthy luminaries who once filled the bass stands at the Cuttyhunk Striper Club.

I thought of Charley Soares, a well-known fisherman on Cape Cod. After his father died, as a teenager he supported his family by selling striped bass in local bars. Today, in lieu of payment, he brings striper steaks to his doctor and takes his dentist on fishing trips—his "wampum," Charley calls it.

Then there was Tom Fote, who'd become the leader of the Jersey Coast Anglers Association and successfully pushed through the legislation making stripers a gamefish in New Jersey. The son of a newspaper deliveryman, Fote had grown up fishing the docks and piers of Brooklyn. "Who was the hero?" he said to me once. "It wasn't the guy who had the fanciest rod or dressed the best, but the guy who could pull the striper onto the dock. It was the great equalizer. This was harvesting to take home to eat."

Fote had spent eighteen months in hospitals after being wounded by artillery fire during the war in Vietnam. Despite a permanently injured leg, he found that he could still fish from the surf. After receiving a disability pension, he'd ended up devoting much of his life to attending meetings, serving as a citizens' representative on numerous fishing boards, and speaking out on behalf of striped bass and its environment. "Now I'm dealing with sewer plants, junkyards, getting the mercury out of the water," he explained. "Striped bass expanded my horizons, made me a different person."

More than twenty years after we met for the first time, at the public hearing in Buzzards Bay, I pulled into the driveway of a modest one-story home with yellow trim in North Attleboro, Massachusetts. A garage door opened and Bob Pond walked out to greet me. Avis Boyd was, as always, at his side. Bob was wearing a plaid shirt, with suspenders holding up his white slacks. The man who'd invented the Atom lure, first sounded the alarm about the striped bass, and founded the first conservation organization devoted specifically to protecting them—Stripers Unlimited, in 1965—still had a wavy mane of white hair at the age of eighty-seven. We greeted each other warmly, like veterans of the same platoon, and walked into the living room.

I looked around. The walls were covered with paintings and prints of the outdoors. A little coffee table was speckled with shells, along with sculptures of a whale and a seal. On the couch, an embroidered pillow bore this inscription: "Men, coffee, chocolate—the richer the better." Avis told me, shyly, that she and Bob had finally gotten married three years ago. Between them, they had seven great-grandchildren. One grandson, an avid fisherman, was studying literature in college; he wanted to become "a fishing writer."

In 1997, they'd sold the fishing tackle company. The new manufacturer, Bob said, "was trying to copy our lures using injection-molding plastics. But we'd always used Styrofoam. We could put the lead where we wanted and the beads in, and hit it with steam, and we could get a lure that was strong enough and light enough to work really well. But they've lost the balance completely."

Could people still buy Atom plugs? I asked. "Well, they still make them," Avis said, "but not many people buy them. They're hard to find now." Bob hadn't fished, she added, since he sold the company.

Nor did they travel much anymore, as Bob couldn't see well enough to drive and Avis had never learned how. But they sure had the memories. They reminisced about the time John Cole came with them to watch the Chesapeake spawning. "He couldn't believe the change in the smell of the river," Bob recalled. "It has a definite odor when the bass are spawning. It's very strong." They reflected too, on how the entire town of Edenton, North Carolina, would turn out to watch the spawning on the Roanoke River, as Bob and others picked up fish in dip nets and raced them to the hatchery up above to strip the eggs.

"It was an interesting time," Bob said.

"It was a battle," Avis said. "So many scientists would say to us, 'We can't talk about what's going on, but you do it!'"

"They didn't dare get involved. They couldn't keep their funding if they started talking about Monsanto or any of the big companies," Bob said, referring to the chemical giant that invented PCBs and sold them to General Electric and others.

There were exceptions—Joe Boone and Jim Uphoff in Maryland, for example. They were "the greatest," Avis said. Bob was glad to hear that Uphoff was still with the Department of Natural Resources and still raising hell. "There's too few of that kind," he said.

Did Pond have concerns about what was happening to the striper population today? "They probably need to shut the fishery down again," he replied. "Why was that?" I asked. He proceeded to lay out a strange theory, one I'd not heard from anyone else. In Pond's view, the Chesapeake stripers had managed to adapt to the PCBs that they'd absorbed along their migration. And now that PCB contamination along the striper's migrating route off Cape Cod has been largely cleaned up, Pond speculated that the adapted larvae are in a way over-reacting to their new circumstances—using up their yolk material too soon and perhaps starving to death.

While there was no evidence for this that I was aware of—indeed, according to one marine scientist, if anything detoxification usually *increased* metabolism, which would require greater activity and foraging for survival—Bob Pond had always been ahead of the curve. It appeared that stripers had managed to adapt to PCBs in the heavily polluted Hudson River, and there was perhaps another precedent for his oddball idea.

Bob Boyle had told me of comparative health studies conducted by two NMFS scientists, Jeannette Whipple and Mickey Eldridge, on West Coast striped bass during the 1980s. All the diversions of fresh water, Whipple had informed Bob, concentrated the pollutants emanating from a number of oil refineries in the San Francisco Bay area. The scientists began finding striped bass that had holes in their sides. They'd ingested a smaller fish that contained a parasite. Because of pollution, the bass had a heightened immune system, one that overreacted and, as a result, opened up a hole in its body in an apparent effort to expel the parasite. These always showed up on the right side, because the fish's intestines coil in that direction.

"Then a curious transformation occurred," Boyle said. "The only striped bass that were surviving didn't have straight stripes. They were broken into a checkerboard pattern. Jeannette told me that they were genetically coded to resist the chronic petrochemical pollution. They became dominant in the population."

So, who knew? Pond continued, "There is so much genetic diversity in any stock that natural selection should eventually straighten out the problem that adaptation caused. But you've got to give it time." Had the pendulum swung back to 1984, when the coastal states took drastic action on the striper's behalf? "Well, the Chesapeake moratorium," Pond said, "was what really saved the fish."

Should such a course be needed again, I thought to myself, overfishing on the bass would be far from the only reason. The issues today encompassed the ecosystem as the death of a thousand cuts—newborn fish being sucked into the maw of power plants and water pumps, juvenile fish being overwhelmed by pollutants in their nursery areas, adult fish suffering from stress-related disease because their primary food source is being fished out.

In an adjacent bedroom, Avis showed me Pond's honorary doctorate from a Boston college, his Chevron Conservation Award and David Belding Award, a mounted striper that was the first one Avis ever caught—"and I never got one much bigger either."

As we stood by the door saying goodbye, Pond said: "A lot of people didn't recognize the problem or want to face it, but we stayed with it."

"Did you ever feel like giving up?" I asked.

Bob smiled, and Avis jumped in with the answer. "No, not Bob! He took the slings and arrows and just kept going."

"Do you have any advice for people fighting for ocean conservation today?"

"Yes," Pond said. "They should really be watching new birth: how are the young of the year coming?"

Bob Pond had seen it all. And the future, then as now, he knew was dependent upon how we cared for the latest arrivals to our ecosystem.

At the end of my journey, sitting at a little desk on Martha's Vineyard looking out upon Menemsha Pond, a long-ago memory drifts in as I watch the outgoing tide. It begins with an image of a striped bass that had been in our boat for some hours before we brought it home. I'd presumed we would have the fish for dinner the next day. It was late on a full-moon Vineyard night when, inside the small outdoor studio where we kept our rods and reels, my mentor held up the fish for a photograph. It wasn't heavy, perhaps eight pounds, but as we stood there, slowly, inexplicably, the bass began to move. It arched its tail in one direction and lifted its head in the other.

"Put some more water in that fish box," my friend said, his voice quavering. Carefully he placed the bass inside, and we watched as, to our amazement, the fish began moving to and fro.

"He wants to live. Shall we let him go?"

We must have been a curious sight, returning the several miles to Menemsha Harbor in two cars, perhaps a half dozen of us. Most of us stood on the dock as my friend hopped into our boat, lifted the bass from the tank, and placed it in a small net. Setting the net slowly into the water, he held the fish's tail for a few minutes, while the rest of us waited to see what would happen. As we watched, life came back to the fish. My friend tipped the net, released the tail, and the bass was free. I can picture it still, in the rays of our flashlights, beginning its journey home.

Acknowledgments

So many remarkable people have been involved in the battle to save the striped bass. I call them heroes, and you have met many in these pages. Would that I had space to provide proper tribute to them all. Let me begin here by thanking three of the great women in my life. My wife, Alice, for the third book in seven years, once again read every word and served as my first editor, while taking care of me in several different locations. Heidi Keegan was always there in the beginning, at all the meetings, petition drives, and our national conference. Jessie Benton, then and now, has always pushed and inspired me to fight for what I love.

My longtime literary agent, Sarah Jane Freymann, remains a dear friend as well as a wonderful adviser/negotiator. My editor on this latest book, Jonathan Cobb, is the best I could possibly have hoped for. Many thanks to David Policansky of the National Research Council and another scientific reviewer of the manuscript who wishes to remain anonymous for their meticulous reading and important corrections and suggestions. My writer friend Ross Gelbspan, ordinarily no fan of fish tales, offered great encouragement and advice in reading the chapters as they unfolded.

It seems fitting that Anthony Benton Gude, a terrific artist (and fisherman) who appears several times in *Striper Wars*, was chosen by Island Press to do the cover art as well as the illustrations inside. Also, that my great friend and adviser during the first round of the fight, George Peper, offered his astute comments on the final editing phase.

This book also honors the memory of my close friend Brian Keating, who piloted our boat through many a storm and who forever tried to teach me how to tie those pesky fishing knots; he will be deeply missed by all who knew him.

The members of our "children's crusade," who braved the cold winters of New England to petition for the bass, are all grown up now—lawyers, contractors, teachers, politicians, mothers, and more. This book is for them too: the Lymans (Norma Lynn, Jackie, Ray, Abra, Lincoln, Daria, Valery, Lynne, Obray, and Tucia); the Pepers (Cybele, Gabriel, and Padrick); the Kweskins (Aaron, Bellina, and Corrina); the Guerins (Bing, Rose, and Marlon); the Gudes (Geordie, Heather, Ruby, and Riley); the Givens (Saskia, Ruby, and Jesse); Deirdre and Kat Goldfarb; Dean and Leelia Franck; Peter and Tom DeWan; Tamerlane and Samoa Wilson; Irene Lanier, Erica Glynn, and Matthew Keegan.

Many others in my extended family were instrumental in the fight. Kurt Franck and Mark Spector accompanied me on crucial trips to the Chesapeake Bay region. Eben Given provided powerful and artful drawings for the campaign. Richie Guerin, Geoff DeWan, Randy Foote, Joey Goldfarb, Carol Franck, Jeremy Greenwood, Mary Curtis, Richard Herbruck, Kay Rose, Jonny Gude, Marilyn Kweskin, Susan DeWan, Laura Given, Candy Guerin, Gale Lyman, David Gude, Nell Foote, Terry Bernhard, and others all hit the streets and spoke out at meetings on the striped bass's behalf.

There was also Horacio Malonson, a Wampanoag Indian born and raised on the cliffs of Gay Head. Horacio was deaf. He died in 1984, shortly before his thirtieth birthday. He wrote a poem about the striped bass that said, in part:

> We will save the bass under the ocean and deeper. . . .
> When a small bass dies, heaven cries. . . .

Through the years, my friends Tom Langman, David Bramhall, and Packy Parris have always been there to offer fishing trips and to grace our dinner table with bass and blues.

Chris Weld, founder of the National Coalition for Marine Conservation and a longtime leader in striper preservation efforts, was kind enough to put the Striped Bass Emergency Council under its non-profit umbrella. His associate Ken Hinman remains a staunch advocate for our marine fisheries.

I've been privileged to have a number of fine editors at magazines and newspapers, skilled professionals who understand the desperate need for conservation of our ocean resources. Peter Borrelli, editor of the NRDC's award-winning *Amicus Journal* (now called *OnEarth*), was the first. I also thank the editors who provided space for me to write numerous articles as the striper wars progressed: Tim Coleman at the *New England Fisherman*, Fred Golofaro and Dr. Bill Muller at the *Long Island Fisherman*, Peter Barrett at the *New Jersey Fisherman*, and Eric Burnley at the *Maryland Delmarva Fisherman*. Thanks also to Jerry Kirshenbaum, editor of the "Scorecard" section at *Sports Illustrated*, and, later, Jim Motavalli at *E* magazine.

Many outdoor writers played a huge role in calling attention to the striper's plight: Tony Chamberlain of the *Boston Globe*; Les Boyd of the *Providence Journal*; Nelson Bryant of the *New York Times*; Bill Burton and Lefty Kreh of the *Baltimore Sun*; Angus Phillips and Dennis Collins of the *Washington Post*; Michael Globetti of the *Boston Herald*; Nick Karas of *Newsday*; and Al Ristori, Charley Soares, and Dusty Rhodes in *The Fisherman* and other publications. At *Salt Water Sportsman*, Hal Lyman and Frank Woolner were early crusaders to protect the striper, later joined by Spider Andresen, Barry Gibson, Rip Cunningham, and Whit Griswold. Thanks also to George Reiger and Duncan Barnes at *Field & Stream*, Skip Fleet with *Commercial Fisheries News*, and my good friend Dan Levin (along with Robert H. Boyle) at *Sports Illustrated*. In the news sections of the paper, Tom Horton of the *Baltimore Sun* and Ry Ryan of the *Boston Globe* editorial page provided immeasurable contributions, as did Dick Reston and Mark Alan Lovewell at the *Vineyard Gazette*. My thanks, too, to Chris Knight of the New Film Company for his hour-long PBS documentary *Striped Bass: Mysterious Disappearance*.

Fishermen all across the Atlantic coast came together to push the various states toward implementing tougher regulations. In my home state of Massachusetts, special thanks must go to John Cherico and his fellow members of the Massachusetts Striped Bass Association (including William Tilley, who made those wonderful bass pins); Conrad Smith of United Mobile Sportfishermen; Daniel Kelliher and others with the Nantucket Anglers' Club; Joe Lawler of Boston Whaler; Vineyard fishing giants Danny Bryant, Cooper Gilkes, and Jack

Koontz; Conrad Smith of the Massachusetts Wildlife Federation; also, Les Smith, Joe Juffre, Dennis Sabo, Bill Sargent, the Cape Cod Salties, the Massachusetts Beach Buggy Association, and the Cape and Island Sportsmen's Association. Others in Massachusetts, especially the Sierra Club's Cynthia Knuth, Richard and Gail Ivker, the Atlantic Salmon Federation's Howard Brown, and then state legislator Robert Durand, offered consistent support to the cause.

In Rhode Island, besides Jim White and Will Barbeau, the stalwart efforts of Fred Thurber were paramount in the fight. Others who played an important role were Bob Greene, Bob Randall, Dave Decker, Charley Matley, John Jolley, Bob Jagolinzer, Rhode Island Marine Fisheries Council member Louis Othote, charter captain Al Anderson, State Senator John Sabatini, and "Salty" Stryker.

In Connecticut, thanks to Shaw Mudge, Lou Taborey, and Art Glowka for their work for the bass.

In New York, besides Fred Schwab, those instrumental in defending the striper included Bob Buss, Bob Rantz, Vic Rivas, and Don Palmer at Save Our Stripers; Charlie Johnson of the New York Sportfishing Federation; and Lou Figurelli of the Staten Island chapter of the Coastal Conservation Association. Marcy Benstock of the New York City Clean Air Campaign spearheaded the fight against the Westway. Carl Safina, a striper fisherman on Long Island, not only has been a fierce conservationist but has also become our most eloquent writer on the oceans. Thanks, too, to Stephen Sloan for his consistent support and to biologist Byron Young.

In Maryland, colleagues in the fight included Mike Pivec of the Maryland Saltwater Sportfishermen's Association, Jack Barnhart of the Maryland Sport Fishing Advisory Commission, Jim Gracie of Trout Unlimited, Bob Spore, and State Senator Gerald Winegrad. In Virginia, thanks to Jim McHugh.

Among national organizations, important roles were played by Gil Radonski and Bob Martin of the Sport Fishing Institute; Donal C. O'Brien, board chairman of the National Audubon Society; Lonnie Williamson, Wildlife Management Institute; John Gottschalk, Alliance for Chesapeake Bay; Lynn Greenwalt and Rudy Rosen, National Wildlife Federation; Amos Eno, National Audubon Society; Carl Sullivan, American Fisheries Society; Liz Raisback, Friends of the Earth; Dery Bennett, American Littoral Society; Steve McAllister, Greenpeace; Jack Lorenz, Isaak Walton League of America; and Dick Schaefer, National Marine Fisheries Service. Thanks also to congressional staff aides Chuck Swezey (with Representative Claudine Schneider), Andy Schwarz (with Representative Gerry Studds), and Steve Shimberg (with Senator John Chafee).

In the course of my recent travels, I received insights and assistance from renowned conservation writer Ted Williams; marine scientist Phil Goodyear; Rolland Schmitten, former director of the National Marine Fisheries Service; Dr. Serge Doroshov of the University of California at Davis; Joe Shelton, Craig Callahan, Rudy Haney, and the Fleeton Fields bed and breakfast in Reedville, Virginia; commercial fisherman Everett Mills; Chesapeake Bay columnist Will Bartlett; and Bill Goldsborough, Alan Girard and Bill Street of the Chesapeake Bay Foundation. Thanks, too, for hosting me on several occasions, to Mary Kennedy and her family, as well as Dianne Taylor-Snow and her husband, Pepper; Jim Price and his wife, Henrietta; and Tom Horton and his wife, Jenny. Thanks also to Mary Beth Postman. And to Robert Lifset for reviewing my chapter on Storm King Mountain.

And much appreciation to those who have carried on the work for our oceans and fisheries: John Bello and Robert Pride of the Coastal Conservation Association; Bill Windley of the Maryland Saltwater Sportfishermen's Association; Jim Donofrio, Gary Caputi, and Michael Doebley of the Recreational Fishing Alliance; Mike Nussman of the American Sportfishing Association; Andy Palmer of the American Oceans Campaign; Peter Shelley of the Conservation Law Foundation; Sarah Chasis, Lisa Speer, Joel Reynolds, and Hal Candee of the Natural Resources Defense Council; Matt Gianni, Niaz Dorry, and Kenny Bruno of Greenpeace; David Rockland; Gerry Leape; the Center for Marine Conservation; and Oceana. If I have inadvertently omitted anyone from this list, my apologies.

Notes

Prologue: Origins of a Fish Story

6. POPULATION FIGURES: "2003 Atlantic Striped Bass Advisory Report," ASMFC Striped Bass Technical Committee Report #2003-03, p. 34.

6. SAFINA QUOTE: "The World's Imperiled Fish," by Carl Safina, *Scientific American*, November 1995, p. 53.

Background on the striped bass's life cycle was drawn from numerous sources, including the following books: *The Complete Book of the Striped Bass*, by Nicholas Karas, Winchester Press, New York, 1974; *Striped Bass Fishing: Salt Water Strategies*, by the editors of *Salt Water Sportsman*, Creative Publishing International and Times Mirror Magazines, 2000; *Striper: A Story of Fish and Man*, by John N. Cole, Atlantic–Little, Brown, 1978; *On the Run: An Angler's Journey Down the Striper Coast*, by David DiBenedetto, William Morrow, 2003; *L. L. Bean Fly Fishing for Striped Bass Handbook*, by Brad Burns, The Lyons Press, 1998; *The Striped Bass Chronicles: The Saga of America's Great Game Fish*, by George Reiger, Lyons & Burford, 1997; and *Chesapeake Stripers*, by Keith Walters, Aerie House, 1990. Also, "Synopsis of Biological Data on Striped Bass, *Morone saxatilis* (Waldbaum)," NOAA Technical Report NMFS Circular 433, June 1980.

11. LAPTEW QUOTE: "The Striper's World," by Mike Laptew, *Salt Water Sportsman*, October 1995, p. 57.

Chapter 1: A Tale of Two Fishes

13. SMITH ON CHESAPEAKE: "The Chesapeake: Who Will Save the Bay?" (newspaper format), by Jack Bowie, Chesapeake Publishing Group, Easton, Maryland, December 7, 1983.

13. SMITH ON BASS: *Striper: A Story of Fish and Man*, by John N. Cole, Atlantic–Little, Brown, 1978, p. 32.

13. NATIVE AMERICANS: "Some Early Notes on Striped Bass," by D. B. Fearing, *Transactions of the American Fisheries Society* 32:90–98, 1903.

13. NARRAGANSETT INDIANS: "The Golden Era of Surfcasting in Rhode Island," by Fred Thurber, unpublished paper, 1983.

13–14. DE RASIERES QUOTE: Cole, *Striper*, p. 32.

14. PLYMOUTH SETTLERS: *History of New England, from the Year 1620 to the Year 1680*, by William Hubbard, quoted in Thurber, "The Golden Era of Surfcasting."

14. WOOD QUOTE: Cole, *Striper*, p. 32.

14. FIRST FISHING REGULATION: *The Complete Book of the Striped Bass*, by Nicholas Karas, Winchester Press, New York, 1974, p. 101.

14. PUBLIC SCHOOLS: "The Life History of the Striped Bass, or Rockfish,

Roccus saxatilis (Waldbaum)," by J. C. Pearson, *U.S. Bureau of Fisheries Bulletin* 49:825–860, 1938.

14–15. GEORGE WASHINGTON: *This Was Potomac River*, written and published by Frederick Tilp, Alexandria, Virginia, 1978.

15. MASSACHUSETTS AND NEW YORK LEGISLATURES: "Tall Tales," by J. D. Bayless, *Wildl. N.C.* 28(5):4–6, 1964.

15. "EMINENT AMONG THE CELEBRATED FISHERMEN": *The Striped Bass Chronicles: The Saga of America's Great Game Fish*, by George Reiger, Lyons & Burford, 1997, pp. 25–26.

16. INDUSTRIAL REVOLUTION: "Decline of the Striped Bass," by Nick Karas, *Newsday*, April 12, 1982.

16. 1896 CATCH: "Make the Striped Bass a Game Fish," ed. by Kip Farrington, *Field & Stream*, April 1941, quoting from *American Food and Game Fishes*, by David Starr Jordan and Barton Warren Evermann, Doubleday, New York, 1902.

16. CHESAPEAKE & DELAWARE CANAL: Karas, *The Complete Book of the Striped Bass*, p. 32.

17. "GREAT AWAKENING OF PUBLIC INTEREST": "The Striped Bass Problem," by William C. Neville, September 1942 presentation provided to the author by Fred Schwab.

17. SEINERS CATCH: "SOS from a Seaside Slaughter," by Robert H. Boyle, *Sports Illustrated*, November 6, 1972.

18. 1981 SURVEY: "Fisheries: An Offspring of the Ice Age, the Striper May Not Survive Nuclear Age," by John N. Cole, *Audubon*, March 1984, p. 113.

19. "CHUNKER FISHING" STORY: A version of this story first appeared in "A Vanishing Treasure," by Dick Russell, *Boston Globe Magazine*, July 29, 1984, p. 12.

21. COMMERCIAL LANDINGS: Ibid., p. 14.

23. MENHADEN FIGURES: "Atlantic Menhaden 2003 Stock Assessment Report," Atlantic States Marine Fisheries Commission (ASMFC), October 2003, p. 25.

24. COLE QUOTE: Cole, *Striper*, p. 9.

25. CATCH FIGURES: "2003 Atlantic Striped Bass Advisory Report," ASMFC Striped Bass Technical Committee Report #2003-03, p. 24.

25. COMMERCIAL INDUSTRY: "Conflicts of Interest, Lack of Diverse Perspectives and Flawed Institutional Design Plague Management of U.S. Fishing Industry" (press release), by the Pew Oceans Commission, November 11, 2003.

25. FISH STOCKS: "Has the Sea Given Up Its Bounty?" by William J. Broad and Andrew C. Revkin, *New York Times*, July 29, 2003, p. D2.

25. OVERFISHING: "The State of America's Marine Fisheries" (report), Oceana, 2003.

25. NATURE STUDY: "Commercial Fleets Slashed Stocks of Big Fish by 90%, Study Says," by Andrew C. Revkin, *New York Times*, May 15, 2003, p. 1.

26. PEW REPORT: "America's Living Oceans: A Course for Sea Change"

(report), Pew Oceans Commission, June 2003, www.pewoceans.org. The fundamental message underlying the 145-page report is that oceans are ecosystems and must be managed as such.

27. MOUNTFORD: Author's interview, July 30, 1998.
27. SHEPHERD: Author's interview, July 1998.
29. NCMC PETITION: "NCMC Launches New Campaign to Protect Menhaden, Save Striped Bass" (press release), by the National Coalition for Marine Conservation, June 13, 2003, www.savethefish.org.
29. OMEGA PLANT: "Omega Protein to Build New Fish Oil Processing Facility" (press release), by Omega Protein, April 15, 2003, www.omega proteininc.com.

Chapter 2: Storm over the Hudson

31. SCULLY QUOTE: *The Hudson River: A Natural and Unnatural History*, by Robert H. Boyle, W. W. Norton, New York, 1969, p. 172.
32. RATHJEN-MILLER STUDY: Ibid., pp. 137–138.
32. "SIX BILLION GALLONS": Ibid., p. 154.
33. "SIX TIMES HIGHER": Ibid., p. 159.
33. ROCKEFELLER STATEMENT: *The Riverkeepers*, by John Cronin and Robert F. Kennedy Jr., Scribner, New York, 1997, p. 28.
33. PERLMUTTER TESTIMONY: Cited in ibid., pp. 29–30.
34. BOYLE: Author's interview, February 12, 2004, Cooperstown, New York.
35. PIRONE QUOTE: "A Stink of Dead Stripers," by Robert H. Boyle, *Sports Illustrated*, April 26, 1965, p. 81.
36. BOYLE QUOTE: Boyle, *The Hudson River*, p. 160.
36. YELLOT QUOTE: Cronin and Kennedy, *The Riverkeepers*, p. 32.
38. GARRISON: "E-Law: What Started It All?" by the Natural Resources Defense Council (NRDC), www.nrdc.org/legislation/helaw.asp.
38. "OUTSTANDING ICHTHYOLOGIST": Boyle, *The Hudson River*, p. 165.
38. "FISHERIES QUESTION": Cronin and Kennedy, *The Riverkeepers*, p. 33.
38. STORM KING DOCTRINE/NEPA: Ibid., p. 36.
38. HUDSON DEBRIS: "At Age 73, Advocate Refuses to Quit Fight to Keep the Hudson River Free of Pollution," by Michael Virtanen, Associated Press, March 22, 2002.
38. BOYLE QUOTE: Author's interview, 2004.
39. BUTZEL QUOTES: Author's interview, February 15, 2004, New York.
40. CLARK TESTIMONY: Cronin and Kennedy, *The Riverkeepers*, p. 34.
40–41. NRDC BEGINNINGS: NRDC, "E-Law."
41. CLARK/INDIAN POINT: Cronin and Kennedy, *The Riverkeepers*, p. 35. See also "The Nukes Are in Hot Water," by Robert H. Boyle, *Sports Illustrated*, January 20, 1969.
41–42. BOREMAN QUOTE: Author's interview, January 9, 2004, Woods Hole, Massachusetts.
42. YOST QUOTE: Cronin and Kennedy, *The Riverkeepers*, p. 38. Judge Yost

wrote about the case in "Science in the Courtroom," *American Fisheries Society Monograph* 4:294–301, 1988.

42. FOUNDATION ESTABLISHED: "Long-Abused Hudson Thrives Again," by Sara Rimer, *New York Times*, November 6, 1986.

43. INDIAN POINT FISH KILLS: "Study Finds Power Plants Harm Fish in the Hudson," by Al Baker, *New York Times*, July 11, 2003.

44. STRIPED BASS KILLS: "Lawsuits Force State to Reconsider Power Plant Fish Kills," *Riverkeeper*, Fall 2003.

44. DEC ORDER: "Plan Would Reduce Fish Deaths Caused by Nuclear Plant," by Lisa W. Foderaro, *New York Times*, November 13, 2003, p. A28.

44. RIVERKEEPER COURT VICTORY: "Fisheries and Power Plant Project Campaign Updates," by Reed Super, *Riverkeeper*, Spring 2004, p. 15.

44. MATTHIESSEN QUOTE: Author's interview, April 2004.

45. PCBs: "Muddy Waters," by Michael A. Rivlin, *The Amicus Journal*, Winter 1998.

45. RISEBROUGH QUOTE: "Poison Roams Our Coastal Seas," by Robert H. Boyle, *Sports Illustrated*, October 26, 1970.

46. BOYLE SEINING QUOTE: Author's interview, 2004.

47. BOYLE, *SPORTS ILLUSTRATED*, AND PCBs: Drawn from the author's interview with Boyle as well as from Boyle's landmark 1970 article.

47. BASS AND PCBs: One striped bass showed levels of 350 parts per million, "a new record for PCB contamination of freshwater fish": Cronin and Kennedy, *The Riverkeepers*, p. 60.

47. GE'S PCB DISCHARGES: "Toxics on the Hudson," by Charlie Cray, *Multinational Monitor*, July/August 2001, p. 12.

47. PCB BAN: Cray, "Toxics on the Hudson." The congressional ban applied except in "totally enclosed" systems.

48. AL REINFELDER: The entire May 31, 1973, issue of *Long Island Fisherman*, in which his letter appears, is devoted to Reinfelder.

Chapter 3: The Conscience of a Lure Maker

49–50. CATCH FIGURES/JUVENILE INDEX: "Update: Striped Bass," by Hal Lyman, *Salt Water Sportsman*, February 1981, p. 83.

50. FLORENCE QUOTE: "Official Warns on Demise of Striped Bass in 5 Years," by Nelson Bryant, *New York Times*, February 19, 1978.

51. BASS SALE FIGURES: "The Venerable Striped Bass," by Dick Russell, *The Amicus Journal*, Fall 1982, p. 35.

52. SIZE LIMIT CONTROVERSY: "The Battle to Save the Striped Bass," by Dick Russell, *Boston*, July 1982.

53. BOONE QUOTE: "1982 Tidewater Fishing Forecast and Homily," paper by Joe Boone.

55. BOB POND: In addition to private conversations, the story of Pond's work was derived from the following: *Stripers Unlimited 1973–1975 Research*

Guide Book (newspaper format); *Stripers Unlimited 1976 Guide Book* (newspaper format); untitled paper by Pond with charts, Summer 1982; "Bob Pond and Stripers," by Vlad Evanoff, *Striper*, January–February 1984; "A Vanishing Treasure," by Dick Russell, *Boston Globe Magazine*, July 29, 1984; "Striped Bass Egg Quality," paper presented by Bob Pond, March 1994; "Thank You Mr. Striper," by Charley Soares, *The Fisherman*, June 29, 1997; and "The Striper: A Great and Noble Fish," by John Pollock, *The Fisherman*, May 16, 2002. Pond's work is also referenced in *Striper: A Story of Fish and Man*, by John N. Cole, Atlantic–Little, Brown, 1978.

58. MERRIMAN QUOTE: "Studies on the Striped Bass of the Atlantic Coast," by Daniel Merriman, *Fish and Wildlife Service Fisheries Bulletin* 35:11.

63. NEW BEDFORD PCBs: see www.buzzardsbay.org/nbprobs.htm.

64. BOONE QUOTE: Author's interview, 2004.

64. BARBEAU QUOTE: Author's interview, 2004.

65. CHESAPEAKE: "The Chesapeake: Who Will Save the Bay?" (newspaper format), by Jack Bowie, Chesapeake Publishing Corp., Easton, Maryland, December 7, 1983.

68. MAYOR WHITE LETTER: Dated March 26, 1982, the two-page letter called for Massachusetts to adopt the twenty-four-inch limit as "a necessary and even minimal step" toward preserving the striped bass.

69. GUDE LETTER: Reproduced in full in Russell, "The Battle to Save the Striped Bass."

69. MASSACHUSETTS LIMIT: "Stricter Striped-Bass Rules Voted," by William Mills, *Cape Cod Times*, April 3, 1982.

Chapter 4: A Man Named Mendonsa

71. MENDONSA ON LEGISLATURE: Author's telephone interview, January 2004.

71. MENDONSA BACKGROUND: In addition to author's interview, January 18, 2004, in Middletown, Rhode Island, Mendonsa provided the author with the transcript of a 194-page interview he gave the Newport Historical Society, February 11, 1987 ("George Mendonsa: The Fishing Industry in Newport, Rhode Island, 1930–1987").

73. MENDONSA IN SAILOR PHOTOGRAPH: Although controversy continues over whether Mendonsa is indeed the sailor in the Eisenstadt photo (several others have claimed to be), strong evidence exists to support his claim. The story has been recounted in several articles, including "A Magic Moment, a Kiss, a Lawsuit," by David Arnold, *Boston Globe*, August 14, 1988, who reported that Mendonsa was suing Time, Inc., for $100,000 "because the magazine was profiting by selling $1,600 prints" of his picture.

75. JIM WHITE: The author featured White in "Who Will Save the Striped Bass?" *Yankee*, June 1983, p. 53, which also discussed Mendonsa and the Massachusetts effort in Rhode Island.

77. RHODE ISLAND FIGHT: "Mass. Man Brings Striper Fight Here," by Les Boyd, *Providence Evening Bulletin*, April 14, 1982.
79. WASHINGTON, D.C., TRIP: April 28–29, 1982.
81. CONFLICT-OF-INTEREST LAWS: "Conflict Arguments Heard for Fisheries Board Members," *Providence Journal*, May 16, 1982.
82. EDITORIAL: "R.I.'s Turn to Protect the Striped Bass," *Providence Journal/Evening Bulletin*, May 10, 1982.
85. NET BUST: "Bass Netters Arrested off Gay Head," by Dick Russell, *New England Fisherman*, July 8, 1982; "Netting Striper Violators," by Tony Chamberlain, *Boston Globe*, June 1, 1982; "High Speed Sea Chase Snares Fishermen with Gill Net Haul," by Holly Higinbotham, *Vineyard Gazette*, June 1, 1982.
86. RHODE ISLAND VOTE: "Surprising Vote Puts Striped Bass Regs in Place," *Commercial Fisheries News*, October 1982.
87. NETTERS FINED: "New Bedford Men Net Vineyard Fines," *Cape Cod Times*, August 13, 1982.
88. TRIP TO CHESAPEAKE: "The Venerable Striped Bass," by Dick Russell, *The Amicus Journal*, Fall 1982, p. 30.
89. SIMNS QUOTE: Ibid., p. 38.
90. *NEWSWEEK* ARTICLE: "Schemes to Save the Noble 'Striper,'" *Newsweek*, September 13, 1982, p. 17.

Chapter 5: How the Striped Bass Stopped a Highway and Eluded the Mob

91. WESTWAY BACKGROUND: Drawn from several sources, including "Westway Halted—Fish Have Right-of-Way," by Dick Russell, *Sierra*, January–February 1983, p. 21; "Life Amidst the Ruins," by Fred Powledge, *The Amicus Journal*, Summer 1982, p. 44.
91. TRUMP QUOTE: Boyle e-mail to author, January 16, 2005.
93. BOYLE QUOTE: Author's interview, 2004.
93. SILVERSTEIN QUOTE: Author's interview, 2004.
93. BUTZEL QUOTE: Author's interview, 2004.
94. EIS/LMS STUDIES: Russell, "Westway Halted"; "NYC Westway Project Challenged in Court by Striped Bass Advocates," by Bruce Stutz, *National Fisherman*, May 1982.
94. STRIPER MIGRATION: "Pier Area on West Side Called Vital to Survival of Striped Bass," by Richard Severo, *New York Times*, April 25, 1982, p. 50.
94. 15 TO 105 TIMES: Russell, "Westway Halted."
95. "MILD WINTER . . . MANY AVAILABLE HABITATS": Russell, "Westway Halted"; "Indignation over Westway," *Sports Illustrated*, July 12, 1982, p. 11.
95. FLETCHER AND THE *TIMES*: *The Riverkeepers*, by John Cronin and Robert F. Kennedy Jr., Scribner, New York, 1997, p. 160.
95. FLETCHER QUOTE: Ibid., p. 161.
96. KOCH QUOTE: "Save the Bass, Dump the Road," by John N. Cole, *Newsday*, March 29, 1984.
96. BIOLOGIST QUOTE: Severo, "Pier Area on West Side Called Vital," p. 50.

96. BOYLE QUOTE: "Westway Threatens Hudson Stripers," by Dick Russell and Rachel Greenwood, *Long Island Fisherman*, March 15, 1984, p. 14.

97. GRIESA BACKGROUND: "Meticulous Judge in Westway Case," by E. R. Shipp, *New York Times*, April 22, 1982.

97. GRIESA QUOTE: Cronin and Kennedy, *The Riverkeepers*, p. 162.

97. GRIESA STATEMENTS: "U.S. Judge Blocks Westway Landfill as Threat to Bass," by Arnold H. Lubasch, *New York Times*, April 1, 1982, p. 1; "Court Order Spares Hudson Stripers from Westway," by Bruce Stutz, *National Fisherman*, June 1982, p. 27.

98. CAREY QUOTE: Powledge, "Life Amidst the Ruins," p. 44.

98. CONASON QUOTE: Ibid, p. 47.

98. CASE HISTORY: "U.S. Judge Halts Landfill Project for the Westway," by Arnold H. Lubasch, *New York Times*, April 15, 1982, p. B1; "Westway: Third-Rate Cover-Up," by Sydney H. Schanberg, *New York Times*, April 20, 1982, p. A27; "Judge Bars $90 Million U.S. Payment for Westway," by Arnold H. Lubasch, *New York Times*, April 21, 1982, p. 1; "Westway Effects to Be Restudied; Process May Take a Year or More," by Michael Goodwin, *New York Times*, April 29, 1982, p. 1; "Hudson Stripers Have Clout," SFI Bulletin, *New England Fisherman*, July 1, 1982; "A Judge Blocks All U.S. Money for the Westway," by Arnold H. Lubasch, *New York Times*, July 1, 1982, p. 1; "Westway Master Gets Judge's Orders," by Arnold H. Lubasch, *New York Times*, July 24, 1982.

99. BUTZEL QUOTE: Author's interview, 2004.

99. BOYLE QUOTE: "Letter from the Publisher," by Philip G. Howlett, *Sports Illustrated*, April 26, 1982.

100. MASSACHUSETTS LIMIT: "Massachusetts Keeps 24 Inch Bass Limit," by Dick Russell, *New England Fisherman*, May 26–June 1, 1983, p. 8.

101. TIMES AD: "One Man Against the Survival of a Species," *New York Times*, April 1, 1970, p. 31.

102. SEINERS: "SOS from a Seaside Slaughter," by Robert H. Boyle, *Sports Illustrated*, November 6, 1972, p. 24; "3 L.I. Fishing Crews Leave Carolina After Violence Threat," *New York Times*, December 10, 1972.

103. FULTON FISH MARKET AND THE MOB: "New York Fish Market Points Up a Pattern of Extortion and Fear," by Stanley Penn, *Wall Street Journal*, April 14, 1982, p. 1; "Three More Indicted in Fulton Investigation," by Bruce Stutz, *National Fisherman*, June 1982.

104. BOREMAN QUOTE: Author's interview, 2004.

104. LEGISLATIVE DEBATE: "New York Legislature Passes 24 Inch Bill on Striped Bass," by Dick Russell, *New England Fisherman*, July 21–27, 1983, p. 6.

105. SCHWAB STORY: Author's interview, 2004.

106. CUOMO: "Cuomo's Cynical Westway Strategy," by Joe Conason, *Village Voice*, November 15, 1983.

106. "SINCE CUOMO'S ASCENSION": Ibid.

106. BUTZEL QUOTE: Author's interview, 2004.

107. "ENEMY OF THE ENVIRONMENTAL MOVEMENT": Conason, "Cuomo's Cynical Westway Strategy."

108. EDITORIAL: July 25, 1983, cited in "New York Enacts New Striped Bass Law," by Dick Russell, *New England Fisherman*, September 8, 1983.
109. *GLOBE* EDITORIAL: "The Striped Bass Massacre," *Boston Globe*, cited in ibid.
109–113. CUOMO AND THE BASS BILL: "New York Enacts New Striped Bass Law," by Dick Russell, *New England Fisherman*, September 8, 1983.
113. SCHWAB QUOTE: Author's interview, 2004.

Chapter 6: How Rhode Island Changed the World

115–118. COUNCIL MEETING: The account of what transpired is drawn primarily from the author's article "Rhode Island Bans Striped Bass Fishing for One Year," *New England Fisherman*, September 22–28, 1983.
118. MENDONSA QUOTE: Author's interview, 2004.
118. "PERVERSE BOOST": "How're They Spitin'?" *Sports Illustrated*, September 19, 1983, p. 19.
118. EDITORIAL: "Captains Outrageous," *Boston Globe*, September 11, 1983, p. 24.
119. RHODE ISLAND CONTROVERSY: "Compromise Sought on Striped Bass Ban," by Dick Russell, *New England Fisherman*, October 13–19, 1983, p. 10.
120. ACID RAIN: An excellent overview is "A Rain of Death on the Striper?" (special report), by Robert H. Boyle, *Sports Illustrated*, April 23, 1984, p. 40. Also drawn from "Pollution in the Chesapeake: Decline of the Striper," by Stephen Crocker and Dick Russell, *Long Island Fisherman*, March 1, 1984, p. 17.
124–126. THREE-YEAR BAN: "Rhode Island Extends Ban to Three Years," by Dick Russell, *New England Fisherman*, October 27, 1983, p. 8; "Sports, Commercial Polarize on 24" Issue," *Commercial Fisheries News*, October 1983, p. 40.
127. MARYLAND PLAN: "Controversy over Maryland Bass Plan," by Dick Russell, *Long Island Fisherman*, December 1–7, 1983, p. 10.
127. BIOLOGIST QUOTE: Ibid.

Chapter 7: Showdown at Friendship Airport

129–131. BOREMAN QUOTES: Author's interview, 2004.
133–135. BASS CONFERENCE: "Marine Experts Say Urgent Steps Are Required to Aid Striped Bass," by Philip Shabecoff, *New York Times*, December 13, 1983, p. B9; "Parley Looks at Ways to Protect Rockfish," by Tom Horton, *Baltimore Sun*, December 13, 1983; "Rockfish Need Relief, but Who's Going to Provide It?" by Bill Burton, *Baltimore Evening Sun*, December 13, 1983, p. C5; "States Must Act to Save Rockfish, Environmentalists Told," Associated Press, December 13, 1983; "Conservationists Hope This Won't Be the Species That Got Away," *New York Times*, December 18, 1983, News of the Week in Review section; "Battle to Save Rockfish Is Tantamount to Saving the Chesapeake," by Dennis Collins, *Washington Post*, December 18, 1983, p. D9; "Striped Bass, Atlantic Salmon Face

Extinction," by Tony Chamberlain, *Boston Globe*, December 18, 1983, p. 106; "The Striper Emergency: What Is to Be Done?" by Dick Russell, *New England Fisherman*, January 19–25, 1984; "Saving the Bass: Writer Wages 1-Man Crusade to Stem Their Rapid Decline," by Angus Phillips, *Washington Post*, January 16, 1984.

137. EDITORIAL: "Rockfish Are Not Forever," *Baltimore Sun*, December 18, 1983.

139. MARYLAND CONTROVERSY: "State's Rockfish Conservation Claims Called Inaccurate," by Tom Horton, *Baltimore Sun*, December 25, 1983, p. B1. Other articles: "Agency Orders Leaks on Rockfish Stopped," by Peter Jensen, *Easton* [Maryland] *Star-Democrat*, December 28, 1983, p. 1; "Rock Gag Order Irks Workers," by John Hartsock, *Annapolis Capital*, January 13, 1984, p. 1; "Md. Rules Spawn Striper Battle," by Les Boyd, *Providence Journal*, January 15, 1984, p. D12.

139. BIOLOGIST QUOTED: "Politics: Adrift on the Chesapeake," by Tony Chamberlain, *Boston Globe*, January 10, 1984, p. 25.

141. MARYLAND LEGISLATURE: "State Opposes Rockfish-ban Bill," by Tom Horton, *Baltimore Sun*, January 14, 1984.

141. MENDONSA CONFLICT: "Bass Controversy Leads to Big Fine," by Nelson Bryant, *New York Times*, January 29, 1984, sports section; "Coastal Striper Update: Conflicts in Rhode Island, Deception in Maryland," by Dick Russell, *Maryland-Delmarva Fisherman*, February 16–23, 1984. See also "State of Rhode Island Decision and Order," Conflict of Interest Commission, Complaint Nos. 83-10, 83-13, 83-17, 83-16, January 10, 1984, Wilfred H. Barbeau Jr., Charles A. Matley, J. Donald Sylvestre, Complainants, vs. George Mendonsa, Respondent.

144. MORATORIUM BILL: "Bill Would Prohibit Striped-bass Fishing," by Michael F. Bamberger, *Boston Globe*, February 18, 1984, p. 18; "Time to Save Striped Bass," by Michael Globetti, *Boston Herald*, February 24, 1984; "Decline of Celebrated Striped Bass Stirs Congressional Concern," by Bob Drogin, *Los Angeles Times*, March 3, 1984, p. 3; "Striper Ban Bill Introduced in Congress," by Dick Russell, *New England Fisherman*, March 8, 1984, p. 19; "U.S. May Protect Striped Bass in East," by Nelson Bryant, *New York Times*, March 18, 1984, section 5.

Chapter 8: Revolt of the Biologists

146. EDITORIAL: "A Debt to the Striped Bass," *Boston Globe*, March 19, 1984, p. 12.

146. RHODE ISLAND MEETING: "Rhode Island Will Keep Striper Ban . . . for Now," by Dick Russell, *New England Fisherman*, March 29, 1984, p. 19.

147–148. CONGRESSIONAL HEARING: "Administration Opposes Ban on Fishing for Bass," by Martin J. Funke, *Providence Journal*, March 21, 1984, p. 2; "Proposed Ban on Stripers Gets a Hearing," by Pamela Glass, *Cape Cod Times*, March 21, 1984; "House Holds Hearing on Striper Moratorium; Senate Bill Coming," by Dick Russell, *Maryland-Delmarva Fisherman*,

April 12, 1984, p. 16; "Rockfish Question Starts Bubbling Up in Congress," by Alison Muscatine, *Washington Post*, April 14, 1984.

149. STUDDS QUOTE: Author's interview, 1984. See also "Striped Bass Caught in Net of Political Wrangling," by Pamela Glass, *Cape Cod Times*, March 24, 1984, p. 1; "Saving the Striped Bass" (op-ed), by Gerry Studds, *Boston Globe*, March 31, 1984, p. 14.

149. ACID RAIN: "A Rain of Death on the Striper?" (special report), by Robert H. Boyle, *Sports Illustrated*, April 23, 1984, p. 40.

150. UPHOFF: "Rockfish Aide Alleges Harassment," by Tom Horton, *Baltimore Sun*, April 22, 1984, p. C1; "*Fisherman* Article Subject of Maryland Controversy," by Dick Russell, *Maryland-Delmarva Fisherman*, May 3, 1984.

150. *SUN* EDITORIAL: "Fingering the Whistleblower," *Baltimore Sun*, May 5, 1984, editorial page.

150–151. SWEZEY LETTER: "Congress Blasts State Striper Plans," by Dick Russell, *New England Fisherman*, June 21, 1984, p. 16.

151. RESTAURANTS: "Cape Eateries Take Bass off Menus," by Michael F. Bamberger, *Boston Globe*, April 29, 1984; "To the Rescue of the Striped Bass," by Michael F. Bamberger, *Boston Globe*, May 20, 1984.

151. TOURNAMENTS: "Noble Gestures Made on Behalf of Stripers," by Tony Chamberlain, *Boston Globe*, August 12, 1984, p. 67; "Fate of the Chesapeake Striper Up to Congress," by Dick Russell, *Long Island Fisherman*, June 21–27, 1984, p. 16.

152. RHODE ISLAND RELAXES: "Rhode Island Adopts New Bass Regulations," by Dick Russell, *New England Fisherman*, July 19, 1984, p. 12.

153. CORPS STUDY: "Army Engineers to End Fish Study a Year Early," by Jane Perlez, *New York Times*, December 16, 1983, p. B3.

153. "SIGNIFICANT ADVERSE IMPACT": "Corps Finds Hudson Bass Habitat Would Be Damaged by Westway," by Sam Roberts, *New York Times*, May 16, 1984, p. 1.

153. WESTWAY OP-EDS: "Instead, Call It Wasteway," by John B. Oakes, *New York Times*, June 2, 1984; "Corruption Ignored," by Sydney H. Schanberg, *New York Times*, June 9, 1984.

155. WESTWAY HEARINGS: "Hearings Open on Whether Westway Is 'Imperative' or Is a 'White Elephant,'" by Sam Roberts, *New York Times*, June 27, 1984, p. B1; "Westway Hearings Raised Dozens of Questions, Army Corps Official Says," by Sam Roberts, *New York Times*, June 28, 1984; "Rockfish May Spawn Westway's End," by Tom Horton, *Baltimore Sun*, July 2, 1984. See also "Bass: Why Is Hudson So Important?" by Sam Roberts, *New York Times*, June 26, 1984, Science Times section.

155. QUOTES ON WESTWAY: "There Are Too Some New Things to Say about the Westway," *New York Times*, July 1, 1984, News of the Week in Review section.

155–156. FDA RULING: "Tougher Controls Limiting PCB's," by Nelson Bryant, *New York Times*, June 3, 1984, section 5.

156. SCHWAB QUOTES: Author's interview, 2004.

156. FISHERMEN AND FDA RULING: "Industry Gloomy over Lowered PCB Tolerance Levels," by Bruce Stutz, *National Fisherman*, August 1984, p. 2.

157. SCHNEIDER STATEMENT: "New FDA Rules and the Striped Bass," *Congressional Record*, May 31, 1984.

157. STRIPER SURVEY: "Rockfish Spawning Again Low," by Tom Horton, *Baltimore Sun*, August 19, 1984, p. C1.

158. POND PETITION: "Bass Endangered Species?" by Michael Globetti, *Boston Herald*, August 12, 1984, p. 73.

158–161. MARYLAND AND BASS: Much of this story originally appeared in the author's article "Fisheries Management in the Chesapeake," *Amicus Journal*, Fall 1984, p. 20.

161–162. VINEYARD DERBY CONTROVERSY: "Petition Seeks an End to Derby Bass Fishing," by Mark Alan Lovewell, *Vineyard Gazette*, August 21, 1984; "Fishermen Back Derby Position on Catching the Bass," by Mark Alan Lovewell, *Vineyard Gazette*, August 24, 1984; "Putting the Rod to Vineyard Derby," by Michael Globetti, *Boston Herald*, August 27, 1984, p. 46; "Martha's Vineyard Striped Bass Derby Draws Fire," by Gary Ghioto, *Boston Globe*, August 28, 1984, p. 19; "Chamber Backs Bass Tourney," by Mark Alan Lovewell, and "The Hunt Must End," editorial, *Vineyard Gazette*, August 31, 1984; "Rally Round the Bass, Boys," editorial, *Boston Globe*, September 2, 1984, p. 6; "Stripers to Stay in Bass Derby," *Martha's Vineyard Times*, September 6, 1984; "The Striped Bass: A Plea for Release," full-page advertisement, *Vineyard Gazette*, September 14, 1984, p. 9.

162. MARYLAND MORATORIUM: "Maryland to Ban Fishing for Rockfish After Jan. 1," by Tom Kenworthy, *Washington Post*, September 12, 1984, p. 1, with accompanying stories; "State Bans Harvest of Rockfish in 1985," by John W. Frece, *Baltimore Sun*, September 12, 1984, p. 1; "Hold the Rockfish," editorial, *Washington Post*, September 15, 1984, p. 18; "Rockfish: Why Md. Decided on the Ban," by John W. Frece, *Baltimore Sun*, September 16, 1984, section C; "Maryland Tries to Save the Striped Bass," by Michael Wright, *New York Times*, September 30, 1984, p. E8; "Maryland Enacts Striper Moratorium," by Dick Russell, *New England Fisherman*, October 4, 1984, p. 12; Russell, "Fisheries Management in the Chesapeake."

163. HUGHES QUOTES: Author's interview, 2004.

163. BOONE QUOTE: Author's interview, 2004.

164. HARRISON QUOTE: Author's interview, 2004.

Chapter 9: Striper Magic

165–167. The story of congressional passage of the striped bass bill is drawn primarily from the author's article "Congress Passes Striper Protection Bill," in *New England Fisherman*, November 8, 1984, p. 8. Also, "Studds Bill for Stricter Fishing Protection of Striped Bass Wins Unanimous House Vote," by Tony Chamberlain, *Boston Globe*, October 5, 1984, p. 16; "Striper Bill Gasps for Air in Senate," by Michael Globetti, *Boston Herald*, October 5, 1984, p. 61; "U.S. Senate Approves Bill to Lessen Fishing of Depleted Striped Bass," by Tony Chamberlain, *Boston Globe*, October 12, 1984, p. 67; "Striped Bass Hooks Federal Protection," by Michael Globetti, *Boston Herald*, October 12, 1984.

166. WILL QUOTE: "Save the Rockfish: Make Maryland's Moratorium National," by George F. Will, *Washington Post*, October 7, 1984, Opinion/Commentary section.

167. NEW WESTWAY REVIEW: "Army Determines Building Westway Could Harm Bass—But Report by Engineers Calls Impact 'Minor'—Ruling Expected in January," by Sam Roberts, *New York Times*, November 28, 1984, p. 1.

167–168. LANDFILL PERMIT/QUOTES: "Westway Landfill Wins the Support of Army Engineer," by Sam Roberts, *New York Times*, January 25, 1985, p. 1.

168. EPA MEMORANDUM: "Rockefeller Pressured EPA to Back Westway," by Arthur K. Lenehan, *Star-Ledger* [Newark, N.J.], June 25, 1985; "Getting 'the Message,'" by Sydney H. Schanberg, *New York Times*, July 2, 1985, editorial page.

168. ROCKEFELLER LETTER: Lenehan, "Rockefeller Pressured EPA."

168. "FALLACIOUS": Ibid.

169. DOVEL TESTIMONY: "Biologist Says Westway Poses No Threat to Bass," by Arnold H. Lubasch, *New York Times*, June 30, 1985, p. 28; "The Testimony Doesn't Wash," by Sydney H. Schanberg, *New York Times*, July 6, 1985, p. 21; "Documents Trace Rockefeller Role in Financing Westway Bass Study," by Arthur K. Lenehan, *Star-Ledger* [Newark, N.J.], July 3, 1985, p. 1.

169. BOREMAN QUOTE: Author's interview, 2004.

170. GRIESA QUOTE: Schanberg, "The Testimony Doesn't Wash."

170. GRIESA, "MATTER TO AN END": *The Riverkeepers*, by John Cronin and Robert F. Kennedy Jr., Scribner, New York, 1997, p. 163.

170. *NEWSWEEK*: "The Death of a 'Boondoggle'?" *Newsweek*, August 19, 1985, p. 28.

170. KEMPTON: "It Was a Fish Story That Tied a Noose Around Westway," by Murray Kempton, *Newsday*, August 9, 1985.

170. WESTWAY DEMISE: "New York Leaders Give Up Westway and Seek Trade-In," by Michael Oreskes, *New York Times*, September 20, 1985, p. 1; "Westway Sleeps with the Fishes," by Joe Conason, *Village Voice*, October 1, 1985, p. 13. Also, "Editorial: Westway, 1971–1985," *Village Voice*, p. 3.

170–171. HUDSON RIVER PARK: Brochure, "Friends of Hudson River Park," provided to the author by Al Butzel, 2004.

171. BUTZEL QUOTES: Author's interview, 2004.

171. NEW YORK PCB STUDY: "Dealing with the Sensitive Issue of PCB-Contaminated Striped Bass," by Edward A. Gargan, *New York Times*, March 27, 1985; "Koch Studies a Ban on Sale of Striped Bass," by Joyce

Purnick, *New York Times*, March 30, 1985; "East End Baymen's Way of Life Hinges on Striped Bass," by John Rather, *New York Times*, March 31, 1985.

172. PCB BAN: "State Broadens Ban on Striped-Bass Fishing," by Edward A. Gargan, *New York Times*, April 1, 1985; "Ban on Striped Bass— Environmentalists Say Albany Acted on Basis of Politics Instead of Science," by Edward A. Gargan, *New York Times*, April 3, 1985.

172. HALPIN QUOTE: "Cuomo Issues Limited Bass Ban," by Geraldine Baum, *Newsday*, April 1, 1985.

172. "STING" OPERATION: "Sting Operation Hits Illegal Bass Sales," by Susan Kellam, *National Fisherman*, April 1985, p. 10.

172. SURE ADVERTISEMENT: *Boston Herald*, May 3, 1985, p. 72; "Detergent Company Lands in Dirty Water—Again," by Michael Globetti, *Boston Herald*, May 6, 1985, p. 54; "Bass Ad: P&G Won't Change Stripes," by Michael Globetti, *Boston Herald*, May 8, 1985, p. 75.

173. VINEYARD DERBY CONTROVERSY: *Vineyard Gazette* articles by Mark Alan Lovewell: "Chamber Throws Weight Behind Bass Tourney; Blasts Gazette Coverage of Fishing Controversy," May 10, 1985, p. 1; "Selectmen Attack Bass Prizes; Sponsors of Derby Withdraw," May 17, 1985, p. 1; "Derby Drops Striped Bass from Tournament in a Dramatic Reversal of Previous Position," May 24, 1985, p. 1; "Derby Support Gains Strength and Sponsors," May 31, 1985, p. 1; "An Act of Vision," editorial, May 31, 1985, p. 16. Also, "Bass from the Brink," editorial, *Boston Globe*, May 24, 1985, p. 20; "Salvation for Stripers Left on Hook by Official," by Michael Globetti, *Boston Herald*, June 24, 1985, p. 54.

175. RHODE ISLAND PCB BAN: "R.I. Bans Sale of PCB-Laden Stripers," by Ken Weber, *Providence Journal*, April 8, 1986, p. 1.

175. NEW YORK PCB BAN: "State to Ban All Commercial Striped Bass Fishing," by Harold Faber, *New York Times*, April 22, 1986, p. A1.

175. EDITORIAL: "Piscatory PCBs," *Boston Globe*, April 28, 1986, p. 18.

175. CHAFEE: "Striper Regulations Scrutinized by a Congressional Committee," by Mark Alan Lovewell, *Vineyard Gazette*, July 26, 1985.

175. 1985 INDEX: "Hopes for Big Rockfish Crop Dashed," by Tom Horton, *Baltimore Sun*, July 19, 1985, p. C1. See also "Trying to Save the Striped Bass," by Nelson Bryant, *New York Times*, July 14, 1985.

176. BASS STUDY: "Handful of Rockfish Given Credit for Bumper '82 Spawn," by Tom Horton, *Baltimore Sun*, January 19, 1986.

176. "ALMOST VIRGIN STOCK": "Stripers: Running the Numbers," by Keith Walters, *Salt Water Sportsman*, April 1987, p. 88.

176. STRIPER SPAWNING: "Bay Scientists Collect More Data on Rockfish, but Big Question Still Unanswered," by Tom Horton, *Baltimore Sun*, August 24, 1986, p. E3.

177. STRIPER REGULATIONS: "Striped Bass, in Comeback, Spur Debate over Fishing," by Philip Shabecoff, *New York Times*, May 24, 1988, p. C1.

177. MAFIA/FULTON: "Mafia Runs Fulton Fish Market, U.S. Says in Suit to

Take Control," by Arnold H. Lubasch, *New York Times*, October 16, 1987, p. 1.

177. BAY CLEANUP: "Chesapeake Cleanup Pact Is Signed," by B. Drummond Ayres, *New York Times*, December 16, 1987.

178. FAULTY SURVEY: "Doubts Raised on Accuracy of Chesapeake Bass Survey," by Dick Russell, *New England Fisherman*, October 12, 1989, p. 20.

178. BOYLE ARTICLE: "A Hasty Call on a Fast Count," by Robert H. Boyle, *Sports Illustrated*, March 19, 1990.

178–179. STRIPER HEARING: "Congress Holds Striped Bass Inquiry," by Dick Russell, *New England Fisherman*, February 22, 1990, p. 32.

180–184. WHITE STORY: Author's interview, 2004; "More Than Just a Lucky Rod," by Jim White, *New England Fisherman*, May 14, 1987, p. 32.

Chapter 10: The Double-Edged Sword of "Full Recovery"

186. "FULLY RECOVERED": "Look but Don't Touch," by Kirk Moore, *National Fisherman*, August 1996, p. 15; "U.S. Atlantic Coast Striped Bass: Issues with a Recovered Population," by K. J. Hartman and F. J. Margraf, *Fisheries Management and Ecology* 10:309–312, 2003.

186. 43 PERCENT INCREASE: Stripers Forever, www.stripersforever.org; Jersey Coast Anglers Association newsletter, January 2003, p. 5.

186. BOREMAN BACKGROUND: "The Art of Fisheries Science," by Dexter Van Zile, *National Fisherman*, October 2003, p. 26. This article includes Boreman's reaction to "Trawlgate," problems with trawl surveys conducted by the National Marine Fisheries Service.

188. FOX QUOTES: Author's interview, 2003.

188. ATLANTIC COASTAL ACT: www.nmfs.noaa.gov/irf/acfcmaac.html.

188. WEAKFISH: Author's interview with Ted Williams, 2004.

190. FOX QUOTE: Author's interview.

190. SECOR STUDY: "Rocks in Fish's Head Tell Tales," *National Geographic* (Earth Almanac), October 1995.

190. WALDMAN QUOTE: Author's interview, 2004.

191. SSB MODEL: "The State of the Striped Bass," by John Merwin, *Field & Stream*, April 1995, p. 65.

191. KARAS: Cited by Jim White in "Stripers, a Looming Problem?" Part 2, *New England Fisherman*, September 30, 1999, p. 15.

192. BOONE QUOTE: Author's interview, 2004.

193. MAY QUOTE: Author's interview, 2004.

193. CATCH/FINANCIAL FIGURES: "Striper Recovery—Not," by Ted Williams, *Fly Rod & Reel*, 2003, www.flyrodreel.com.

193. WILLIAMS QUOTE: Ibid.

193. "F" TARGET: Author's notes from the ASMFC Striped Bass Board meeting, New York, December 2003; "2003 Review of the Atlantic States Marine Fisheries Commission's Fishery Management Plan for Atlantic Striped Bass (*Morone saxatilis*)," prepared by the Striped Bass Plan Review Team, November 2003.

195. BOREMAN QUOTE: Author's interview, 2004.
197. UPHOFF QUOTES: Author's interview, 2003.
198–199. CLIMATE AND CHESAPEAKE: "Climate Forced Changes in the Striped Bass Forage Base within Chesapeake Bay," research paper presented by Bob Wood, NOAA Chesapeake Bay Program and Cooperative Oxford Lab, at the 60th Annual Northeast Fish and Wildlife Conference symposium. See also "Future Consequences of Climate Change for the Chesapeake Bay Ecosystem and Its Fisheries," by Robert J. Wood, Donald F. Boesch, and Victor S. Kennedy, *American Fisheries Society Symposium* 32:171–184, 2002.
199. ECOSYSTEM MANAGEMENT: "Ecosystem-Based Fishery Management," (multiple authors) *Science* 305, July 16, 2004.
199. KAUFMAN QUOTE: "The Next Wave," by Charles A. Radin, *Boston Globe Magazine*, October 26, 1997, p. 38.

Chapter 11: The Myco Mystery

203. SOARES: Author's interview, 2004.
203–204. PRICE/BOYLE: "Bringing Back the Chesapeake," by Robert H. Boyle, *Audubon*, May–June 1999, p. 78.
204. VOGELBEIN: This statement, and all subsequent quotations unless otherwise noted, is from a telephone interview with the author, Summer 2004. Vogelbein, David Gauthier, Martha Rhodes, Howard Kator, Rob Latour, and Chris Bonsak, all of the Virginia Institute of Marine Science, and Christopher Ottinger, U.S. Geological Survey, Leetown Science Center, presented a research paper, "Mycobacteriosis in Striped Bass (*Morone saxatilis*) from Chesapeake Bay," at a symposium of the 60th Annual Northeast Fish and Wildlife Conference, April 27, 2004, Ocean City, Maryland.
204. "53 PERCENT": "The Prevalence of Mycobacterial Infections in Striped Bass in Chesapeake Bay," by A. S. Overton, F. J. Margraf, C. A. Weedon, L. H. Pieper, and E. B. May, *Fisheries Management and Ecology*, 10:301–308, 2003.
205. CALIFORNIA: Ibid.
205. FIRST INCIDENCE: Ibid.
205. SPREAD OF DISEASE: Author's interview with Eric May, 2004.
206. DISEASE RATE INCREASE: "Mycobacteriosis Infection Rate in Bay's Striped Bass Increasing," by Karl Blankenship, *Bay Journal*, June 2004, p. 10. See also "Scientists Studying Bay's Sick Striped Bass," Associated Press, November 18, 2002, and "Delaware's Striped Bass Studied for Infection," by Molly Murray, *Wilmington News-Journal*, December 29, 2003.
206. STRIPER PERCENTAGES: Blankenship, "Mycobacteriosis Infection Rate in Bay's Striped Bass Increasing."
206. "FISH HANDLER'S DISEASE": "Striped Bass Illness Baffles Bay Scientists," by Karl Blankenship, *Bay Journal*, April 2002; "Handling Fish? No Problem.

Common Sense and Continued Awareness about Myco-bacteria," by Mary Madison, *Waterman's Gazette*, July 2002, p. 3.

206–207. STUDY: "Infection Attacking Rockfish in Chesapeake," by Anita Huslin, *Washington Post*, April 4, 2002, p. B3.

209. CRECCO STUDY: "Method of Estimating Fishing (F) and Natural (M) Mortality Rates from Total Mortality (Z) and Exploitation (U) Rates for Striped Bass," by Victor Crecco, Connecticut Division of Marine Fisheries, September 30, 2003.

209. KAHN STUDY: "Tag-Recapture Data from Chesapeake Bay Resident Striped Bass Indiates That Survival Has Declined," research paper presented by Desmond Kahn, Delaware Division of Fish and Wildlife, at a symposium at the 60th Annual Northeast Fish and Wildlife Conference.

209. KAHN QUOTE: Author's interview, 2004.

210. MAY STUDY: "Potential Impacts of Mycobacteriosis in Striped Bass on Chesapeake and Atlantic Stocks," research paper presented by Eric May and V. Pernell Lewis, University of Maryland Eastern Shore (UMES); Anthony Overton, East Carolina University; John Jacobs, NOAA Cooperative Oxford Lab; and Larry Alade, UMES, at a symposium at the 60th Annual Northeast Fish and Wildlife Conference.

211. HATCHERIES AND DISEASE: In Vogelbein's interview with the author, he said he understood mycobacteriosis to be a "major problem" in operations that culture hybrid striped bass for food. In terms of hatchery-raised fish, Vogelbein added that he could imagine "a hatchery manager sitting on 100,000 fingerlings to be released, who starts to experience chronic fish losses over a period of time from disease. What's he going to do? Well, maybe release those fish to get them out before they die, so he can claim to have released 98,000 fish that year." That would potentially send the disease out into the wild population. Vogelbein cautioned that the disease most likely emanates from "a multitude of things."

212. JACOBS STUDY: "Nutritional Health of Chesapeake Bay Striped Bass, *Morone saxatilis*, in Relation to Disease," research paper presented by John Jacobs, NOAA National Ocean Service, Cooperative Oxford Lab; Hobart Rogers and William Van Heukelem, University of Maryland Horn Point Lab; and Brett Coakley, Charles Gieseker, and Mark Matsche, Maryland DNR Cooperative Oxford Lab, at a symposium at the 60th Annual Northeast Fish and Wildlife Conference.

214. MENHADEN IN BASS DIET: "Predatory Demand and Impact of Striped Bass, Bluefish, and Weakfish in the Chesapeake Bay: Applications of Bioenergetics Models," by K. J. Hartman and S. B. Brandt, *Canadian Journal of Fisheries and Aquatic Sciences* 52:1667–1687, 1995.

214. MENHADEN IN BASS DIET, 1998–99: "Striped Bass Predator-Prey Interactions in the Chesapeake Bay and Along the Atlantic Coast" (report), by A. S. Overton for the Maryland Cooperative Fish and Wildlife Research Unit, 2003. See also "The Diet of Chesapeake Bay Striped Bass in the Late 1950s," by J. C. Griffin and F. J. Margraf, *Fisheries Management and Ecology* 10:323–328, 2003, and "Food Habits

of Large Striped Bass in the Lower Chesapeake Bay and Its Tributaries" (final report), by Herbert M. Austin and John Walter, Virginia Institute of Marine Science, submitted to the Marine Recreational Fisheries Advisory Board and Commercial Fisheries Advisory Board, Virginia Marine Resources Commission.

214. OVERTON QUOTE: Cited in Karl Blankenship, "Striped Bass Illness."

214. UPHOFF REPORT: "Predator-Prey Analysis of Striped Bass and Atlantic Menhaden in Upper Chesapeake Bay," by J. H. Uphoff Jr., *Fisheries Management and Ecology* 10:313–322, 2003. Uphoff also presented an updated paper, "Striped Bass and Atlantic Menhaden: Is There a Predator-Prey Imbalance in Chesapeake Bay?" at a symposium at the 60th Annual Northeast Fish and Wildlife Conference.

214. JIM PRICE: "Chesapeake Bay Forage Base Collapse and Interactions of Striped Bass and Atlantic Menhaden" (report), by the Chesapeake Bay Ecological Foundation, presented to the Atlantic Menhaden Management Board, New York, December 17, 2003, www.chesbay.org.

215. ANCHOVIES: Overton, "Striped Bass Predator-Prey Interactions."

215. GIBSON QUOTE: Author's interview, 2004.

215. PETERSON QUOTE: Author's interview, 2004.

215. PRICE STUDY: Author's interview, 2004.

216. MENHADEN LANDINGS: "Atlantic Menhaden 2003 Stock Assessment Report," Atlantic States Marine Fisheries Commission, October 2003, p. 25.

Chapter 12: The Town That Menhaden Built

217. PER CAPITA INCOME: "Reedville 1874–1974," by Miriam Haynie, booklet published by the Greater Reedville Association, Reedville, Virginia.

217. PORT/POUNDS: National Marine Fisheries Service, www.st.nmfs.noaa.gov/st1/index.html.

218. ELIJAH REED: "Time and Tidewater," by Larry Chowning, *National Fisherman*, April 1999, p. 30.

218. "MUNNAUWHATTEAUG": "Mad About Menhaden," by Charlie Coates, *Tide* (Coastal Conservation Association), Spring 2004.

218. JOHNSON, 1628: Exhibits, Reedville Fishermen's Museum, Reedville, Virginia.

219. MENHADEN: "Atlantic Menhaden 2003 Stock Assessment Report," Atlantic States Marine Fisheries Commission (ASMFC), October 2003. See also "Ecological Role of Atlantic Menhaden (*Brevoortia tyrannus*) in Chesapeake Bay and Implications for Management of the Fishery," by Sara Jean Gottlieb, University of Maryland at College Park, master's thesis, 1998.

219. CARSON QUOTE/MENHADEN USES: Exhibits, Reedville Fishermen's Museum, Reedville, Virginia.

220. BAIT SALES: "Menhaden in the Middle," by Kirk Moore, *National Fisherman*, April 2004, p. 19.

221. SEINING OPERATION: Industry videotape viewed at Reedville Fishermen's

Museum, Reedville, Virginia; "The Most Important Fish in the Sea," by H. Bruce Franklin, *Discover*, September 2001; "Purse Seining for Pogies," by Kirk Moore, *National Fisherman*, November 1997.

221. NEW FACTORY: "Omega Protein to Build New Fish Oil Processing Facility" (press release), by Omega Protein, April 15, 2003, www.buyomega protein.com.

222. FDA APPROVALS: ASMFC, "Atlantic Menhaden 2003 Stock Assessment Report," p. 28.

222. NEW FACILITY: "Omega Protein Announces Completion of New Health and Science Center in Virginia" (press release), by Omega Protein, October 21, 2004, www.buyomegaprotein.com.

222. MEAL/OIL PRODUCTION: Coates, "Mad About Menhaden," p. 28.

222. 2002 SALES: Moore, "Menhaden in the Middle," p. 18.

223. Bush/Zapata: "Adios, Zapata!" by Monica Perin, *Houston Business Journal*, April 26, 1999.

223. SEC FILINGS: Zapata Corp., June 15, 1998, www.themotleyfool.com.

223–224. Background on Glazer and Zapata was drawn from several sources, including "A Sports Mogul and a Mystery," *BusinessWeek*, March 15, 2004; "Another Mysterious Investor Bids for Glazer Company," by Scott Barancik, *St. Petersburg Times*, September 5, 2003; and "Little-known Firm Makes Bid for Zapata," by Thor Valdmanis, *USA Today*, www.usa today.com/money/industries/food/2003-03-06-zapta_x.htm. In a letter to *BusinessWeek* from Avram Glazer, president and CEO of Zapata Corporation, posted on www.zapatacorp.com, it was stated that the Glazer family "has never had any contact with Mr. Rexford" and that allegations about Malcolm Glazer selling Zapata stock or owning shares of Omega Protein "are false." The article, according to Avram Glazer, was "filled with incorrect facts, unfounded accusations, and unsubstantiated speculation."

225. MENHADEN LANDINGS: National Marine Fisheries Service figures.

226. MENHADEN RECRUITMENT: ASMFC, "Atlantic Menhaden 2003 Stock Assessment Report."

226. MENHADEN BIOMASS: "Chesapeake Bay Forage Base Collapse and Interactions of Striped Bass and Atlantic Menhaden" (revised report), by the Chesapeake Bay Ecological Foundation, submitted to the Menhaden Ecological Workshop for the Atlantic Menhaden Management Board, Atlantic States Marine Fisheries Commission, October 12–14, 2004, Alexandria, Virginia, www.chesbay.org.

226. MAY QUOTE: Author's interview, 2004.

227. WALDMAN QUOTE: Author's interview, 2004.

227. SPITZER BIRD FIGURES AND QUOTES: Franklin, "The Most Important Fish in the Sea."

227. MENHADEN IN MAINE: Author's interview with Barry Gibson, 2004.

228. SCIENCE BLOCKED: Author's interview with Bill Goldsborough, 2004; "Controversy Sidelines Proposed Study of Bay Menhaden," *Bay Journal*, January–February 1995, p. 5; "Unpopular Science," commentary by

Donald Boesch, *Bay Journal*, March 1995; "The Menhaden Fishery Controversy: A Convening Analysis," by Richard C. Collins, Institute for Environmental Negotiation, University of Virginia, published by Virginia Sea Grant College Program; "The Menhaden Controversy: A Wrong Turn for Fisheries Management," commentary by the Chesapeake Bay Foundation, November 17, 1994.

228. ASMFC BOARD: "At Issue: ASMFC Menhaden Board Membership," *Commercial Fisheries News*, December 1999, p. 8B.

229. UPHOFF QUOTE: Author's interview, 2004.

229. JUVENILE AGGREGATION: Ibid.

229. KAUFMAN QUOTE: Author's interview, 2004.

230. SCIENTISTS PANEL: See National Coalition for Marine Conservation, www.savethefish.org.

230. UPHOFF PAPER: "Striped Bass and Atlantic Menhaden: Is There a Predator-Prey Imbalance in Chesapeake Bay?" by James H. Uphoff Jr., Fisheries Service, Maryland Department of Natural Resources.

230. HARTMAN PAPER: "Increases in Coastal Striped Bass Predatory Demand and Implications of Declines in Atlantic Menhaden Populations," by Kyle J. Hartman, Wildlife and Fisheries Resources Program, West Virginia University Division of Forestry.

230–231. MENHADEN WORKSHOP: Author's interviews with Bill Goldsborough, Ken Hinman, Jim Price, Jim Uphoff, and Toby Gascon, 2004.

231. CONSERVATIONISTS ALLIANCE: See Menhaden Matter, www.menhaden matter.org.

231. INDUSTRY WEBSITE: See Menhaden Facts, www.menhadenfacts.org/the_facts.html.

232. SOYBEAN ALTERNATIVE: Franklin, "The Most Important Fish in the Sea."

232. OMEGA-3: See also "Health Rewards from Omega-3 Fatty Acids Aren't Just Fish Tails," by Louis B. Parks, *Houston Chronicle*, August 12, 2003; "The Fats of Life: The Role of Omega-3 Fatty Acids in the Prevention of Coronary Heart Disease," by Charles R. Harper, M.D., and Terry A. Jacobson, M.D., *Archives of Internal Medicine*, Vol. 161, October 8, 2001.

232. MENHADEN CAP PROPOSAL: "Menhaden Issue Critical to Bay's Future," by Keith Walters, *Easton* (Maryland) *Star-Democrat*, February 13, 2005 (includes Simns quote); "Governor Ehrlich Supports Cap on Menhaden Fishing in Chesapeake Bay," press release, Maryland governor's office; "Another Victory for Menhaden Conservation," Conservation Coast Association Maryland Info Alert, February 11, 2005; "Commission Votes to Limit Menhaden Catch in Chesapeake Bay," National Coalition for Marine Conservation, February 10, 2005, http://www.savethefish.org/action_items_striped_bass_EAN.htm.

Chapter 13: Upriver

235. KING QUOTES: Author's interview, 2004.

236. FISH KILLS: "*Pfiesteria* May or May Not Be Toxic, but the Dispute over the Issue Is," by Karl Blankenship, *Bay Journal*, October 2002. For more on

the *Pfiesteria* controversy, see also "Deadly or Dull? Uproar over a Microbe," by Nicholas Wade, *New York Times*, August 6, 2002.

238. BAY FIGURES: "Watershed Blues," by Dexter Van Zile, *National Fisherman*, May 2004, p. 26.

238. 40 PERCENT: "Manure's Impact on Rivers, Streams, and the Chesapeake Bay" (report), by the Chesapeake Bay Foundation, July 28, 2004.

238–239. NUTRIENT STATISTICS: "Heavy Rains Poured Nutrients, Sediment into the Chesapeake," *Bay Journal*, June 2004, citing U.S. Geological Survey figures.

239. BAY GRASSES: "Study Says Bay Grasses Declined Steeply in '03," by Rona Kobell, *Baltimore Sun*, May 19, 2004; "Bay's SAV Fell Off Almost 30% in 2003," by Karl Blankenship, *Bay Journal*, June 2004.

239. CHESAPEAKE "DEAD ZONES": "Bay's 'Bad Water' Churns Unease," by David A. Fahrenthold, *Washington Post*, August 16, 2004, p. B1. General: "Sea 'Dead Zones' Threaten Fish," by Alex Kirby, BBC News, March 29, 2004.

239–240. VIRGINIA "DEAD ZONES"/FISH REQUIREMENTS: "'Dead Zone' a Threat to Bay," by Lawrence Latane III, *Richmond Times-Dispatch*, July 15, 2004.

240. WATER CLARITY IMPROVEMENT: "Cleaner Chesapeake Bay Waters Mystify Scientists," by Alex Dominguez, *Washington Times*, August 11, 2004.

240. ERIC MAY: Author's interview, 2004.

243. "FLUSH TAX": Van Zile, "Watershed Blues."

244. CBF REPORT: "Sewage Plants Fail to Reduce Nitrogen," by Dennis O'Brien, *Baltimore Sun*, October 29, 2003.

244. POST INVESTIGATION: "Bay Pollution Progress Overstated," by Peter Whoriskey, *Washington Post*, July 18, 2004, p. 1.

244. HOWARD ERNST: For the latest in-depth books on the Chesapeake situation, see *Chesapeake Bay Blues: Science, Politics, and the Struggle to Save the Bay*, by Howard R. Ernst, Rowman & Littlefield, Lanham, Maryland, 2003, and *Turning the Tide: Saving the Chesapeake Bay*, by Tom Horton, Island Press, Washington, D.C., 2003.

244. SENATORS' LETTER: "3 Senators Seek GAO Review of Bay Cleanup," by Peter Whoriskey, *Washington Post*, August 12, 2004, p. B1.

244. CBF LAWSUIT: "Advocates for Bay to Sue the EPA," by David A. Fahrenthold, *Washington Post*, November 10, 2004, p. B1. See also "CBF Takes Legal Action against EPA" (press release), by the Chesapeake Bay Foundation, November 2004, www.cbf.org.

244. WATER GROUPS' LAWSUIT: "4 Water Watchdog Groups Sue EPA over Levels of Pollution in Maryland," by Rona Kobell, *Baltimore Sun*, December 10, 2004.

244. EPA/STATE PACT: "Pact Will Limit Sewage in Bay," by Lawrence Latane III, *Richmond Times Dispatch*, January 4, 2005.

245. CBF MANURE QUOTE: Chesapeake Bay Foundation, "Manure's Impact," p. 1. See also "Manure Cited as Key Pollutant," by David A. Fahrenthold, *Washington Post*, July 29, 2004.

245. DELMARVA/POULTRY FIGURES: "The Economic Contribution and Long-Term Sustainability of the Delmarva Poultry Industry" (report), prepared

for the Maryland Agro-Ecology Center, Queenstown, Maryland, April 2003, MCAE Pub-2003-02, pp. 1 and 14.

245. PERDUE FIGURES: Corporation brochures and author's interview with Jim Perdue, 2004.

245. MANURE FIGURES: Cited in Chesapeake Bay Foundation, "Manure's Impact," p. 3.

246. GOVERNORS AND RULES: "Manure into Money," by Anita Huslin, *Washington Post*, August 3, 2003.

247. OYSTERS: "Can Introducing This Oyster Save the Chesapeake? Or Will It Take over the Bay and Hurt the Native Species?" by Beth Daley and Gary Cately, *Boston Globe*, August 19, 2003, p. E1. See also "Plan for Asian Oysters Worries Delaware and New Jersey," by David A. Fahrenthold, *Washington Post*, December 30, 2004, p. 87.

247. BAY CLEANUP COST: "Panel Brings Bay Cleanup Cost into Focus," by David A. Fahrenthold, *Washington Post*, October 28, 2004, p. 10.

247–248. PENNSYLVANIA: Chesapeake Bay Foundation, "Manure's Impact," p. 7. See also "PA's Draft Agricultural Regulations Fall Short," *Save the Bay* (Chesapeake Bay Foundation), Spring 2004. Chesapeake Bay Program officials, however, have pointed to Pennsylvania's phosphorus reduction efforts as playing a role in the resurgence of aquatic grasses in the upper bay. See "Aquatic Grasses Bounce Back," by Rona Kobell, *Baltimore Sun*, August 25, 2004.

248. PENNSYLVANIA HOG FARMS: "State Proposal Aims to Curb Pollution from Factory Farms," by Don Hopey, *Pittsburgh Post-Gazette*, August 11, 2004.

248–249. PERDUE QUOTE: Fahrenthold, "Manure Cited."

249. PFIESTERIA AND REGULATIONS: "Fears of Deadly Organism Cast Shadow on Chesapeake," by Michael Janofsky, *New York Times*, April 26, 1998, p. 18.

250. PERDUE AND MANURE LIABILITY: Huslin, "Manure into Money."

251. PELLET USES: Ibid.; author interviews with Jim Perdue and Tom Ferguson, 2004; company brochures. Also, "Perdue Markets Manure Pellets," by Luladey B. Tadesse, *Wilmington News Journal*, September 3, 2003.

252. ARSENIC: Author's interviews with Bill Goldsborough and Tom Horton, 2004. Another potential problem is the widespread use of antibiotics in poultry drinking water and its impact on human health. See "Poultry Farms' Use of Antibiotics Raises Concerns about Drug-Resistant Germs," by Tom Pelton, *Baltimore Sun*, August 31, 2004, and "In Soil, Water, Food, Air," by Amy Ellis Nutt, *Star-Ledger* [Newark, N.J.], December 8, 2003.

253. WATERMEN'S LAWSUIT: "Watermen Consider Suing for Cleaner Bay," by Tom Pelton, *Baltimore Sun*, July 28, 2004, p. B1.

253. PRICE QUOTE: "Bringing Back the Chesapeake," by Robert H. Boyle, *Audubon*, May–June 1999, p. 84.

Chapter 14: California Stripers

254. SPANISH EXPLORERS: "Tapped Out," *The Amicus Journal*, Fall 1998, p. 17.

255. STRIPED BASS COMING TO CALIFORNIA: "History and Status of Introduced Fishes in California, 1871–1996," by William A. Dill and Almo J.

Cordone, *Fish Bulletin* (California Department of Fish and Game) 178:136–137, 1997; *Striper: A Story of Fish and Man*, by John N. Cole, Atlantic–Little, Brown, 1978, pp. 33–35.

255. BASS AND GOLD MINING: Author's interview with Peter Moyle, 2004.

256–257. STRIPED BASS DISTRIBUTION: *Inland Fishes of California*, by Peter B. Moyle, University of California Press, Berkeley, 2002, pp. 365–366.

257. COOS BAY BASS: For a fascinating analysis of hermaphroditism in that fish population, see "Multiple Population Bottlenecks and DNA Diversity in Populations of Wild Striped Bass, *Morone saxatilis*," by John R. Waldman, Reese E. Bender, and Isaac I. Wirgin, *Fishery Bulletin* 96:614–620, 1998.

257. COMMERCIAL BAN: "Management School," by Robert H. Boyle, *Outdoor Life*, April 1995, p. 58.

257. DIMAGGIO: Ibid.

257. DROUGHT AND BASS: "Status and Protection of the San Francisco Bay–Sacramento–San Joaquin Delta Striped Bass Population" (report), by Thomas Cannon, Four Pumps Committee, for the California Striped Bass Association, October 2004. Cannon also told the author that often there is less water diverted in droughts because there is less to release from reservoirs. However, during the first year of drought, they divert, thinking the drought will end. In 1976 there was a large diversion, but 1977's was a low diversion, while the proportion diverted remained high.

258. LANDMARK STUDY: "Distribution and Abundance of Young-of-the-Year Striped Bass, *Morone saxatilis*, in Relation to River Flow in the Sacramento–San Joaquin Estuary," by J. L. Turner and H. K. Chadwick, *Transactions of the American Fisheries Society* 101:442–452, 1972.

258. FLOW DIVERSION: "Where Have All the Fishes Gone? Interactive Factors Producing Fish Declines in the Sacramento–San Joaquin Estuary," by William A. Bennett and Peter B. Moyle, from *San Francisco Bay: The Ecosystem*, J. T. Hollibaugh, ed., American Association for the Advancement of Science, San Francisco, 1996, p. 523; "Brother, Can You Spare a Dam?" by Robert H. Boyle, *Amicus Journal*, Fall 1998, p. 20.

258. EXPORTS DOUBLING: Cannon, "Status and Protection," p. 6.

258. BOYLE QUOTE: Boyle, "Management School," p. 60.

258. STRIPER POPULATION/RECRUITMENT FIGURES: California Department of Fish and Game figures, http://www.delta.dfg.ca.gov/data/townet/indices.asp?species=0. Also, Dill and Cordone, "History and Status of Introduced Fishes," p. 142.

260. TOM CANNON: Cannon, "Status and Protection."

261. TOXICS/COPEPOD DECLINE: Moyle, *Inland Fishes*, p. 371.

261. ALIEN SPECIES: Ibid.; see also "Dietary Shifts in a Stressed Fish Assemblage: Consequences of a Bivalve Invasion in the San Francisco Estuary," by Frederick Feyrer, Bruce Herbold, Scott A. Matern, and Peter B. Moyle, *Environmental Biology of Fishes* 67:277–288, 2003; and "Native and Alien Fishes in a California Estuarine Marsh: Twenty-One Years of Changing Assemblages," by Scott A. Matern, Peter B. Moyle, and Leslie C. Pierce, *Transactions of the American Fisheries Society* 131:797–816, 2002.

261. DAVE KOHLHURST: Author's interview, 2004.

262. PERIPHERAL CANAL: *The King of California: J. G. Boswell and the Making of a Secret American Empire*, by Mark Arax and Rick Wartzman, Public Affairs, New York, 2003, p. 351.

262. REISNER QUOTE: Ibid., p. 352.

262. 1986 AGREEMENT: *Four Pumps Agreement*, California Department of Water Resources, http://des.water.ca.gov/mitigation_restoration_branch/fourpumps/.

263. BASS EATING SALMON: Moyle, *Inland Fishes*, p. 372.

263. MOYLE ON DELTA SMELT: Author's interview, 2004.

263. HATCHERY AND SALMON IMPACT: Moyle, *Inland Fishes*, p. 369.

264. BASS STAMP: Author's interviews with California Striped Bass Association (CSBA) members; see also the CSBA website, www.striper-csba.com.

264–265. BENNETT HYPOTHESIS: "El Niños and the Decline of Striped Bass," by W. A. Bennett and E. Howard, *IEP Newsletter* 10(4):17–21, Autumn 1997; "Climate Change and the Decline of Striped Bass," by Bill Bennett and Liz Howard, *IEP Newsletter* 12(2): Spring 1999. (IEP stands for Interagency Ecological Program for the Sacramento–San Joaquin Estuary.)

265. STUDY FUNDING: Author's interview with William Bennett, 2004. Interestingly, in their coauthored 1996 paper "Where Have All the Fishes Gone?" Bennett and Moyle write in their conclusion: "the best evidence available to date indicates that the single most important factor associated with recruitment failure in fishes of the upper Sacramento–San Joaquin estuary is the reduction of outflowing freshwater through the estuary. This is produced by water diversions both within and upstream of the Delta and exacerbated by drought."

265–266. OSTRACH OTOLITH STUDY: Author's interview, 2004.

267. TOXIC RUNOFF: "The Killing Fields," by Robert H. Boyle, *Sports Illustrated*, March 22, 1993.

267. BOSWELL: "USA," by L. G. Carter, *Penthouse*, January 1999, p. 64.

267. DOUGLAS QUOTE: Ibid.

268. HOMELAND CANAL: Arax and Wartzman, *The King of California*, pp. 388–394.

268. KINGS COUNTY FIGURES: Author's interview with Lloyd Carter, 2004.

269. CORCORAN: Arax and Wartzman, *The King of California*, p. 414.

272. OSTRACH STUDY: Author's interview, 2004. Also, "Maternal Transfer of Xenobiotics and Effects on Larval Striped Bass in the San Francisco Estuary," abstract provided to the author by David J. Ostrach and coauthored with J. M. Low, S. J. Whiteman, and J. G. Zinkl, Department of Pathology, Microbiology and Immunology, School of Veterinary Medicine, University of California, Davis. The *Los Angeles Times* reported on nutrient pollution "suffocating some of California's most popular lakes and streams" in "Muck Runs Amok," by Marla Cone, September 8, 2004.

272. HEALTH ADVISORY: "Striped Bass Information Page," California Department

of Fish and Game, Central Valley Bay–Delta Branch, www.delta.dfg. ca.gov/stripedbass/health_advisory.asp.

273. FRIANT DAM RULING: *Nature's Voice* (Natural Resources Defense Council), November/December 2004.

273. WHITEY RASMUSSEN: Author's interview, 2004.

273–274. CANNON REPORT: Cannon, "Status and Protection."

274. ECOSYSTEM RESTORATION: "Water Accord Said to Be in Peril," by Bettina Boxall, *Los Angeles Times*, November 7, 2004, p. B1. Restoration programs include dam removal, installation of fish screens and ladders, "levees set back and tons of gravel laid to create salmon spawning grounds." See also, "In Fish vs. Farmer Cases, the Fish Loses Its Edge," by Dean E. Murphy, *New York Times*, February 22, 2005, p. 15.

274. PUMPING INCREASE FIGURES: "Thirst of Growers, Developers Huge Threat," by Lloyd G. Carter, *Fresno Bee*, March 22, 2004, p. B9. Author's interview, Barry Nelson.

274. NELSON BACKGROUND: Author's interview, 2004.

275. CARTER COLUMN: Carter, "Thirst of Growers, Developers Huge Threat."

275. FISHERIES REPORT ALTERED: "Rewrite Softens Report on Risks to Fish," by Stuart Leavenworth, *Sacramento Bee*, October 2, 2004.

275. CALIFORNIA DEPARTMENT OF FISH AND GAME FIGURES: http://www.delta. dfg.ca.gov. Author's conversation with Marty Gingras, senior biologist (Fisheries).

276. ENVIRONMENTAL WORKING GROUP: "Farmers Fill Up at Federal Water Trough," by Mark Arax, *Los Angeles Times*, December 15, 2004; "Big Farms Found to Get Most Water Subsidies," by Glen Martin, *San Francisco Chronicle*, December 15, 2004.

Chapter 15: Keepers of the River

279. JACK NEWFIELD: Author's interview, 2002.

279. DRUG ADDICTION: *The Riverkeepers*, by John Cronin and Robert F. Kennedy Jr., Scribner, New York, 1997, p. 90.

280. ON ROBERT BOYLE: Ibid., p. 92.

280. RIVERKEEPER LEGAL ACTIONS: "A Guardian for Troubled Waters," by Robert F. Kennedy Jr. with Dick Russell, *Conservation Matters* (Conservation Law Foundation), Winter 2002–2003, p. 24.

280. CHELSEA PUMP STATION: Cronin and Kennedy, *The Riverkeepers*, p. 204.

283. ESOPUS CREEK LAWSUIT: "Legal Eagle," by Dick Russell, *E*, November/ December 2003, p. 35.

284. PCBs AND STRIPERS: "Did PCB's Save the Stripers? A Fish Story," by James Gorman, *New York Times*, March 25, 2003, p. D3; "PCB's in Hudson Striped Bass at Safe Levels, State Says," by Andrew C. Revkin, *New York Times*, February 23, 1999.

284. GE DREDGING: "E.P.A. to Proceed on Dredging Plan for Hudson PCB's," by Kirk Johnson, *New York Times*, August 1, 2001, p. 1; "G.E. Is Accused of Trying to Undercut Order to Clean Up the Hudson River," by

Katharine Q. Seelye, *New York Times*, October 1, 2001, p. A19; "Dredging the Hudson Without Slinging the Mud," by Kirk Johnson, *New York Times*, April 21, 2002, p. 30; "PCB Cleanup in Upper Hudson River Is Delayed for Another Year," by Kirk Johnson, *New York Times*, March 11, 2003. See also "Is GE Mightier Than the Hudson?" by Richard Pollak, *The Nation*, May 28, 2001, p. 11; "The Case Against GE," *Multinational Monitor*, July/August 2001 (several articles). Conversely, a National Research Council report, "A Risk-Management Strategy for PCB-Contaminated Sediments," basically concluded that in most cases a careful analysis was needed and that the risks of trying to remove the contaminated sediments could exceed the risks of leaving them there.

285. SUPERFUND BANKRUPT: *Crimes against Nature*, by Robert F. Kennedy Jr., HarperCollins, New York, 2004, p. 113.

286. WASHINGTON IRVING: "20 Years of Stewardship: Riverkeeper Interview with Robert F. Kennedy Jr.," by Alex Matthiessen, *Riverkeeper*, Spring 2004, p. 9.

286–287. ZEBRA MUSSELS: "Muscling Their Way into the Food Chain: Zebra Mussels Alter Fish Populations in the Hudson River" (press release), by the Institute of Ecosystem Studies, www.ecostudies.org/press/zebra_mussels.html.

287. WATERSHED AGREEMENT: "The Kennedy Who Matters," by Pat Wechsler, *New York*, November 27, 1995, cover story.

289. KENNEDY QUOTE: Cronin and Kennedy, *The Riverkeepers*, p. 276.

290. POWER PLANTS: "Fisheries and Power Plant Project: Campaign Updates," by Reed Super, *Riverkeeper*, Spring 2004, p. 15; author's interview with Reed Super, 2004.

291. DESMOND KAHN STUDIES: "Mortality of Delaware River Striped Bass from Entrainment and Impingement by the Salem Nuclear Generating Station" (report), by Desmond Kahn, Delaware Division of Fish and Wildlife, March 30, 2000; "Assessment of the Impact of the Salem Nuclear Generating Station on Weakfish and Striped Bass," Desmond Kahn, March 8, 2001. Both papers provided to the author by Kahn.

Chapter 16: Stripers Forever?

295–296. KENNEBEC STRIPER RESTORATION: "The Big Ones That Always Get Away," by John N. Cole, *Down East*, May 1985, p. 47. See also "Rebound on the Kennebec," by Terry Tessein, *Tide* (Coastal Conservation Association), 1996. Cole's obituary appeared in *New York Times* on January 10, 2003.

296. PRESUMPSCOT RIVER: "Presumpscot River Facts," www.presumpscotriver.org/Text/RiverFacts.html.

298. OVERFISHING: For more details, see Stripers Forever, www.stripersforever.org.

298. "GROWTH OVERFISHING": "Can the Striper Fishing Last?" by Ted Williams,

Tide, July/August 1998, p. 13; "Striper Recovery—Not," by Ted Williams, *Fly Rod & Reel*, 2003, www.flyrodreel.com.

298. TOM FOTE: Author's interview, 2003.

298. OPERATION BACK DOOR: "Fish-market Sting Results in 14 Arrests," by Lawrence Latane III, *Richmond Times-Dispatch*, April 6, 2004.

299. TALLMAN & MACK BUSTED: "Three Firms Fined for Violating Fishing Rules," by Karen Lee Ziner, *Providence Journal*, September 21, 2003.

299. ECOSYSTEM AND BYCATCH: "Ecosystem-Based Fishery Management," *Science*.

300. CHARLES WITEK: Author's interview, 2004.

300. STRIPER BYCATCH: "Is Striped Bass Bycatch a Looming Problem?" by Charley Soares, *New England Fisherman*, January 8, 1998, p. 12; "How Many Dead Bass? Striper Bycatch Issue," by Charley Soares, *New England Fisherman*, February 18, 1999, p. 26.

300. OCEANA AND BYCATCH: Author's interview with Gib Brogan, 2004; "Striped Bass: Trawl Bycatch; Health Worries," by Lorelei Stevens, *Commercial Fisheries News*, October 2004, p. 13B. See also www.oceana.org. Late in 2004, the National Marine Fisheries Service notified Oceana "that there was insufficient justification to take emergency action. . . ." (*Commercial Fisheries News*, January 2005, p. 22A).

301. GOLDSBOROUGH ON MENHADEN BYCATCH: Author's interview, 2004. On December 28, 2004, the North Carolina Marine Patrol cited the entire crew of a menhaden seining boat from Louisiana for illegally netting and filleting thirteen striped bass.

302. EEZ OPENING: "NMFS Considers Opening EEZ Striper Fishery," by Lorelei Stevens, *Commercial Fisheries News*, December 2003, p. 21A. Due to "uncertainty" in the latest striped bass stock assessment, the NMFS has decided to postpone a decision on whether to reopen federal waters (*Commercial Fisheries News*, "No EEZ Striper Decision," January 2005, p. 22A).

304–305. CIRCLE HOOKS: "1999 Striped Bass Circle Hook Study," by Rudy Lukacovic, www.dnr.state.md.us/fisheries/recreational/crsb.html; "Spotlight on Circle Hooks," by Mark Fulton, *New England Fisherman*, July 1, 1999, p. 14; "Circle Hooks: A Conservation Tool," by Captain Al Anderson, *New England Fisherman*, September 9, 1999, p. 10.

305. CATCH-AND-RELEASE FISHING: "Catch-and-Release Recreational Fishing: a Historical Perspective," by David Policansky, in *Recreational Fisheries: Ecological, Economic and Social Evaluation*, T. J. Pitcher and C. E. Hollingworth, eds., Blackwell Science, Oxford, U.K., 2002, pp. 74–94; "A Bioenergetic Evaluation of the Chronic-Stress Hypothesis: Can Catch-and-Release Fishing Constrain Striped Bass Growth?" by Jason D. Stockwell, Paul J. Diodati, and Michael P. Armstrong, *American Fisheries Society Symposium* 30:144–147, 2002; "The Role of Catch-and-Release Mortality Estimates in the Assessment of Marine Fish Populations," by Gary R. Shepherd and Mark Terceiro, *American Fisheries Society Symposium* 30:230–233, 2002; "Mortality Associated with Catch and

Release of Striped Bass in the Hudson River," by M. J. Millard, S. A. Welsh, J. W. Fletcher, J. Mohler, A. Kahnle, and K. Hattala, *Fisheries Management and Ecology* 10:295–300, 2003. See also "Study Cites Sport Fishermen's Impact on Species," by Rick Weiss, *Washington Post*, August 27, 2004, p. A02.

306. CLIMATE CHANGE IN CHESAPEAKE: "Future Consequences of Climate Change for the Chesapeake Bay Ecosystem and Its Fisheries," by Robert J. Wood, Donald F. Boesch, and Victor S. Kennedy, *American Fisheries Society Symposium* 32:171–184, 2002.

307. DOUGLAS RADER: Author's interview, 2003.

307. BOB WOOD: Author's interview, 2004. For more on Wood's climate work regarding the bay, see "Whether a Fish Has a Hot Spawning Season May Depend on Weather," by Karl Blankenship, *Bay Journal*, October 2004, www.bayjournal.com/04-10/spawn.htm.

307. WARMING AND STRIPER MIGRATION: "Final Fish," by Peter Barone, *The Fisherman*, October 22, 1998, p. 10; "Stripers Get the Message: Global Warming Not Just Hot Air," by Tony Chamberlain, *Boston Globe*, October 23, 1998.

308. LES KAUFMAN: Author's interview, 2004.

308. DELAWARE RIVER STRIPERS: "Restoration of the Delaware River Stock of Striped Bass," by Desmond M. Kahn, Roy W. Miller, and Craig A. Shirey, Delaware Division of Fish and Wildlife, and Stephen Grabowski, U.S. Fish and Wildlife Service, July 1998; "Origin of the Present Delaware River Striped Bass Population as Shown by Analysis of Mitochondrial DNA," by John R. Waldman and Isaac I. Wirgin, *Transactions of the American Fisheries Society* 123:15–21, 1994.

308. OIL SPILL: Author's interview with Kahn, 2004.

308. BOYLE: "Management School," by Robert H. Boyle, *Outdoor Life*, April 1995, p. 60.

311. ERIC MAY: Author's interview, 2004.

311. DAVID POLICANSKY: E-mail to author, December 7, 2004. See "Fishing as a Cause of Evolution in Fishes," by David Policansky, pp. 2–18 in T. K. Stokes, J. M. McGlade, and R. Law (editors), *The Exploitation of Evolving Resources*, Springer, 1993a; "Evolution and Management of Exploited Fish Populations," by David Policansky, pp. 651–664 in D. M. Eggers, C. Pautzke, and T. J. Quinn II (editors), *Management Strategies for Exploited Fish Populations*, University of Alaska Sea Grant Program, Fairbanks, 1993b.

314. POND THEORY: However, a recent study has found no relationship between PCB exposure and any measures of striped bass abundance: "The Effects of Historic PCB Exposures on Reproductive Success of the Hudson River Striped Bass Population," by L. W. Barnthouse, D. Glaser, and J. Young, Oak Ridge National Laboratory, *Environmental Science and Technology* 37:223–228, 2003.

Index

Abranowicz, Bill, 292
acid rain, 120–21, 166
adaptation, 308–9, 314–15
AgriRecycle, 246, 250–52
alewives, 293
algae as vitamin source, 232
algae blooms, 26–27, 65, 233, 235–42
Anadromous Fish Conservation Act (1965),
    78, 188
anadromous fishes, 7
anchovy, 195, 215
Andresen, Spider, 19, 106, 151, 173
Animal Waste Technology Fund, 252
anoxic conditions, 239–40
Antioch, California, 269
Army Corps of Engineers, 95–98, 107,
    153–55
arsenic, 252
Asian clam, 261
Atlantic Coastal Act (1993), 188
Atlantic menhaden. See menhaden
Atlantic States Emergency Conference
    on the Future of the Striped Bass,
    129–35
Atlantic States Marine Fisheries Commission
    (ASFMC)
  1984 annual meeting, 165
  catch data, recreational, 303, 305
  compliance claim by, 175
  ecosystem management and, 196, 197
  management plan of, 50–52, 79
  on menhaden, 29, 228–32
  moratorium issue debate, 135–40
  on recovery of stripers, 186, 191–92,
      297–98
  regulatory authority, lack of, 133
  Studds Amendment and, 78–79
  transition plan, 178–79
Atlantic Striped Bass Restoration Act (1984),
    142–45, 165–67, 188

Atomic Energy Commission (AEC), 41
Atom plug lure, 55, 56–58
Azores-Bermuda high pressure systems,
    198

Baird, Spencer, 15
Baker, Kenny, 66–67
Baker, Will, 162, 244
Barbeau, Will, 64, 78, 81, 119, 126, 143
bay anchovy, 215
Baynard, Sherman, 28
Beaufort Fisheries, 30
Bendick, Robert, 116–19, 124
Bennett, William (Bill), 264–66
Benstock, Marcy, 96
Beulah Winds poultry farm, 250
biomass approach, 191–92
black market sales, 298–99
Boesch, Donald, 306, 307
Bohemia River, 21, 88
Boone, Joe
  at Chesapeake young-of-the-year survey,
      87–90
  on ecological problems, 194–95
  Edenton Hatchery and, 60
  on hot spots, 178
  on malnutrition, 212–13
  on menhaden, 232–33, 234
  on modeling, 192
  on organizing, 89–90
  on Pond (Bob), 64
  on population trends, 100, 121–23
  reunion with, 20–23
  at scientific conclave, 119
  on size limits, 52–83
  on state efforts as farce, 175
  on state vs. federal regulation, 148
  Tidewater Administration and, 158–60,
      163–64
Boozer, Brenda, 286

Boreman, John
    at Atlantic States Emergency Conference,
        134–35
    background of, 129–31
    at Buzzards Bay hearing, 54
    on commercial vs. sports catch, 80
    on ecosystem management, 197
    on Fulton Fish Market, 104
    lessons learned by, 185–90
    on modeling, 191, 192
    moratorium debate and, 125, 136, 140,
        143, 147–48
    on power plants, 41–42
    on Westway Project, 153, 169–70
Boreman Model, 170
Boswell, J. G., II, 262, 267–69
Boyd, Avis, 60–61, 87, 313–16
Boyle, Robert H,
    on adaptability, 308–9, 315
    on California stripers, 254, 256–58
    closed-cycle cooling campaign, 290
    Con Ed dispute and, 32–42
    environmental reporting introduced by, 38
    Hudson River Fishermen's Association and,
        279
    Kennedy (Robert F., Jr.) and, 279–280,
        292
    on lifting of moratorium, 178, 179
    NRDC and, 280
    PCBs exposed by, 45–47, 284
    Pond (Bob) and, 60–61, 64
    on Price, 204
    Riverkeeper movement and, 43–45
    Striped Bass Fund and, 48
    Westway Project and, 91, 93–99, 153, 155
Bratton, J. K., 153
Breaux, John, 148–49, 152
Briggs, Fred, 107–8, 110–12
Brogan, Gib, 300–301
Brown, Torrey, 140–41, 162–64
Bryant, Nelson, 108
Bund, Emmanuel, 40
Burns, Brad, 295–97, 303–5, 309, 311
Bush, George H.W., 29, 223
Bush, George W., 4, 285, 289–90
Buss, Bob, 105
Butzel, Al
    Cuomo and, 106–7, 153
    Friends of Hudson River Park and, 171
    Storm King controversy and, 39–40, 42
    Westway Project and, 93–95, 99
Buzzards Bay hearing (Mass.), 49–50, 52–55

bycatch, 299–301
Byrd, Robert, 166
Byrne, John, 147

Cal-Fed Bay-Delta Program, 273–75
California striped bass
    agribusiness and, 266–69, 276
    Cal-Fed Bay-Delta Program and Napa
        Agreement, 273–76
    de-emphasis of stripers, 262–66
    introduction and history of, 254–58
    map of Bay–Delta, 256
    pollution and bioaccumulation, 271–73
    population surveys, 261–62, 269–72, 275
    water diversion and, 257–62, 271, 274
California Striped Bass Association (CSBA),
        264, 273, 276
Cannon, Tom, 260, 273–276
Cape Cod Canal Salties club, 58
Cape Cod Commercial Hook Fishermen's
        Association, 300
Carey, Hugh, 91, 98, 102, 103
Carson, Rachel, 46, 61, 219
Carter, Lloyd, 266–68, 274, 275
catch-and-release fishing
    in California, 259
    Fisherman Tournament and, 151
    Martha's Vineyard Striped Bass and
        Bluefish Derby and, 161–62
    mortality in, 304–5
    as norm or ethic, 179, 202, 288–89, 303–4
Catskill watersheds, 287–88
Chafee, John, 50, 79, 120, 142, 149, 163,
        165–67, 175
Chamberlain, Tony, 86
Chapman, Jack, 264
Chapman, Robert W., 176
Chelsea pumping station (NY), 280–82
Chesapeake and Delaware Canal, 16–17
Chesapeake Bay Acid Rain Foundation, 22
Chesapeake Bay and River system
    algae blooms in, 26–27, 65, 233, 235–42
    DDT in, 62
    disease in, 203–5, 209
    human land use, impact of, 237–42
    map, 236
    menhaden decline in, 23, 214–15, 222–32
    migratory vs. in-situ bass, 9
    nutrient pollution in, 243–53
    population forecasts, 306
    regional clean-up effort, 177
    size limits in, 100

spawning location in, 16
spawning-season closure to bass fishing, 50, 89
See also entries at Maryland; young-of-the-year survey
Chesapeake Bay Commission, 247
Chesapeake Bay Ecological Foundation, 21
"Chesapeake Bay Forage Base Collapse" (Price), 28
Chesapeake Bay Foundation (CBF), 243–48, 252
Chesapeake Bay Program, 243, 245
Chesapeake strain of striped bass, uniqueness of, 59
chicken manure, 245–53
Chinook salmon, 262–63, 275
Choptank River, 20–24, 121–23, 157, 178, 227
chunker fishing, 19
Cieri, Matt, 29–30
clam, Asian, 261
Clark, John, 36, 40, 41, 45–46
Clean Water Act (1972), 39–40, 41, 97, 244
Clifton Forebay, 263
climate change, 26, 265, 306–8
closed-cycle cooling systems, 44, 290–91
Coastal Migratory Fish Conservation Act, 149
Coates, Phil, 68, 74, 136, 179
Cole, John, 24, 50, 142, 161, 295–96, 311, 313–14
Cole, Kenneth, 288
colonial American bass fishing, 14
commercial catch value decline, 51
Conason, Joe, 98, 106, 107
Condon, Rae, 78, 85
Conservation Department, New York, 36–37
conservation equivalency, 187–88
Consolidated Edison (Con Ed), 32–42, 290
Corcoran, California, 268–69
Cordts, Alan, 53–54
cotton industry, 268
Crecco, Victor, 209
Creek Chub lure, 57
Cronan, John
  moratorium debates and, 146
  on NBC, 110
  size limits debate and, 74–75, 78, 79, 86, 105–6, 114–15, 117
Cronin, John, 43, 95, 132, 135, 280
Croton River, 289–94
Cuomo, Mario, 102

PCBs and, 172, 175
striper protection law and, 105–13
Westway Project and, 106–7, 152–53, 168, 170
Cuttyhunk Striper Club (Massachusetts), 15, 16

Daggett, Christopher, 168
D'Amato, Alfonse, 168
dams, in Industrial Revolution, 16
David, Laurie, 289
Dawson, Joe, 117, 125–26
DDT, 45–46, 60–62
dead zones, 239, 306
Delaware River, 308
Delmarva Peninsula, 245
de Long, Owen, 68, 77–78
delta smelt, 262–63
De Rasieres, Isaack, 13–14
development of eggs. *See* embryonic development
diet and feeding
  communal vs. solitary feeding, 11
  during development, 9
  energy requirements, 213
  of larvae, 8–9
  malnutrition, 27–29, 211–16
  migration and, 11
  predator-prey imbalance, 227–28
  prey population patterns, 194–95
  See also menhaden
DiMaggio, Joe, 257
Diodati, Paul, 302
disease
  fish handlers disease, 206
  lesions, 202–4, 206, 208
  malnourishment and, 27–29
  mycobacteriosis, 27–28, 203–11
dominant year class, 5, 51, 59–60, 176–77
Douglas, William O., 267
Dovel, William, 168–69
Duggan, Stephen, 40
Dukakis, Michael, 109
Duryea, Perry B., Jr., 101–2

ears, 12, 190, 265–66
ecosystem management, 26, 195–99, 299, 307–8
Edenton National Fish Hatchery, 59–60
eggs. *See* embryonic development; spawning
Ehrlich, Robert J., 232, 246
Eldridge, Mickey, 314–15

embryonic development, 61–64, 87–88,
   119–20, 272
   See also larvae; spawning
Emergency Striped Bass Study
   Boreman and, 54
   Chafee and, 50
   on economic loss, 106
   mandate for, 18
   moratorium issue and, 140, 143
   scientific conclave, 120
   See also Boreman, John
Endangered Species list
   Chinook salmon and delta smelt on,
      262–63
   Pond's petition for, 65, 140, 158
energy requirements, 213
environmental policy development, influence
   on, 4–5
Environmental Protection Agency (EPA), 42,
   168, 244
environmental reporting, introduction
   of, 38
environmental review, 38
environmental standing, 37–38
Environmental Working Group, 276
Esopus Creek, 283–84
exclusive economic zone (EEZ), 301–3
eyes, 12

Fairbanks, Randy, 68
Federal Highway Administration (FHA), 95,
   98
Federal Power Commission (FPC), 33, 36,
   37, 41
Federal Refuse Act (1899), 38
Federated League of Sportsmen's Clubs, 80
feeding. *See* diet and feeding
Ferguson, Tom, 251–52
fingerlings, development of, 9
fins, 11–12
Fisherman Tournament, 151
Fishermen's Association. *See* Hudson River
   Fishermen's Association
fish handlers disease, 206
fish oil, from menhaden, 219, 222, 223
fish stocks, depletion of, 25–26
Flaherty, Mo, 288
Fletcher, Ian, 95–97, 281
Florence, Ben, 50, 158–60, 162
Flower, Ken, 270–71
flush tax, 243–44
F-max, 193

Food and Drug Administration (FDA),
   156–57
Fote, Tom, 298, 312
Fox, Bill, 188, 190
Friends of Hudson River Park, 171
Fulton Fish Market (NY)
   Cuomo and, 110
   federal regulatory control of, 177
   lobbying by, 103–5, 111–13
   monopoly of, 73–74
   undersized bass at, 90, 103

gamefish status, 310–11, 312
Garrahy, J. Joseph, 114, 118, 124, 174
Garrison, Lloyd, 37
Gascon, Toby, 222
General Electric, 47, 284–85
genetic change, 120
genetic lineage study, 176
Gibbons, Boyd, 263
Gibson, Barry, 215, 227–28
gillnetting
   bycatch and, 300
   in Chesapeake, 17
   Martha's Vineyard bust, 82–85
   method of, 75
   in Rhode Island, 74–75
gill system, 12
Girard, Alan, 246
Giuliani, Rudy, 177
Glazer, Malcolm, 29, 223, 224
Glendening, Parris, 246, 249
global warming, 26, 306–8
Glowka, Art, 35
Goldsborough, Bill, 228, 230–31, 234, 301
Good Day Show, 66–67
Goodyear, Phil, 41, 120, 130–31, 143, 148
Gordon, William, 151
Greenough, Peter B., 161
Grey, Zane, 304
Griesa, Thomas P., 96–99, 106, 168–70
Griffis, Fletcher, 107, 167
groundfishery, Northeast, 187
growth overfishing, 298
Gude, Anthony Benton, 68–69, 82–84,
   119–23, 136–37
Guerin, Bing, 67
Guerin, Richie, 82–84
Gulf Coast Conservation Association, 76–77

Halpin, Patrick, 103, 104, 105, 108, 156, 172
Harrison, Verna, 163

Hartman, Kyle, 230
hatchery programs
  in California, 263–64
  disease and, 210–11
  Maryland proposal for, 132
hatching studies, 59–66
haul seine fisheries, 101, 102, 104, 155
Haynie, Wendell, 217–20, 225
hearing, sense of, 12
high-grading, 302
Hinman, Ken, 29–30, 195, 231
hook type, 304–5
Horn Point Laboratory, 212
Horton, Tom, 138–39, 150, 160, 175
hot spots, 178
Houde, Ed, 198
House Fisheries Subcommittee, 147–49, 152
Hudson Highlands, 31
Hudson River
  Chelsea pumping station battle, 280–82
  Indian Point Power Plant, 35–36, 42–44,
    290
  Kennedy (Robert F., Jr.) and, 277–82
  map, 154
  PCB contamination on, 45–48, 283–85
  polluters on, 39
  Storm King power plant dispute, 32–42
  watersheds, 287–88
  Westway Project, 91–99, 106–7, 152–53,
    155, 167–70
  See also Riverkeeper movement
*Hudson River: A Natural and Unnatural
  History, The* (Boyle), 32
Hudson River Fishermen's Association
  Chelsea pumping station battle, 281
  formation of, 39, 279
  Silverstein and, 93
  Storm King dispute and, 41, 42
  Westway Project and, 95
Hudson River Foundation, 42, 185
Hudson Riverkeepers. *See* Riverkeeper
  movement
Hudson River Park Act, 170–71
Hudson River Peace Treaty, 42
Hudson River Tribes, 13–14
Hughes, Harry, 124, 132, 163–64, 243, 247
hybrids, hatchery-raised, 121–23
hydroelectric projects, 32–42
hypoxia, 239–42

illegal commercial catches, 298–99
Impervious Surface Project, 235–42

Indian Point Power Plant, 35–36, 41, 42,
  43–44, 290
Industrial Revolution, 16
invasives, 261, 287
inverse catchability, 229–30
Isleton, California, 259

Jacobs, John, 212–13
Jensen, Pete, 89, 158–60
Jewett, Robert, 277
Johnson, Edward, 218
juveniles, development of, 9

Kahn, Desmond, 209, 229, 291
Karas, Nick, 191–92
Kaufman, Les, 199, 229, 308
Kempton, Murray, 170
Kennebec River, 295–96
Kennedy, Brendan, 290
Kennedy, Edward, 149, 166
Kennedy, Robert F., Jr.
  at Croton River, 289–94
  Hudson River and, 277–82, 289
  Magdalen Island camping trip, 282–88
  on Martha's Vineyard, 288–89
  Riverkeeper movement and, 43
  on Westway Project, 95
Kesterson National Wildlife Refuge, 267
King, Howard, 235
Koch, Ed
  on Fulton Fish Market, 177
  Westway Project and, 91, 95, 96, 98, 168,
    170
Kogut, Nina, 270–71
Kohlhurst, Dave, 261–65
Krantz, George, 160, 162
Kweskin, Jim, 68, 81

Lamb, George, 169
land-use development, 237–38, 252–53
Laptew, Mike, 11
larvae, 8–9, 87–88, 272
  See also embryonic development
Lawler, Matusky & Skelly (LMS), 94–95
lawsuits, environmental standing in, 37–38
Leo, Arnold, 134, 135, 155, 172
Leopold, Aldo, 196
lesions, 202–4, 206, 208
life span of striped bass, 27
Lippe, Tricia, 271
live bait and mortality, 304
livestock manure, 245–53

lobster, as bait, 15
loons, 227
Lucas, Sandra, 83–85
Ludwig, Michael, 168
Lukacovic, Rudy, 240–42
Lyman, Hal, 134
Lyman, Lincoln, 20

Magdalen Island, 282–83
Maine rivers, 295–97
malnutrition, 27–29, 211, 212–16
"Managing Our Nation's Fisheries"
    conference, 196
Manchester, Francis, 115–19, 125–26, 143,
    152, 174
Mansuetti, Romeo, 62, 168–69
manure, 245–53
maps
    Chesapeake, 236
    Hudson, 154
    migration, 10
    San Francisco Bay–Delta, 256
Marietta, Martin, 291
Markey, Ed, 68
Martha's Vineyard, gillnetting bust on,
    82–85
Martha's Vineyard Striped Bass and Bluefish
    Derby, 161–62, 173
Maryland
    hatchery program, 132
    interstate management plan, 127
    moratorium bill, 140–41
    See also Chesapeake Bay and River system
Maryland Charter Boat Association, 141
Maryland Department of Natural Resources,
    89
    See also Maryland Tidewater
        Administration
Maryland Fisheries Division, 137–40
Maryland Saltwater Sportsfishermen's
    Association, 89, 141
Maryland Tidewater Administration
    Impervious Surface Project, 235–42
    King as head of, 235
    management vs. biologists in, 135–40,
        149–50, 158–61
    moratorium decision, 162–64
Maryland Watermen's Association, 89, 253
Maryland Wildlife Federation, 140
Massachusetts Bay Colony, 3, 14
Massachusetts Marine Fisheries Commission,
    52, 66–69, 100

Massachusetts Maritime Academy hearing,
    49–50, 52–55
Massachusetts Striped Bass Association, 68
Matthiessen, Alex, 44
May, Eric
    on disease, 210–11
    on energy requirements, 213
    on gamefish status, 310
    on global warming, 307
    on hypoxia, 240
    on menhaden, 216, 226
    on models, 192–93
    on trophy fishing, 311
McGinty, Margaret, 240–42
McKay, Bonnie, 196
McLain, Ferrell, 220–221
Mendonsa, George
    alleged 130,000-pound take of, 77, 80,
        125
    conflict-of-interest ruling on, 82, 85, 86,
        141, 145
    government tagging and, 174–75
    illegal trafficking and, 299
    life of, 71–74
    Marine Fisheries Council and, 71–72, 85,
        86, 173–74
    moratorium debate and, 115–19, 124–26,
        146
    on NBC, 110, 112
    size-limit debate and, 77–78, 82
menhaden
    algae blooms and, 235–36, 242
    as commodity, 219
    controversy over restrictions on, 222–32
    debate over, 28–30
    ecosystem management and, 26, 195–96
    modeling of, 194
    overview of, 218–20
    proposed cap on landings, 232
    scarcity of, 22–23, 213–16
    spawning stock biomass, 226
Menhaden Matter, 231
Menhaden Stock Assessment Committee, 29
Merchant Marine and Fisheries Committee,
    152
Merriman, Daniel, 5, 58, 59, 61
metals, toxic, 252
migration, 9–11, 307
Miles River, 238–42
Miller, Lee, 259, 266, 270
Miller, Lewis, 32, 34
Minner, Ruth Ann, 251

models and modeling, 170, 191–95
Mollica, Joe, 118, 119, 126–27, 141, 152
monofilament nets, 89
Monsanto Chemical, 45, 47
moratorium on bass fishing
  Atlantic States Emergency Conference and,
    131–35
  in EEZ (federal waters), 301–3
  in Maryland, 140–41, 177–79
  Rhode Island controversy over, 115–19,
    124–27, 141–45, 152
"More Than Just a Lucky Rod" story, 179–84
Morone saxatilis, meaning of name, 8
mortality rates, 208–9
mortality target, 193
Moser, Ron, 163
Mountford, Kent, 27
Mowrer, Jim, 240–42
Moyle, Peter, 261, 263, 265
Mudd, Roger, 112
mussels, zebra, 286–87
mycobacteriosis, 27–28, 203–11

Nanticoke River, 60, 250
Napa Agreement, 274–75
nares (nostrils), double, 12
Narragansett Indians, 13
National Coalition for Marine Conservation,
  29, 231
National Environmental Policy Act (NEPA),
  38
National Marine Fisheries Service, 93–94,
  151, 300–301
National Oceanic and Atmospheric
  Administration (NOAA), 147
Natural Resources Defense Council (NRDC),
  40–42, 273, 278–80
Native Americans, 13–14
NBC Nightly News, 110–12
Nelson, Barry, 274, 275–76
Neville, William C., 17
New Bedford Harbor (Mass.), 63
Newfield, Jack, 279
New Jersey, gamefish status in, 312
New York
  PCBs and, 45–48, 171–72, 175, 283–85
  pressure from commercial interests in, 90
  striper protection law, 101–13
  Westway Project, 91–99, 106–7, 152–53,
    155, 167–70
  See also Hudson River
New York Central Railroad, 39

New York Department of Conservation
  (DEC), 43–44
New York Fish and Game Journal, 32
New York Rivers and Harbors Act (1888), 38
New York Sportfishing Federation, 103
New York Times, 108
nitrogen excess, 26–27, 244–53, 306
North Carolina, colonial settlement in, 15
nursery areas, 9
nutrient overload, 26–27, 65, 233–34,
  237–42

Oakes, John B., 153
Oceana, 300–301
Office of Information and Regulatory Affairs,
  290–91
Ohio River Valley low pressure systems, 198
Olney, Charles, 61
omega-3 fatty acids, 222, 223, 232
Omega Protein Corporation, 23, 29, 217,
  220–25, 228, 231, 301
Operation Back Door, 298–99
ospreys, 227
Ostrach, David, 265–66, 271–73
Othote, Louis, 116, 117, 124–26
otoliths, 190, 265–66
Overton, Anthony, 214
oxygen levels, 239–42
  See also algae blooms
Oyster Recovery Project, 247

Pace Environmental Litigation Clinic,
  283–84, 289
Pallone, Frank, 310
Parascondolo, Michael, 85, 86, 117, 125–26,
  146
PCBs (polychlorinated biphenyls)
  adaptation to, 314–15
  in California, 272
  FDA ruling on, 156–57
  Hudson River and, 45–48, 155–57,
    283–85
  in Massachusetts, 63
  in New York Harbor and off Long Island,
    171–72
  state sale bans and, 175
Pennsylvania, animal waste from, 247–48
Peper, George, 78, 132–35, 161
Perdue, Jim, 246–53
Perdue Farms, Inc., 245–52
Perlmutter, Alfred, 33, 34, 36, 37–38
pesticides, 45–46, 60–62

Peterson, Alan, 79, 128, 129, 134, 215, 311
Pew Oceans Commission, 26
Pfiesteria scare, 204, 249
phosphorus excess, 26–27, 243–51
phytoplankton, 219
Pikitch, Ellen, 199
Pilgrims, 14
Pirone, Dominick, 35, 38–39
Plymouth Colony, 3, 14
"Poison Roams Our Coastal Seas" (Boyle), 46
Policanksy, David, 305, 311
pollution
    acid rain, 120–21
    DDT, 45–46, 60–62
    Kesterson National Wildlife Refuge,
        266–67
    nutrient overload, 26–27, 65, 233–34,
        237–42
    polybrominated diphenyl ethers (PDBEs),
        271–72
    in San Francisco Bay–Delta, 260–61, 272
    See also PCBs (polychlorinated biphenyls)
polybrominated diphenyl ethers (PDBEs),
    271–72
Pond, Bob
    Atom plug lure and, 56–58
    at Buzzards Bay hearing, 54–55
    on eggs and larvae research, 89
    endangered-species petition, 65, 140, 158
    reunion with, 313–16
    Rhode Island debate and, 75, 76, 81
    at scientific conclave, 119–20
    spawning and hatching studies by, 58–66,
        178
    young-of-the-year survey and, 87, 90
poultry manure, 245–53
power plants
    closed-cycle cooling systems, 44, 290–91
    Indian Point, 35–36, 42–44
    nitrogen and, 306
    Pacific Gas & Electric, 260
    Storm King dispute, 32–42
Presumpscot River, 295, 296–97, 309
Price, Jim
    at Choptank River, 122–23, 178, 233
    on disease, 27–29, 203–4, 206
    on illegal fishing, 302–3
    menhaden analyses, 214, 225, 226
    monofilament nets and, 89
    on political mismanagement, 253
    reunion with, 20–24
    threatened-species petition, 131–32

menhaden, 233
    on weak Maryland plan, 127
Providence Journal, 82

Rader, Douglas, 307
Range, Jim, 133–35
Rasmussen, Whitey, 273
Rathjen, Warren, 32, 34
recreational fishing, history of, 15
red drum moratorium, 189
Reed, Elijah, 218
Reedville, Virginia, 217–20, 224–25
Reinfelder, Al, 48
Reinfelder, Jesse, 83–84
Reisner, Marc, 262
Reston, Dick, 133, 161
Reston, James, 133–34
"restoration measures," 290
Rhode Island Conflict of Interest
    Commission, 78, 82, 85, 141
Rhode Island Department of Environmental
    Management (DEM), 81
Rhode Island League of Federated Sportsmen,
    77
Rhode Island Marine Fisheries Council
    creation of, 71
    Mendonsa and, 71–72, 85, 86, 173–75
    moratorium debates, 115–19, 124–27,
        146, 151–53
    New York law and, 105–6
    size limits regulation debate and, 69–70,
        71–87, 100–1
Rhode Island public health officials, 175
Rhode Island Saltwater Anglers Association,
    200–201
Rich, Tim, 83–84
Risebrough, Robert, 45, 46
Riverkeeper movement
    Chelsea pumping station battle, 280–82
    closed-cycle cooling and, 290–91
    Indian Point power plants and, 43–45
    Kennedy (Robert F., Jr.) and, 280
    moratorium proposal and, 132
Robinson, Bill, 282–83, 286
Rockefeller, David, 91, 107, 168
Rockefeller, Nelson, 33
"rock fights," 8
Roosevelt, Robert B., 16
Roosevelt, Theodore, 16
Roxford, Theodore, 224
Ruckelshaus, William, 168
runoff, 26–27, 65, 233–34, 237–42